Combustion Processes in Propulsion

Combustion Processes in Propulsion

Control, Noise and Pulse Detonation

Edited by Gabriel D. Roy

ELSEVIER
BUTTERWORTH
HEINEMANN

AMSTERDAM • BOSTON • HEIDELBERG • LONDON
NEW YORK • OXFORD • PARIS • SAN DIEGO
SAN FRANCISCO • SINGAPORE • SYDNEY • TOKYO

Elsevier Butterworth–Heinemann
30 Corporate Drive, Suite 400, Burlington, MA 01803, USA
Linacre House, Jordan Hill, Oxford OX2 8DP, UK

 Recognizing the importance of preserving what has been written, Elsevier prints its books
on acid-free paper whenever possible.

Library of Congress Cataloging-in-Publication Data
Application submitted

British Library Cataloguing-in-Publication Data
A catalogue record for this book is available from the British Library.

ISBN 13: 978-0-12-3693945
ISBN 10: 0-12-369394-2

For information on all Elsevier Butterworth–Heinemann publications
visit our Web site at www.books.elsevier.com

Printed and bound by CPI Group (UK) Ltd, Croydon, CR0 4YY

Transferred to Digital Print 2012

In loving memory of my uncle

Mr. Arulappa Elias

Contents

SECTION TWO
HIGH-SPEED JET NOISE   211

SECTION THREE
PULSE DETONATION ENGINES   271

Foreword

The Office of Naval Research (ONR), from its inception in 1946, has been sponsoring research efforts in various fields of science and technology to support the U.S. Navy and Marine Corps. One of the areas is Energy Conversion–Propulsion that has been providing significant contribution in advancing the fundamental knowledge in combustion and contributing to the literature. Various topics ranging from the preparation of fuel and its distribution in a combustion chamber to the control of the resulting exhaust in terms of emission and noise have been identified and managed by Dr. Gabriel Roy of the ONR. Annual meetings are conducted to review and share the accomplishments of the scientists and engineers participating in his program and the propulsion research and development community and to disseminate the knowledge base for further follow up and development.

I had the opportunity to participate in review meetings and to meet with the researchers working in this program. Their accomplishments are presented in this book. It covers the entire spectrum of combustion applied to air breathing propulsion. Analysis, computation, experimentation, and diagnostics are blended logistically to maximize the utilization of research to practical applications. Fundamental aspects such as the detonation cell structures to applied topics such as multiaxis fluidic thrust vector control are covered in this book.

I would like to congratulate all those who have contributed their time and efforts to get this volume published, and I am pleased to write the foreword to this book. I am sure that this book will be a valuable contribution to the combustion literature and a reference to the research and educational community.

Dr. Jane A. Alexander

Executive Director
for Science and
Technology (2002–2003)
Office of Naval Research
January 2005   Arlington, VA 22217

The Editor

Gabriel D. Roy received his B.S. and M.S. in Mechanical Engineering, did his graduate study at the University of Tennessee Space Institute (UTSI) in Tullahoma and received his Ph.D. degree in engineering science. He served as a faculty member of the Mechanical Engineering department in India and the U.S. and as a major thesis advisor for over a dozen graduate students, besides managing the Heat Engines Laboratory projects. His early research involved hydrodynamic air bearings, crack propagation, and fatigue failure. As a Senior Research Engineer at UTSI, where he developed the diffuser and heat transfer diagnostics, he conducted heat transfer studies on the magnetohydrodynamic (MHD) power train and also generated pressure recovery performance curves for slagging MHD diffusers. He was responsible, as Project Leader in Heat Transfer, for all heat transfer aspects of the U.S. DoE sponsored MHD project. Later, he joined the industry (TRW, Inc.), where he received the TRW Roll of Honor Award and patents on combustor and atomizer. He was responsible for the development of the pulverized and slurry fuel injector and the high-voltage slag isolation and removal system.

Currently, Dr. Roy manages the Energy Conversion–Propulsion Program for the U.S. Navy at the Office of Naval Research (ONR). He also manages the Pulsed Power Program for the Ballistic Missile Defense Organization. In addition to the fundamental combustion program, he also envisioned and monitored focused multiyear programs such as High-Energy Strained Hydrocarbon Fuels and Air-Emission Control of Navy Marine Engines, and he presently manages research programs on High-Speed Jet Noise Control, Pulse Detonation Engines for Propulsion, Nano-Metallic Catalytic Fuels, and Combustion Control. With more than 35 years of government, industry, and university experience, he is recognized worldwide in magnetohydrodynamics, combustion, propulsion, and fuels research and management. His research interests include control of combustion processes, detonation, electromagnetic propulsion, electrorheology, advanced fuels, and jet noise control. He has organized several national and international conferences and workshops on innovative aspects of the combustion field such as electrorheological fuels, electromagnetic propulsion, pulse power, pulse detonation engines, combustion and noise control, and atmospheric pollution.

Dr. Roy is the recipient of the ASME Jean F. Lewis Energy Systems Award, the JANNAF Combustion Award, and the AIAA Energy Systems Award. He has edited over a dozen books and has over 100 publications and research reports to his credit. He is listed in the Marquis' Who's Who in Science and Engineering and the Who's Who in the South and Southwest. He has been a member of various technical committees of the American Institute of Aeronautics and Astronautics (AIAA) and a number of government review panels. He served as an associate editor of the AIAA Journal of Propulsion and Power and is a Fellow of AIAA. Dr. Roy is also an artist and has received national awards in painting. As a college student, he received the Duthie Memorial Award for 2 years.

Contributors

S. ABRAHAM

Department of Aerospace Engineering
& Engineering Mechanics
University of Cincinnati
Cincinnati, OH 45221

S. ACHARYA

Mechanical Engineering Department
Louisiana State University
Baton Rouge, LA 70803

A. AFSHARI

Department of Mechanical
Engineering
Michigan State University
East Lansing, MI 48824-1226

R. C. ALDREDGE

Department of Mechanical
& Aeronautical Engineering
University of California
Davis, CA 95616-5294

M. R. AMIN

Department of Mechanical
& Industrial Engineering
Montana State University
Bozeman, MT 59717

S. L. ANDERSON

Department of Chemistry
University of Utah
Salt Lake City, UT 84112

A. M. ANNASWAMY

Massachusetts Institute of Technology
Cambridge, MA 02139

V. ARAKERI

Department of Mechanical
Engineering
Florida A&M University
and Florida State University
Tallahassee, FL 32310

S. ARCHER

Department of Mechanical
Engineering
University of Maryland
College Park, MD 20742

A. A. ATCHLEY

The Pennsylvania State University
University Park, PA 16802

A. J. BARRA

University of Texas
Austin, TX 78712

W. T. BAUMANN

College of Engineering
Virginia Polytechnic Institute
and State University
Blacksburg, VA 24061

A. A. BEHRENS

University of Minnesota
Minneapolis, MN 55455

A. A. BORISOV

Institute of Chemical Physics
Russian Academy of Sciences
Moscow 119991, Russia

C. M. BROPHY

Naval Postgraduate School
Monterey, CA 93943

J. BRUMBERG

Department of Mechanical
& Nuclear Engineering
The Propulsion Engineering
Research Center
The Pennsylvania State University
University Park, PA 16802

J.-L. CAMBIER

Department of Mechanical
& Industrial Engineering
Montana State University
Bozeman, MT 59717

A. CHANDY

University of Florida
Gainesville, FL 32611

S. CHEATHAM

Center for Reactive Flow
& Dynamical Systems
Laboratory for Computational
Physics & Fluid Dynamics
Naval Research Laboratory
Washington, DC 20375-5344

J. D. CHENOWETH

Combustion Research
& Flow Technology, Inc.
Dublin, PA 18917

J. Y. CHOI

The Pennsylvania State University
University Park, PA 16802

S. CIPOLLA

Department of Aerospace Engineering
University of Maryland
College Park, MD 20742

C. CONRAD

Department of Mechanical
& Nuclear Engineering
The Propulsion Engineering
Research Center
The Pennsylvania State University
University Park, PA 16802

S. C. CREIGHTON

Pratt & Whitney
East Hartford, CT 6108

S. M. DASH

Combustion Research
& Flow Technology, Inc.
Dublin, PA 18917

D. F. DAVIDSON

High Temperature Gasdynamics
Laboratory
Department of Mechanical
Engineering
Stanford University
Stanford, CA 94305

G. DIEPVENS

University of Texas
Austin, TX 78712

C. F. EDWARDS

Department of Mechanical
Engineering —
Thermosciences Division
Stanford University
Stanford, CA 94305-3032

J. L. EDWARDS

College of Engineering
Cornell University
Ithaca, NY 14853

J. L. Ellzey
University of Texas
Austin, TX 78712

D. J. Forliti
University of Minnesota
Minneapolis, MN 55455

S. H. Frankel
School of Mechanical
Engineering
Purdue University
West Lafayette, IN 47907

S. M. Frolov
N. N. Semenov Institute
of Chemical Physics
Russian Academy of Sciences
Moscow 119991, Russia

T. B. Gabrielson
The Pennsylvania State University
University Park, PA 16802

Z. Gao
Department of Mechanical
& Industrial Engineering
University of Illinois at Chicago
Chicago, IL 60607

A. F. Ghoniem
Massachusetts Institute
of Technology
Cambridge, MA 02139

J. P. Gore
School of Mechanical
Engineering
Purdue University
West Lafayette, IN 47907

F. C. Gouldin
College of Engineering
Cornell University
Ithaca, NY 14853

R. J. Green
Department of Chemistry
University of Utah
Salt Lake City, UT 84112

B. Greska
Department of Mechanical
Engineering
Florida A&M University
and Florida State University
Tallahassee, FL 32310

F. F. Grinstein
Laboratory for Computational
Physics & Fluid Dynamics
Naval Research Laboratory
Washington, DC 20375-5344

M. Gundersen
Department of Electrical
Engineering — Electrophysics
University of Southern California
Los Angeles, CA 90089-0271

A. K. Gupta
Department of Mechanical
Engineering
University of Maryland
College Park, MD 20742

E. J. Gutmark
Department of Aerospace
Engineering & Engineering
Mechanics
University of Cincinnati
Cincinnati, OH 45221

B. Habibzadeh
Department of Mechanical
Engineering
University of Maryland
College Park, MD 20742

R. K. HANSON

High Temperature Gasdynamics
Laboratory
Department of Mechanical
Engineering
Stanford University
Stanford, CA 94305

O. HSU

Department of Aerospace Engineering
University of Maryland
College Park, MD 20742

K.-S. IM

Mechanical Engineering Department
Wayne State University
Detroit, MI 48202

F. A. JABERI

Department of Mechanical
Engineering
Michigan State University
East Lansing, MI 48824-1226

G. B. JACOBS

Department of Mechanical
& Industrial Engineering
University of Illinois at Chicago
Chicago, IL 60607

B. J. JANSEN

National Center for Physical
Acoustics
The University of Mississippi
Oxford, MS 38677

K. KAILASANATH

Center for Reactive Flow
& Dynamical Systems
Laboratory for Computational
Physics & Fluid Dynamics
Naval Research Laboratory
Washington, DC 20375-5344

C. KANNEPALLI

Combustion Research
& Flow Technology, Inc.
Dublin, PA 18917

D. C. KENZAKOWSKI

Combustion Research
& Flow Technology, Inc.
Dublin, PA 18917

N. J. KILLINGSWORTH

Department of Mechanical
& Aeronautical Engineering
University of California
Davis, CA 95616-5294

B. M. KNAPPE

Department of Mechanical
Engineering —
Thermosciences Division
Stanford University
Stanford, CA 94305-3032

E. KOC-ALKISLAR

Fluid Mechanics Research Laboratory
Department of Mechanical
Engineering
Florida A&M University
and Florida State University
Tallahassee, FL 32310

E. KORUSOY

Imperial College of Science,
Medicine & Technology
London SW7 2BX, UK

A. KROTHAPALLI

Department of Mechanical
Engineering
Florida A&M University
and Florida State University
Tallahassee, FL 32310

M. Krstić

Department of MAE
University of California at San Diego
La Jolla, CA 92093-0411

A. Kuthi

Department of Electrical
Engineering — Electrophysics
University of Southern California
Los Angeles, CA 90089-0271

N. M. Kuznetsov

N. N. Semenov Institute
of Chemical Physics
Russian Academy of Sciences
Moscow 119991, Russia

L.-C. Lee

Department of Electrical
Engineering — Electrophysics
University of Southern California
Los Angeles, CA 90089-0271

S.-Y. Lee

Department of Mechanical
& Nuclear Engineering
The Propulsion Engineering
Research Center
The Pennsylvania State University
University Park, PA 16802

J. C. Legros

Free University of Brussels
Brussels 1050, Belgium

C. Li

Center for Reactive Flow
& Dynamical Systems
Laboratory for Computational
Physics & Fluid Dynamics
Naval Research Laboratory
Washington, DC 20375

G. Li

Department of Aerospace Engineering
& Engineering Mechanics
University of Cincinnati
Cincinnati, OH 45221

M. Linck

Department of Mechanical
Engineering
University of Maryland
College Park, MD 20742

J. Liu

Department of Electrical
Engineering — Electrophysics
University of Southern California
Los Angeles, CA 90089-0271

L. Lourenco

Fluid Mechanics Research Laboratory
Department of Mechanical
Engineering
Florida A&M University
and Florida State University
Tallahassee, FL 32310

J. A. Lovett

Pratt & Whitney
East Hartford, CT 6108

F. H. Ma

The Pennsylvania State University
University Park, PA 16802

L. Ma

High Temperature Gasdynamics
Laboratory
Department of Mechanical
Engineering
Stanford University
Stanford, CA 94305

F. Mashayek

Department of Mechanical
& Industrial Engineering
University of Illinois at Chicago
Chicago, IL 60607

D. W. Mattison

High Temperature Gasdynamics
Laboratory
Department of Mechanical
Engineering
Stanford University
Stanford, CA 94305

D. Mikolaitis

University of Florida
Gainesville, FL 32611

L. Mongeau

School of Mechanical
Engineering
Purdue University
West Lafayette, IN 47907

S. Nakra

Department of Chemistry
University of Utah
Salt Lake City, UT 84112

V. Nenmeni

Department of Aerospace
Engineering
University of Maryland
College Park, MD 20742

D. W. Netzer

Naval Postgraduate School
Monterey, CA 93943

V. F. Nikitin

M. V. Lomonosov Moscow State
University
Moscow 119899, Russia

E. Noordally

Combustor & Heat Transfer
Technology Group
School of Engineering
Cranfield University, Cranfield
Bedfordshire, MK43 0AL, UK

S. Pal

Department of Mechanical
& Nuclear Engineering
The Propulsion Engineering
Research Center
The Pennsylvania State University
University Park, PA 16802

R. V. R. Pandya

Department of Mechanical
& Industrial Engineering
University of Illinois at Chicago
Chicago, IL 60607

B. Pang

Department of Aerospace
Engineering
University of Maryland
College Park, MD 20742

S. Park

Massachusetts Institute
of Technology
Cambridge, MA 02139

G. D. Roy

Office of Naval Research
Arlington, VA 22217

S. T. Sanders

High Temperature Gasdynamics
Laboratory
Department of Mechanical
Engineering
Stanford University
Stanford, CA 94305

R. J. Santoro
Department of Mechanical
& Nuclear Engineering
The Propulsion Engineering
Research Center
The Pennsylvania State University
University Park, PA 16802

S. Saretto
Department of Mechanical
& Nuclear Engineering
The Propulsion Engineering
Research Center
The Pennsylvania State University
University Park, PA 16802

W. R. Saunders
College of Engineering
Virginia Polytechnic Institute
and State University
Blacksburg, VA 24061

D. P. Schmidt
University of Massachusetts
Amherst, MA 01003

C. Segal
University of Florida
Gainesville, FL 32611

J. M. Seiner
National Center for Physical Acoustics
The University of Mississippi
Oxford, MS 38677

V. M. Shevtsova
Free University of Brussels
Brussels 1050, Belgium

B. Shotorban
Department of Mechanical
& Industrial Engineering
University of Illinois at Chicago
Chicago, IL 60607

N. Sinha
Combustion Research
& Flow Technology, Inc.
Dublin, PA 18917

J. O. Sinibaldi
Naval Postgraduate School
Monterey, CA 93943

N. N. Smirnov
M. V. Lomonosov Moscow State
University
Moscow 119899, Russia

P. J. Strykowski
University of Minnesota
Minneapolis, MN 55455

B. A. Tang
University of Minnesota
Minneapolis, MN 55455

J. H. Uhm
Mechanical Engineering Department
Louisiana State University
Baton Rouge, LA 70803

L. Ukeiley
National Center for Physical
Acoustics
The University of Mississippi
Oxford, MS 38677

U. Vandsburger
College of Engineering
Virginia Polytechnic Institute
and State University
Blacksburg, VA 24061

D. Wee
Massachusetts Institute
of Technology
Cambridge, MA 02139

J. H. WHITELAW
Imperial College of Science,
Medicine & Technology
London SW7 2BX, UK

J. J. WITTON
Combustor & Heat Transfer
Technology Group
School of Engineering
Cranfield University, Cranfield
Bedfordshire, MK43 0AL, UK

R. D. WOODWARD
Department of Mechanical
& Nuclear Engineering
The Propulsion Engineering
Research Center
The Pennsylvania State University
University Park, PA 16802

V. YANG
The Pennsylvania State University
University Park, PA 16802

T. YI
Massachusetts Institute
of Technology
Cambridge, MA 02139

C. YOUNG
Department of Electrical
Engineering — Electrophysics
University of Southern California
Los Angeles, CA 90089-0271

T. R. YOUNG
Laboratory for Computational
Physics & Fluid Dynamics
Naval Research Laboratory
Washington, DC 20375-5344

K. YU
Department of Aerospace Engineering
University of Maryland
College Park, MD 20742

S.-T. J. YU
Mechanical Engineering Department
Wayne State University
Detroit, MI 48202

Preface

Conversion of energy stored in a fuel for practical purposes has been known to the human race from ancient times. Chemical energy conversion for propulsion is motivated by the need to move around and travel, explore, defend, and conquer. Chemical propulsion is a multidisciplinary science and requires sophisticated experimentation, diagnostics, and computational tools for its advancement. The fundamentals of propulsion and the progress made in specific areas are covered by a number of books, monographs, and archive publications. However, books presenting the worldwide state of the art written by the performers and compiled into one volume are very limited. This book is an attempt to provide such a source of reference, for practicing engineers and graduate students, or as a text book for a graduate course in Advanced Topics in Combustion or Propulsion.

Propulsion systems of today and the future require increased and more rapid energy-release rates from smaller and more efficient combustors, in order to meet the demand for increased speed and range and for operational flexibility. Additional information on the fundamentals of combustion is needed to accomplish this. Environmental regulations and community concerns call for substantial reductions in combustion-generated noise from propulsion engines. Further, it is about the right time to explore new propulsion engine concepts. To address these issues, this book is organized under the following three sections: (1) Control of Combustion Processes, (2) High-Speed Jet Noise, and (3) Pulse Detonation Engines.

The chapters are written by those who have actually conducted the research, either independently or as part of a team, on research projects envisioned, initiated, and monitored by the editor. They describe the important findings and what is required and what can be expected in the future. Comments or questions raised by the chapter authors and the editor's comments are added as footnotes.

Section 1 is aimed at providing new fundamental and practical data for the development of efficient combustion control strategies. It includes development of novel diagnostic techniques to provide *in situ* measurement capabilities, novel computational tools to acquire deeper insight into the fundamental mechanisms, and interactions in two-phase flows and flames relevant to realistic combustors as well as their operating conditions. Further, detailed experimental and numerical studies of the physical and chemical phenomena in turbulent flames, and the novel passive and active control methodologies to control combustion instabilities, and to improve efficiency, from laboratory-scale models to actual combustor configurations are presented.

Reduction of combustion-generated noise from high-speed jet flows is the topic discussed in Section 2. Fundamental research and novel noise-reduction techniques are presented. These include detailed experimental and computational studies of jet-plume noise, as well as precise measurements. A narrow-band acoustic database generated from actual aircraft landing practice is used as a benchmark for the development of the effective noise-suppression technology.

The pulse detonation engine (PDE) is considered as a candidate for future propulsion systems due to its higher thermodynamic efficiency, simplicity, and ease of scaling. Section 3 is devoted to the advances made in understanding the fundamental features of propagating detonations and various means of detonation initiation for air-breathing PDEs. Topics include PDE fuels and fuel compositions, novel diagnostics with higher temporal and spatial resolutions and simultaneous multiparameter measurement capability, and component and system computational codes to enable parametric studies and performance optimization of PDEs.

Though minor details are limited, each chapter depicts completely the issues addressed and the solutions obtained. The large number of references given at the end of each chapter provides the reader the required source for further information. The editor has taken particular care to maintain uniformity of the individual chapters and to maintain clarity and continuity.

Any effort of this nature takes time, and I thank my wife Vimala for her understanding and cooperation during the long hours spent during this undertaking. My daughter Sitara and my daughters-in-law Vino and Rita continue to be my sources of inspiration, and their support to my efforts is acknowledged. The assistance given by Jennifer Doerman during the manuscript preparation is appreciated.

The contents of this book have been taken from the research sponsored by the Office of Naval Research. The valuable support provided by RADM Jay Cohen, Chief of Naval Research, for the editor's program, is acknowledged.

The editor acknowledges his gratitude to Joel Stein and Dawnmarie Simpson of Elsevier Science for their timely assistance in getting this book published.

Of course, this endeavor would not be possible without the contribution of the chapter authors. Their contribution is greatly appreciated. It is hoped that this book will provide a single source for up-to-date information on Chemical Propulsion for practicing engineers and researchers in combustion and propulsion.

Gabriel D. Roy

Office of Naval Research
Arlington, VA 22217

January 2005

Introduction

G. D. Roy

The past five years have seen a shift to new emphasis in propulsion research. This is due to the increased demand towards faster, yet smaller, propulsion systems with increased range. Compliance with regulations and environmental constraints on pollutants and noise have also contributed to this shift. As in the past, propulsion science and technology (S&T) efforts have been sponsored by various government agencies worldwide and also by the industry. S&T sponsored by the Office of Naval Research (ONR), while pursuing the area of combustion control, focused also on pulse detonation phenomena applied to propulsion and high-speed jet noise mitigation.

Much of the research on advanced fuel synthesis and characterization, fundamentals of combustion, and early work on combustion control and emissions and plumes were covered in the recent book, "Advances in Chemical Propulsion: Science to Technology" [1]. Some of the accomplishments that were not included, the progress made in Combustion Control since this publication and in the new areas of pulse detonation engines and high-speed jet noise control (over the past five years), have been reviewed. A number of these research efforts and the new scientific accomplishments and contributions are presented in this book.

CONTROL OF COMBUSTION PROCESSES

Combustion control, both active and passive, has been used as a tool to minimize combustion-generated oscillations to obtain stable operation, maximize combustion efficiency to improve performance, control temperature distribution, minimize emissions, and improve mixing and residence times to reduce combustor size. Techniques previously used with gaseous fuel combustion have been extended to combustion of liquid fuel of interest. Porous media inserts, countercurrent combustors, etc. have been investigated to control flow and temperature within the combustion chamber, aiming at compact combustors with reduced emission.

Though the measured velocity field provides a fairly complete description of the flow physics in incompressible, nonreacting flows, information on an additional variable is usually required to understand and control compressible or reacting flows. To this effect, *Lourenco* and his colleagues at Florida State University made simultaneous measurements of velocity and temperature fields in a premixed jet flame using particle image velocimetry (PIV) and laser speckle displacement (LSD) techniques (Chapter 1). As a nonintrusive diagnostics tool, *Gouldin* and his coworkers at the Cornell University (Chapter 2) completed the analysis of tomographic CO_2 measurements of forced jets. They refined the tomographic reconstruction technique, adaptive Finite Domain Direct Inversion (FDDI), and used Proper Orthogonal Decomposition (POD) analysis both to assess the possibility for reduced-order models of the vorticity and concentration fields and to examine the correlation between vorticity and scalar mixing in reacting rectangular jets.

Several computational efforts focused on predicting and controlling turbulent reacting flows encountered in combustors to perform parametric evaluations. Numerical simulations using deterministic and probabilistic approaches to predict two-phase turbulent flows in liquid-fueled combustion are presented by *Mashayek*'s group at the University of Illinois at Chicago (Chapter 3). They provided for the first time the basic understanding of the countercurrent shear flow phenomenon occurring in *Strykowski*'s countercurrent combustor (Chapter 8). Such computations will help in obtaining the optimal geometry and flow rate ratios (countercurrent vs. primary flow) in this combustor and assist in formulating new experiments. This can lead to the development of optimized combustion based on this concept.

At Michigan State University, *Jaberi*'s group (Chapter 4) has extended their work on the Large-Eddy Simulation (LES) method to two-phase flows, and some preliminary simulations of nonreacting particle dispersion in a turbulent jet have been conducted. Direct numerical simulation (DNS) was also performed for certain turbulent flames, in order to utilize the DNS-generated data for systematic assessment and validation of the LES closures. Detailed study of the dense spray region in an atomizer flow field has been a subject of speculation, rather than systematic evaluation. *Schmidt*'s group at the University of Massachusetts (Chapter 5) investigated the initial spray breakup process. The formation of fuel drops was simulated based on first principles in order to offer detailed insight into primary atomization. The three-dimensional (3D), transient calculation tracked the interface evolution through droplet formation and breakup. This general approach will lead to better models for engine-design applications.

At the Imperial College of Science in London, *Whitelaw* and his researchers (Chapter 6) conducted detailed experiments with turbulent premixed opposed-jet flames relevant to modern gas-turbine technologies based on lean premixed combustion. Using photography and chemiluminescence imaging, they examined the effects of equivalence ratio, bulk flow velocity, and flow separation on flame extinction and relight dynamics in both unforced and acoustically forced opposed-jet flames.

It was shown that mixtures with similar flame speeds can have very different behaviors due to their level of susceptibility to stretch effects as characterized through the Markstein number by *Aldredge*'s group at the University of California, Davis (Chapter 7).

Strykowski and his coworkers at the University of Minnesota pursued further work on the countercurrent combustion concept proposed by them [2] that involved studies to evaluate the fundamental nature of a planar nonreacting turbulent countercurrent shear layer (Chapter 8). Experiments were also conducted using premixed, prevaporized JP-10 fuel in a dump combustor, modified to accommodate countercurrent flow control at the dump plane.

Mixing by active feedback control was investigated by *Krstić*'s group at the University of California, San Diego (Chapter 9). A two-dimensional (2D) jet and simple control strategies were considered and proposed that a small control effort can generate a large increase in turbulent kinetic energy of the jet flow.

Practical configuration, such as a swirl-stabilized spray combustor, and a multiswirl spray combustor, has been explored by *Gutmark*'s group at the University of Cincinnati (Chapter 10). In a joint effort (with General Electric and Goodrich), they experimented with a multiswirl spray combustor closely mimicking industrial design. Among others, PIV has been used as a diagnostic tool. Mapping the flow in different cross-sectional planes and along a centerline, streamwise plane in the triple annular swirling flow with co-swirling and counter-swirling cases uncovered numerous significant flow structures. *Grinstein* of the Naval Research Laboratory provided computational support to this effort (Chapter 11) using measurements for the time-averaged or steady flow field and instantaneous or unsteady flow field to benchmark LES.

Gupta and his coworkers at the University of Maryland obtained for the first time 3D PIV images of a flame and used these images to interpret airflow characteristics associated with the high-shear regime of the flow (Chapter 12). These features — flame plume and spray flame — are considered to manipulate passive flame control of swirl-stabilized combustion systems.

Combustion in porous media offers some advantages for propulsion devices, notably, wide range, low emissions, and good spatial temperature control. The study at Cranfield University by *Witton* and his coworkers (Chapter 13) focused on these characteristics at operating conditions of pressure and air-inlet temperature representing typical high-performance propulsion devices. A significant concern in these burners is the stabilization of the flame within the matrix. *Ellzey*'s group at the University of Texas at Austin (Chapter 14) provides computational support to *Witton*'s experiments. A potential design that consists of two sections of porous medium with different characteristics has been identified and examined.

Acharya and his group at the Louisiana State University are focused on active control of combustion processes in a 150-kilowatt swirl-stabilized spray combustor (Chapter 15). Effective model-based controllers with feedback are tested for reducing pressure oscillations and improving pattern factor in the combustor. Phase-locked CH measurements are presented to provide improved understanding of heat-release dynamics.

Ken Yu's group at the University of Maryland extended the previously reported work [3] to include four sets of experimental and analytical studies to identify key physical mechanisms for controlling mixing and combustion processes in ramjet and scramjet combustors (Chapter 16). In a supersonic mixing-enhancement experiment, the practicality of utilizing passive acoustic excitation as a means to enhance mixing in scramjets is assessed.

Lovett's team at Pratt & Whitney (Chapter 17) applied a fuel-control system in an actual aircraft engine combustor to actively control pattern factor. For this purpose, available fuel injectors were properly equipped with miniature valves to make possible spatial control of individual fuel injection sites. Optical temperature sensor probes and a traversing gas-sampling rake were integrated into a test rig to quantify the spatial exit temperature distribution in the combustor. Preliminary results have shown good ability to control pattern factor this way.

Experimental and analytical results for pulsed control of combustion instabilities at both fundamental and subharmonic frequencies were obtained by *Baumann*'s group at the Virginia Polytechnic Institute (Chapter 18). They have also developed two suites of control algorithms. One algorithm is based on least-mean-square (LMS) techniques that are suitable for inner-loop stabilization of combustion instabilities, and the other is based on direct optimization that can be used for either stabilization or outer loop optimization of combustion process objectives, such as flame compactness or emissions.

Active combustion control strategies included model-based, optimal control of liquid-fueled combustion systems (Chapter 19) by *Annaswamy* and *Ghoniem*'s group at the Massachusetts Institute of Technology. A recursive proper orthogonal decomposition algorithm for flow control problems was developed. It showed that an adaptive low-order, posi-cast control is able to effectively control a large-scale combustor model with satisfactory results. The test rig to demonstrate active pattern factor control was completed, and tests were run at higher combustion pressures for better characterization of the spray pattern.

HIGH-SPEED JET NOISE REDUCTION

Decades of research have been conducted by various universities worldwide and agencies, such as NASA, and aircraft engines and airframe companies to mitigate jet noise. As the flight velocities of the jet aircraft increase, combustion-generated wide-band noise of the exhaust jet becomes a concern. In a military aircraft with high flight speed and low bypass ratio, the jet noise is a major issue due to the very high velocities of the exhaust jet. Here, plume-generated noise becomes the predominant factor as compared to commercial aircraft, where noise contribution from the airframe can also be significant. This makes the noise control methodologies development over the past decades for commercial aircraft either not applicable or less effective to military jet aircraft. Environmental concerns and regulations call for a substantial reduction in jet noise, whereas the complexity of the combustion–acoustic interactions makes the problem very challenging. As a result, innovative methods of reducing jet noise are investigated.

A combined experimental and computational research program investigating the role of partial premixing and swirl on pollutant and noise emissions from aeropropulsion gas-turbine engines is being performed by *Frankel* and his coworkers at Purdue University (Chapter 1). The diagnostics include PIV, microphone arrays, and gas analyzers in multiswirl and trapped-vortex combustors.

Accurate measurements are a prerequisite for evaluating the noise reduction concepts developed. The research by *Atchley*'s group at the Pennsylvania State University (Chapter 2) is centered on a thorough study of sensors and techniques for making high-fidelity measurements of very-high-amplitude noise fields, such as might be found in the near-vicinity of high-performance, military jet aircraft.

Experiments have shown that high-speed jet noise can be suppressed by using microjects suitably injected to the jet [4]. At Florida State University, *Krothapalli*'s group (Chapter 3) recorded far-field acoustic measurements from a high-temperature (1033 K), supersonic ($M_j = 1.38$), axisymmetric jet issuing from a 50.8-millimeter converging nozzle and demonstrated the suppression of screech tones, Mach-wave radiation/crackle, and large-scale mixing noise by using microjets. While air can be used for microjets, water microjets are shown to have better noise reduction potential. Addition of certain polymers further reduces the noise levels. With total microjet mass flow of about 1% of the primary jet mass flux, it is projected that a 6-dBA reduction in the peak noise radiation direction can be obtained in a full-scale nozzle.

Due to environmental constraints and community noise issues, reduction of exhaust jet noise from military aircrafts, particularly during field take-off and landing practices, has become a major concern. *Seiner*'s research and development (R&D) work with his colleagues at the University of Mississippi focuses on the demonstration of practical noise-reduction strategies on full-scale jet engines. As an initial study, tests on one-tenth-scale model nozzles operating at realistic engine operating points on the Field/Carrier Landing Practice were performed (Chapter 4). Techniques jointly developed at *Krothapalli*'s laboratory are scaled up for this study. *Dash* and his associates at CRAFT, Inc. (Chapter 5) are utilizing Reynolds-Averaged Navier–Stokes (RANS)-based computational fluid dynamics (CFD) to support *Krothapalli*'s laboratory subscale experiments by establishing optimal geometries and conditions. They have also been investigating microjet injection and Bluebell tab/chevron nozzles (with various designs for divergent flap corrugations) in efforts so suppress jet noise and support *Seiner*'s work. Scale-up studies are also performed to evaluate these concepts on full-size aircraft engines.

ADVANCED PULSE DETONATION ENGINES

An engine that utilizes a more efficient thermodynamic cycle of operation that consumes less fuel and is simple and capable of operation at both subsonic and supersonic speeds would be an attractive alternative for future propulsion systems. Pulse Detonation Engines (PDE), in principle, can provide higher efficiency [5] and better performance over a wide range of operating conditions, with fewer moving parts. The following seven fundamental issues in the development of PDEs have been identified [6]:

1. Understanding the complex physical, chemical, and thermodynamic phenomena associated with liquid-phase injection, mixing, and ignition; those which influence rapid development of detonation waves; and the role of transverse waves in the detonation process.

2. Investigating efficient fuel injection and ignition.

3. Exploring methods of efficiently integrating PDE with mixed compression supersonic inlets and high-performance exhaust nozzles.

4. Understanding the dynamic coupling between multitube detonation chambers.

5. Developing complex diagnostics including semiconductor surface sensors and optical sensors, based on tunable laser diodes for sensing both gaseous and liquid characters.

6. Investigating adaptive, active control to ensure optimal performance while maintaining a margin of stability.

7. Performing mathematical analysis, advanced computational simulation, and modeling of detonation of multicomponent mixtures using real chemistry and molecular mixing.

Participation of a number of universities in the U.S. and foreign technical institutions has enabled significant progress in addressing these issues. Single and multicycle operations of PDEs with gaseous and liquid fuels have been demonstrated. The studies include from the fundamental understanding of the cellular structure of the detonation waves to the multiaxis thrust measurement from PDEs.

A fundamental study at Stanford University by *Edwards* and his coworkers (Chapter 1) in examining the effect of a two-phase mixture state on spray detonation characteristics has been initiated; preliminary results using hexane are presented. *Santoro* and his associates (Chapter 2) have performed a series of studies of the phenomena involved in the transition of a detonation for a geometry in which a significant area change occurs. A coaxial initiator geometry, currently being used on an integrated PDE system, is being characterized by both experimental and computational efforts at the Naval Postgraduate School by *Brophy*'s group (Chapter 3).

Theoretical and experimental investigations of control of the deflagration-to-detonation transition (DDT) processes in hydrocarbon–air gaseous mixtures relative to propulsion applications were performed by *Smirnov*'s group at Moscow State University (Chapter 4). At the University of Southern California, *Gundersen*'s group extended their transient plasma work to the PDE. The design and operation of a pulse generator using an advanced pseudospark device for corona or transient plasma-assisted flame ignition and combustion are presented in Chapter 5. The studies by *Segal*'s group at the University of Florida (Chapter 6) observed shock–droplet interactions for mixtures of JP-10 with ethyl-hexyl nitrate.

The idea of the use of a high-pressure jet produced by self-ignition or burning of a monopropellant in a small closed volume and injected in the main combustion chamber filled with air as a source of a high-intensity reactive shock wave in the chamber carrying a large impulse is tested both experimentally and by numerical modeling by *Borisov* at the Semenov Institute of Chemical Physics (Chapter 7). *Frolov* and his coworkers investigated three approaches to ascertain the total pressure and gas-phase composition in water–hydrogen peroxide (HP) two-phase systems depending on solution composition and temperature. Their results are presented in Chapter 8. *Anderson et al.* at the University of Utah (Chapter 9) used a microflow tube, guided-ion beam, quadruple mass spectrometer instrument to examine the pyrolysis chemistry of exotetrahydrodicyclopentadiene (exo-THDCP or JP-10) and related compounds, adamantane, cyclopentadiene (CPD), dicyclopentadiene, and benzene from room temperature up to > 1700 K in order to provide the fundamental chemical information. *Hanson*'s group at Stanford University (Chapter 10) developed diode-laser based sensors for *in situ* measurements of flow properties. These are used for simulation validation, fuel charge monitoring, active PDE control, and spray characterization. The sensors and instrumentation are portable and have been utilized in facilities at both the Naval Postgraduate School and Stanford University. *Kailasanath* and his coworkers (Chapter 11) assessed the impact of chemical recombination and detonation initiation conditions on the computed performance and estimated the theoretical performance of an ideal PDE through computational studies. *John Yu*'s group at Wayne State University (Chapter 12) performed a high-fidelity simulation of the direct initiation process of cylindrical detonation waves by concentrated energy deposition.

In order to evaluate the future potential of PDEs, verifiable system performance codes and design codes are required. To this effect, research at the Pennsylvania State University by *Yang*'s group is concerned with the system performance and thrust chamber dynamics of air-breathing PDEs. The details are presented in Chapter 13. Further, the new software for PDE configuration design and performance has been evaluated by *Cambier* and his associates at Montana State University (Chapter 14) for user-friendly application.

SUMMARY

The above chapters provide comprehensive information on state-of-the-art technology and research accomplishments in the areas of advanced propulsion with significantly higher efficiency and reduced emissions than ever accomplished before. Novel combustion control techniques have been found to be effective in reducing combustion instability and enhancing performance of liquid-fueled airbreathing propulsion engines. Innovative methodologies show promise in achieving the reduction of jet noise of supersonic aircrafts while improving their IR characteristics.

Detonation combustion phenomenon has been exploited in repetitive pulsed detonations in order to achieve uniform thrust for propulsion engines. Research on PDEs addressed all the important issues, and solutions have been presented. Further research is needed to bring the technology to fruition so that a next-generation propulsion engine will emerge.

The last few years have witnessed substantial progress in the air-breathing combustion field. An attempt is made to give as much comprehensive information as possible for practicing engineers, researchers, and graduate students in a single volume.

REFERENCES

1. Roy, G. D., ed. 2002. *Advances in chemical propulsion: Science to technology.* Boca Raton, FL: CRC Press.
2. Lonnes, S., D. Hofeldt, and P. Strykowski. 2002. Flame speed control using a countercurrent swirl combustor. In: *Advances in chemical propulsion: Science to technology.* Ed. G. D. Roy. Boca Raton, FL: CRC Press. 277–90.
3. Yu, K. H., and K. C. Schadow. 2002. Afterburning characteristics of passively excited supersonic plumes. In: *Advances in chemical propulsion: Science to technology.* Ed. G. D. Roy. Boca Raton, FL: CRC Press. 477–94.
4. Krothapalli, A. 2003. Supersonic jet suppression: A review and an extension. In: *Combustion and noise control.* Ed. G. D. Roy. Cranfield, UK: Cranfield University Press. 159–63.
5. Bussing, T., and G. Pappas. 1994. An introduction to pulse detonation engines. AIAA Paper No. 94-0263.
6. Roy, G. 2002. Chemical propulsion: What is in the horizon? In: *Advances in chemical propulsion: Science to technology.* Ed. G. D. Roy. Boca Raton, FL: CRC Press. 497–508.

SECTION ONE

CONTROL OF COMBUSTION PROCESSES

Chapter 1

SIMULTANEOUS VELOCITY AND TEMPERATURE FIELD MEASUREMENTS OF A JET FLAME

L. Lourenco and E. Koc-Alkislar

Simultaneous measurements of the velocity and temperature fields of an axisymmetric premixed jet flame are carried out using the Particle Image Velocimetry (PIV) and Laser Speckle Displacement (LSD) techniques. The configuration of the premixed propane–air jet flame includes a suction collar. In the proposed arrangement, a single Nd:Yag pulse laser provides the required laser sheet illumination for the PIV and is also used as the coherent light source for the recording of the Speckle patterns. The relative displacement of Speckle patterns is used as the means to evaluate the refractive index changes primarily caused by the temperature gradients in the reacting flow. Image recording is done by two high-resolution charge-coupled device (CCD) cameras: one camera dedicated to the PIV recording, i.e., the images generated by the scattering of seed particles, and the second camera for the simultaneous recording of the Speckle patterns. A common processing algorithm provides the displacement (velocity) of the seeds in the flow as well as the displacement of Speckle pattern. The Speckle displacements are in turn used as the input to a Poisson equation to yield the index values. An equation of state provides the required relationship between the refractive index change and the temperature.

1.1 INTRODUCTION

Whereas in incompressible, nonreacting flows the velocity field provides a fairly complete description of the flow physics, the knowledge of an additional state variable is usually required in compressible or reacting flows. The purpose of the present study is to integrate two complementary optical techniques, PIV and LSD, for complete diagnostics of reacting and/or compressible flow. Since the principle of independent operation for both techniques has been described in previous publications, here it will be described how the methods can be simultaneously applied.

1.2   TEST ARRANGEMENT AND RESULTS

As shown in Fig. 1.1, the beam of a single coherent pulsed laser source is split into unequal intensity beams. The high-intensity beam is shaped into a sheet for the PIV illumination, and the low-intensity beam is filtered, diverged, and collimated into a large-diameter beam that illuminates a region of the flame encompassing about 8 nozzle diameters. The large beam is focused onto a ground glass

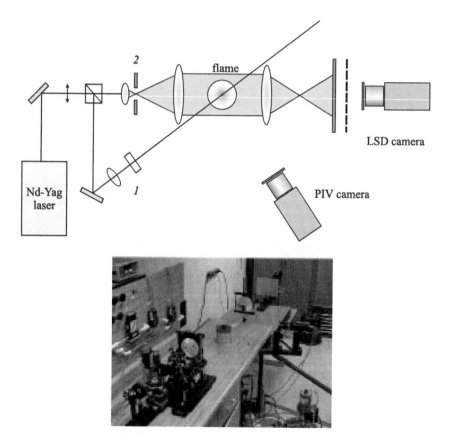

Figure 1.1 Setup for simultaneous PIV/LSD: *1* — laser sheet forming optics; and *2* — collimated beam forming optics.

to produce a Speckle pattern. The position of this pattern shifts as a function of the local beam deflections caused by the index of refraction changes. The laser sheet is a plane that contains the axis of the jet.

The jet flow is issued from the Jet/Counter-current facility at the Fluid Mechanics Research Laboratory at Florida State University (Fig. 1.2). To make the PIV measurement, the flame is seeded with 0.3-micrometer Al_3O_2 particles. To illustrate the use of the technique, the operating conditions for the jet when chosen showed a periodic vortex-ring shedding from the noz-

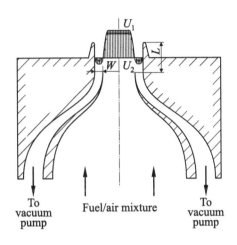

Figure 1.2 Jet flow setup.

zle with a very slight growth rate. The operating parameters for these conditions are: jet exit velocity of 3.7 m/s, equivalence ratio of 5.4, and no suction applied to the counter flow. These conditions produce a diffusion-like flame. Figure 1.3 shows the corresponding Speckle and PIV pattern images for a single realization. Both images depict the strong signature of the periodic vortex shedding present in the flow.

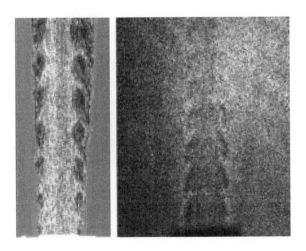

Figure 1.3 LSD (left) and PIV (right) patterns.

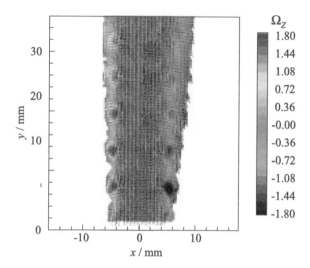

Figure 1.4 Flow field velocity color coded with the vorticity. (Refer color plate, p. I.)

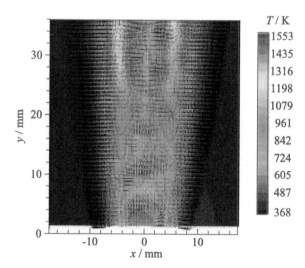

Figure 1.5 Temperature field for propane–air jet flame. (Refer color plate, p. I.)

The velocity field for a single realization presented in the reference frame moving with the convective velocity of the vortices is shown in Fig. 1.4. The vorticity distribution vividly depicts the vortex ring signature. This signature is also put in evidence in Fig. 1.5. This figure shows the Speckle displacement field superposed to the temperature.

The temperature is obtained by converting the refractive index values using the Gladstone–Dale law. The underlying assumption is that the behavior of the gas mixture can be well approximated by ideal gas. The temperature estimate is well within the expected range as the adiabatic flame temperature for a propane–air mixture flame is 1977 °C. This is the maximum temperature that a propane–air flame can reach under atmospheric pressure and for $\Phi = 1$.

1.3 CONCLUDING REMARKS

The simultaneous use of two nonintrusive diagnostic techniques is demonstrated in a reacting flow. This approach provides both the hydrodynamic and temperature fields from which correlations can be easily derived.

ACKNOWLEDGMENTS

This work was performed under ONR contract N00014-98-1-0424.

Comments

Santoro: The index of refraction depends on species, which is a seriously flawed assumption you have made. Also, axisymmetric problems are very sensitive to this approach.

Lourenco: The index of refraction is a much stronger function of temperature than it is of species for the type of chemical species we are dealing with here. Axisymmetry is not a problem and, in fact, is an excellent test for this approach.

Santoro: Our 20 years of experience suggest that this will be a difficult approach.

Lourenco: Our success here indicates that the measurements are indeed accurate and this diagnostic tool is viable.

Whitelaw: It would appear that a complete uncertainty study should be performed before the approach can be used in practical flows.

Lourenco: My experience in these flows is considerable and cannot be summarized in a 15-minute talk. I assure you that I have given this approach serious attention.

Chapter 2

INFRARED ABSORPTION TOMOGRAPHY FOR ACTIVE COMBUSTION CONTROL

F. C. Gouldin and J. L. Edwards

The goals of the research are to develop infrared (IR) absorption tomography as a sensor for closed-loop control of combustion and propulsion systems of interest to the Navy, to apply IR absorption tomography to study mixing in forced-jet actuators, and to use proper orthogonal decomposition (POD) analysis to investigate large-scale structures in reacting jet flows and the potential for reduced-order models of jet actuators. This research is in collaboration with Dr. Grinstein and Dr. Kailasanath of the Naval Research Laboratory (NRL). Over the past year, the authors have completed the analysis of tomographic CO_2 measurements made on forced jets; refined the tomographic reconstruction technique, adaptive Finite Domain Direct Inversion (FDDI); and used POD analysis both to evaluate the potential for reduced-order models of the vorticity and concentration fields and to investigate the relationship between vorticity and scalar mixing in reacting rectangular jets.

2.1 INTRODUCTION

The overall goal of the present research is to develop and apply multiple line-of-sight (LOS) IR absorption tomography for problems of combustion control in naval propulsion systems such as ramjets and turbojets. For IR absorption tomography, the attenuation of radiation along multiple LOSs in a common plane is measured, and the data are used to reconstruct the spatial distribution in the measurement plane of the radiation-absorbing chemical species, e.g., CO_2, CO, and NO. In addition, the authors have collaborated with Dr. Grinstein and Dr. Kailasanath in the study of forced-jet actuators and the use of POD as a tool to characterize large-scale structures in these flows and to evaluate the potential of reduced-order modeling.

9

Many aspects of performance such as efficiency, signature, and pollutant emissions depend in whole or in part on mixing processes. These processes depend on the flow configuration and large-scale flow structures, macromixing, molecular transport, and micromixing. Tomography provides data on the state of macromixing.

For practical reasons, the number of LOS measurements is finite, and the tomographic reconstruction problem is ill-posed. Two reconstruction methods have been developed for cases where optical access is restricted, and the number of measurement LOSs is limited. One method, adaptive FDDI, requires 100 or more LOSs [1–3], while the other method, Tomographic Reconstruction via a Karhunen–Loéve Basis, requires far fewer [4, 5]. Because it requires very few LOSs, the authors believe that this latter method has potential for use in sensing for feedback control of combustion systems where optical access is limited; however, it requires considerable *a priori* information in the form of a set of expected distributions, the training set. This set is analyzed via POD to yield a set of basis functions, the Karhunen–Loéve eigenfunctions, that are used for reconstruction. These training sets could come from measurements on prototype equipment or from computational combustion simulations.

In the past year, the authors have analyzed CO_2 concentration measurements of a square, forced jet collected using the IR laser absorption facility. The absorption facility incorporates a tunable color center laser system and six scanning modules, allowing for measurements to be made along multiple LOSs on a millisecond time scale. Tomographic reconstruction of the concentration field is performed using adaptive FDDI. Techniques for refinement of adaptive FDDI are being evaluated. Proper orthogonal decomposition analyses of reacting, rectangular jets have been performed on a set of CO_2 and vorticity magnitude distributions calculated at the NRL by Dr. Grinstein and Dr. Kailasanath, and the results have been analyzed. In addition, the design and fabrication of linear array detectors are underway for chemiluminescence emission tomography, and collaboration with Pratt & Whitney on a project to design the optical configuration for an IR-laser-based tomographic pattern-factor sensor is underway.

2.2   ABSORPTION TOMOGRAPHY

In absorption tomography, laser-beam transmission measurements are made along many LOSs distributed over several viewing angles, θ. For current measurements, LOSs sharing a common viewing angle are parallel to each other and are defined by their offset, s, Fig. 2.1. For isothermal conditions, line integrals of number density $n(x, y)$ over an LOS can be related to the laser-beam transmission:

$$p(s,\theta) = -\ln\left(\frac{I_\nu^t}{I_\nu^0}\right) = \sigma \int\limits_{-T}^{T} n(x,y)\, dt \qquad (2.1)$$

The projection $p(s,\theta)$ is the line integral of $n(x,y)$ along the LOS, defined by s and θ, times the absorption cross-section σ at the laser source line frequency ν; $\pm T$ are the limits of integration. It can be related to the initial (I_ν^0) and transmitted (I_ν^t) laser-beam intensities as shown.

The practical reconstruction problem is to solve Eq. (2.1) for $n(x,y)$ in cases where $p(s,\theta)$ is known at a discrete set of s and θ. Because the reconstruction problem is ill-posed, the solution or reconstruction method must be tuned to the problem at hand as defined by the character of $n(x,y)$ and the number and distribution over s and θ of the available projections $p(s_i,\theta_j)$.

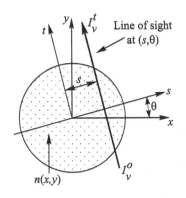

Figure 2.1 Schematic diagram of LOSs in single view and of coordinate system.

As noted, two reconstruction methods have been developed, both of which are series expansion methods. In the case of adaptive FDDI, n is expanded in a set of basis functions, b_k, with arguments $x - x_k$ and $y - y_k$, where x_k, y_k define the points of a triangular grid; $n = \sum f_k b_k$. In the original FDDI, there were 97 basis functions and, hence, 97 unknown weighting factors to be found by solution of Eq. (2.1). For adaptive FDDI, additional basis functions are added to the expansion at grid points located in regions of high gradients determined from the original FDDI solution. In the second reconstruction method, the basis functions are Karhunen–Loéve eigenfunctions of the training set, and in many cases relatively few basis functions are needed for a good representation of the distributions to be reconstructed.

2.3   INFRARED ABSORPTION AND FLOW FACILITY

An IR absorption facility for making LOS measurements on confined and unconfined, reacting and nonreacting flows has been constructed. The facility consists of an Nd:YVO$_4$-pumped, broadly tunable, KCl:Li color center laser (CCL) system and an optical apparatus composed of six scanning modules that permit simultaneous measurement of LOS absorption in a single plane at six different viewing angles, θ_j.

A schematic of the facility is shown in Fig. 2.2. The CCL beam is electro-optically modulated for phase sensitive detection at a frequency of approximately 1 MHz. Each detector is connected to a demodulating "lock-in" circuit which beats the 1-megahertz transmitted beam with a coherent reference signal at 1 MHz to generate a signal that represents the transmitted laser intensity as a function of time. The demodulated detector output is low pass filtered at 50 kHz and read and processed by a fast PC-based data acquisition system. For each scanning module, the time history of the sweep signal is used to determine the transmission of the beam along selected LOSs. It takes approximately 1.5 ms to complete a sweep over the flow region of interest. Measurements have been made on a forced, square jet (8 × 8 mm) with flows of air seeded with CO_2. Gas flows are supplied from high-pressure tanks and controlled by electronic mass-flow controllers; total flow velocities up to 50 m/s are possible. A speaker that is driven by an audio power amplifier provides jet forcing at 30 Hz.

2.4 PROPER ORTHOGONAL DECOMPOSITION

Detection and control of the large-scale features in combustion systems are of primary importance for mixing control and combustion control in general. Reduced-order models for representing the dynamics of a combustion system and large-scale structures are crucial to the development of combustion control. Proper orthogonal decomposition is one method used to develop reduced-order models and to investigate large-scale structures in combusting flows.

For POD, the Karhunen–Loéve procedure decomposes a set of distributions or functions into an optimal, orthonormal set of eigenfunctions able to represent the distributions of interest [4–6]. These eigenfunctions can offer highly efficient representations of important variables and structures in combustion systems. Furthermore, the potential for a reduced-order model utilizing these basis functions can be evaluated from the associated eigenvalue spectrum derived from decomposition.

The snapshot method [7] has been used to obtain eigenfunctions and eigenvalues from CO_2 concentration and vorticity magnitude distribution sets of unsteady, reacting flows. These sets are from numerical computations using Large-Eddy Simulation of turbulent reacting, forced rectangular jets of varying aspect ratios (AR = 1–3) that were performed at the NRL [6, 8, 9]. The simulated (initially laminar) propane–nitrogen jets are issued into a quiescent oxygen–nitrogen background with a Mach number of 0.3 and a Reynolds number greater than 85,000 based on the circular equivalent jet diameter. The jet exit velocity is

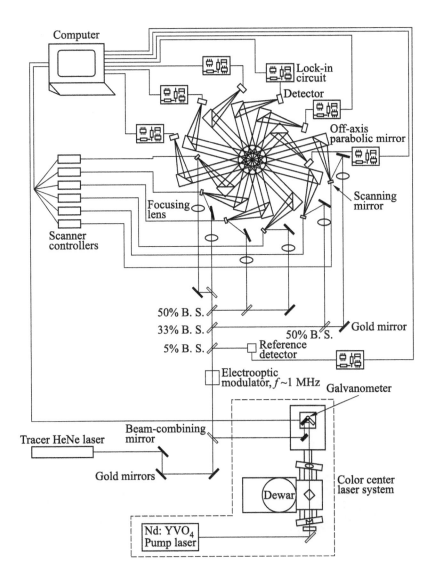

Figure 2.2 Schematic diagram of the IR absorption facility; B.S. — beam splitter.

forced axially by superimposing a single-frequency sinusoidal perturbation to the jet velocity U_0, having a root-mean-squared (rms) level of 2.5% and a Strouhal frequency $St = fD_e/U_0 = 0.5$, where D_e is the circular-equivalent diameter. Multispecies temperature-dependent diffusion and thermal conduction processes are calculated explicitly using central difference approximations and are coupled to chemical kinetics and convection using time step-splitting techniques. A global (single-step irreversible) model for propane chemistry is used [10]. Further details on the simulated transport properties and their validation are discussed in [8, 9] and references therein.

2.5   RESULTS

Square Jet Reconstructions

Inversion of projection data from module absorption measurements was performed using adaptive FDDI. Figure 2.3a shows a reconstruction of experimental CO_2 concentration data from a forced square jet. The structure seen at the top of the peak in Fig. 2.3a is due in part to the shape of the FDDI basis functions and the number of basis functions used in the reconstruction. Refinement of adaptive FDDI is being investigated.

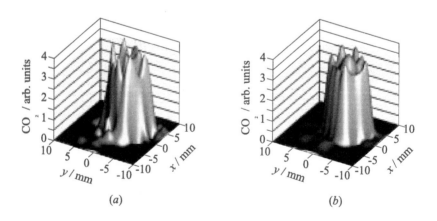

(a)   (b)

Figure 2.3 (a) Adaptive FDDI reconstruction of forced-jet CO_2 concentration distribution from a 3 m/s jet with 25% CO_2; and (b) refined adaptive FDDI reconstruction of the data from (a). Forcing phase angle is 90°.

Refinement of FDDI

The majority of the jet flows the authors reconstructed were square top-hat profiles with steep, smooth sides and relatively flat tops. These features are difficult to accurately reconstruct with tomographic reconstruction methods. Currently, strategies are under investigation for the refinement of the adaptive FDDI method and ways to improve the reconstruction of these features. In the original implementation of adaptive FDDI, 147 basis functions of the same shape were used. These basis functions were sharp, narrow, axisymmetric peaks derived from the Kaiser window function and designed for use in a densely packed grid. A possible refinement of adaptive FDDI can be made by using basis functions of two different shapes, both derived from the Kaiser window function. In Refined Adaptive FDDI (RAFDDI), a specified number of high-aspect ratio basis functions are used in the high-gradient regions, including but not limited to the location of the additional 50 basis functions used in the reconstruction. Low-aspect ratio basis functions are used in regions of lower concentration gradient. The results of performing RAFDDI on the data used to generate Fig. 2.3a are shown in Fig. 2.3b.

Another method of refinement under consideration is the inclusion of a global square top-hat basis function in the FDDI basis set. The size, location, and orientation of the square basis function in the measurement domain are extremely important. Work with phantom data has illustrated the sensitivity of the FDDI reconstruction to changes in the relative positions and orientations of the scalar field being reconstructed and the basis function (Fig. 2.4).

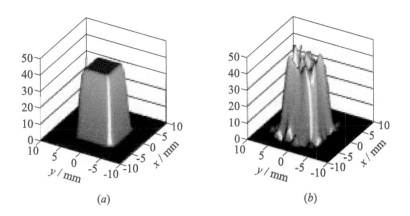

(a)   (b)

Figure 2.4 (a) Adaptive FDDI reconstruction of centered phantom jet incorporating a centered square basis; and (b) reconstruction of a phantom square jet with the center shifted 1 mm in the positive y direction incorporating a centered square basis function.

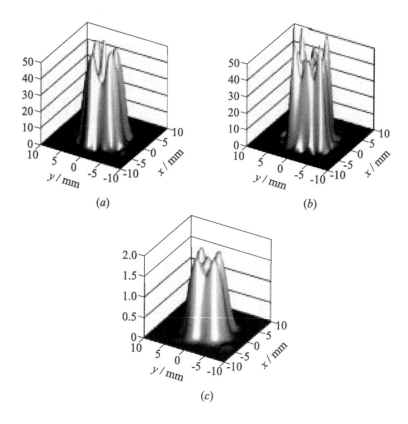

Figure 2.5 (*a*) Adaptive FDDI reconstruction of phantom square jet data; (*b*) adaptive FDDI reconstruction utilizing the new global basis function in (*c*), notice that while the structure on top of the peak has become more complex, the sides more closely approximate that of a square top-hat; and (*c*) normalized global basis function constructed from smoothing the adaptive FDDI reconstruction in (*a*).

A third refinement method includes the development of a global basis function specific to each reconstruction that could be added to the current basis function set. This global basis function can be developed based on preliminary results of an adaptive FDDI reconstruction of the scalar field and included in a subsequent implementation of adaptive FDDI (Fig. 2.5).

Proper Orthogonal Results

Proper orthogonal decomposition has been used to analyze ensembles of concentration and vorticity magnitude distributions in forced reacting rectangular jets

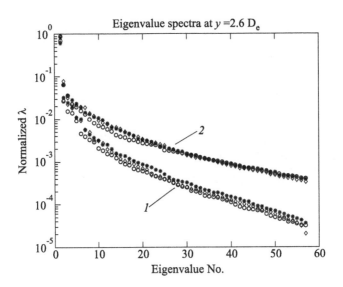

Figure 2.6 Eigenvalue spectra of CO_2 (*1*) and vorticity magnitude (*2*) at $y = 2.6D_e$. The faster decrease of CO_2 eigenvalues than vorticity eigenvalues indicates better potential for reduced-order modeling of the CO_2 field.

of aspect ratios 1 to 3 [6]. Ensembles were constructed at various downstream locations using CO_2 concentration and vorticity data and contained 60 and 63 ensemble members, respectively. Through evaluation of the eigenvalue spectra (Fig. 2.6), it was determined that there is stronger potential for reduced-order modeling based on the CO_2 field than based on the vorticity field, but this potential is approximately the same for the different rectangular jets studied — aspect ratios 1, 2, and 3. A physical insight has been gained through examination of the resulting eigenfunctions, specifically relating to vortex structures and the development of the CO_2 concentration field through mixing and spreading at locations downstream from the jet exit. Axis-switching phenomena and the influence of vortex dynamics on mixing were also evident in the POD eigenfunctions.

Sample POD eigenfunctions are presented in Fig. 2.7. Note that the CO_2 and vorticity eigenfunctions for each jet are closely related in shape and spatial extent, indicating the strong influence of vorticity on mixing.

Chemiluminescence

Linear array equipped 35-millimeter cameras are being designed and built for chemiluminescent emission tomography studies of heat-release dynamics. The

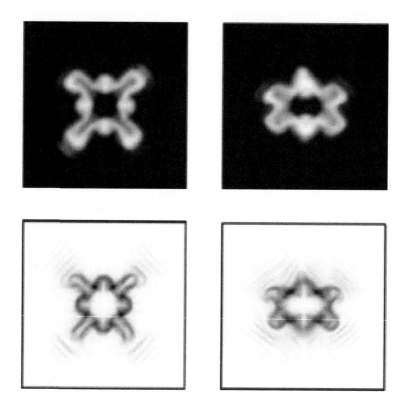

Figure 2.7 First eigenfunctions of CO_2 concentration (top) and vorticity magnitude (bottom) for AR = 1 (left) and AR = 2 (right) jets; $y = 2.6D_e$.

intensity of chemiluminescent emissions is known to be closely coupled to combustion heat-release rates, and for this reason chemiluminescent emission measurements are widely performed. It is planned to use chemiluminescent emission tomography to measure the spatially resolved power spectral density of these emissions as signatures of local heat-release fluctuation dynamics. To perform these measurements, 35-element array detectors will be mounted in the focal plane of six 35-millimeter cameras. After amplification and filtering, the output of each element will be stored on a PC for later data analysis. This analysis will include Fast Fourier Transform (FFT) and spectral analyses of the element signals and reconstruction of interesting frequency components of the FFTs. By mounting the detectors in 35-millimeter cameras, a measurement system that is highly portable and easy to use will become available. Design and fabrication of a prototype camera is currently underway.

2.6   CONCLUDING REMARKS

The current research on tomographic techniques and forced-jet studies has progressed significantly over the past few years. The previously developed adaptive FDDI method is used to determine two-dimensional concentration distributions from the one-dimensional projection data acquired by the absorption tomography facility. Strategies for the refinement of adaptive FDDI are being evaluated and implemented to better reconstruct the top-hat profiles of the square jet. Proper orthogonal decomposition analysis has been completed on computational combustion simulations of forced, rectangular jets with aspect ratios of 1, 2, and 3 performed at NRL.

Results of the analysis indicate that the CO_2 concentration profiles, and to a lesser extent the vorticity profiles, can be well represented by a reduced-order model. This result shows promise for implementing a reduced-order model for the development of a combustion control system.

Finally, linear array cameras for chemiluminescence emission tomography measurements will be used.

REFERENCES

1. Ravichandran, M., and F. C. Gouldin. 1988. Reconstruction of smooth distributions from a limited number of projections. *Applied Optics* 27:4084.

2. Ha, J., M. Feng, and F. C. Gouldin. 1999. Laser tomographic reconstruction in a complex concentration flow field. AIAA Paper No. 99-0444.

3. Feng, M., and F. C. Gouldin. 2000. Experimental evaluation of an adaptive tomographic inversion method. AIAA Paper No. 2000-0949.

4. Torniainen, E. D., A. Hinz, and F. C. Gouldin. 1998. Tomographic analysis of unsteady, reacting flows. *AIAA J.* 36:1270–78.

5. Torniainen, E. D. 2000. Ph.D. Thesis. Cornell University, Ithaca, NY.

6. Edwards, J., F. Gouldin, F. Grinstein, and K. Kailasanath. 2002. Reduced order structure of reacting rectangular jets. AIAA Paper No. 2002-1011.

7. Sirovich, L., and R. Everson. 1992. Management and analysis of large scientific data sets. *Int. J. Supercomputer Applications* 6:50.

8. Grinstein, F. F., and K. Kailasanath. 1996. Exothermicity and three-dimensional effects in unsteady propane square jets. *26th Symposium (International) on Combustion Proceedings*. Pittsburgh, PA: The Combustion Institute. 91–96.

9. Grinstein, F. F. 2001. Vortex dynamics and entrainment in rectangular free jets. *J. Fluid Mechanics* 437:69–101.

10. Westbrook, C. K., and F. L. Dryer. 1981. Chemical kinetic modeling of hydrocarbon combustion. *Combustion Science Technology* 27:31–43.

Chapter 3

DETERMINISTIC AND PROBABILISTIC APPROACHES FOR PREDICTION OF TWO-PHASE TURBULENT FLOW IN LIQUID-FUEL COMBUSTORS

G. B. Jacobs, R. V. R. Pandya, B. Shotorban, Z. Gao, and F. Mashayek

The "deterministic" approach of direct numerical simulation (DNS) and the "probabilistic" approach of probability density function (PDF) modeling are implemented for prediction of droplet dispersion and polydispersity in liquid-fuel combustors. For DNS, a multidomain spectral element method was used for the carrier phase while tracking the droplets individually in a Lagrangian frame. The geometry considered here is a backward-facing step flow with and without a countercurrent shear. In PDF modeling, the extension of previous work to the case of evaporating droplets is discussed.

3.1 INTRODUCTION

The need for optimized and stable operation under various conditions has promoted a growing interest in control of combustion in systems involving liquid fuels. Various control strategies for complex liquid-fuel combustors are usually devised via making several simplifying assumptions and following long and tedious mathematical procedures. As a result, the proposed control schemes must undergo a number of stringent tests under various combustor operation conditions before they can be implemented in practice. The standard procedure until recently has been to implement the proposed strategies in "test combustors" that could duplicate a simplified model of the real combustor. The main difficulty with this procedure is that very often new and complex sensors and actuators must be developed for the purpose of such experiments. Furthermore, to monitor

the performance of test combustors one requires advanced diagnostics which are
still limited in providing detailed information on turbulence, two-phase dynamics, and combustion.

With the rapid growth in the speed and memory of computers in recent
years, computational fluid dynamics (CFD) has emerged as a viable means that
may be used to eliminate some of the expensive steps in the process of development and assessment of control strategies for liquid-fuel combustors. Although
CFD would by no means completely eliminate the need for laboratory experiments, it could be used to significantly reduce the extent of such expensive
practices. Further, the detailed flow information generated by CFD could provide valuable insight that is crucial in the process of development of various
control schemes. In this chapter, recent advances in implementation of the "deterministic" approach of DNS and the "probabilistic" approach of PDF modeling
are discussed.

3.2   DIRECT NUMERICAL SIMULATION
OF COUNTERCURRENT SHEAR FLOW

In previous papers, the authors have reported on progress in various steps in the
development of an advanced multidomain spectral element code for simulations
of turbulent two-phase flows [1, 2]. This code was also applied for simulation of
the flow over a square cylinder and the channel flow for preliminary assessments.

The major task undertaken as a part of the code development efforts during
the past year was implementation of the Message Passing Interface (MPI) such
that the code could be run on parallel computers. This is a very important step,
as the intensity of the planned simulations makes it impossible to generate the
results using only a single processor.

The MPI implementation was performed according to a method that Fischer [3] used in his incompressible Legendre hp-spectral method. The method is
based on the use of three arrays that keep track of the global/local element and
mortar numbering, the processor on which the local element and mortars are
allocated, and the face side of the slave element of a mortar. The mortars are
allocated at the same processor in which the master domain is allocated. The
MPI code was tested for a laminar backward-facing step flow on the local 8-node
(16-processor) Beowulf cluster and showed an efficiency of 90% for large grids
(Fig. 3.1).

More recently, the application of the code for simulation of the flow over a
backward-facing step with a countercurrent flow has been considered. The step
is bounded by two parallel plates at the top and the bottom and resembles the
dump combustor configuration considered by Strykowski's group at the Univer-

sity of Minnesota (UMN) in Minneapolis, MN as part of its involvement in this research (Fig. 3.2).

The UMN group has conducted extensive experimental studies in the geometry shown in Fig. 3.2. This novel design for dump combustor employs countercurrent shear to increase the heat release and controllability of the combustion. Thus far the UMN experiments have been limited to single-phase flows using Particle Image Velocimetry (PIV) [4]. It has been shown that through a relatively small suction flow rate, a large effect on large-scale coherent structures close to the step can be established, which in turn could lead to larger heat releases, without the loss of flame stability due to the increased

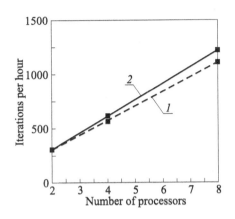

Figure 3.1 Actual (*1*) and linear (*2*) trends of iterations per hour plotted vs. the number of processors for the local 8-node dual PIII/750 MHz-processor Beowulf cluster.

surface of the instantaneous flame sheet. The results have focused on flow visualization and measurements of turbulence kinetic energy for various suction mass flow rates, suction gaps, and trailing edge thicknesses.

The DNS study of the dump combustor will shed light on the details of the flow both with and without the presence of droplets. Three issues will be addressed. First, it is planned to investigate the flow in the dump combustor with and without a countercurrent shear flow. The study will be focused on low-Mach-number flows, as was the case in previous homogeneous simulations, so that singularities such as shocklets are avoided. In analyzing the results, special attention will be paid to flow features that cannot be captured by PIV.

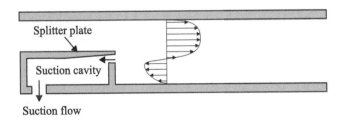

Figure 3.2 Schematic of a dump combustor with countercurrent shear.

The most prominent of these features is the temporal evolution of the flow. Employing PIV, it is not possible to study this feature directly, since PIV takes snapshots of the flow at random times, whereas in the case of DNS the equations are integrated in time. The importance of this study to the dump combustor is that an efficient combustion is highly sensitive to temporal evolution of the fluid structures. For example, the injection frequency can significantly change combustion characteristics.

Second, some of the issues pertaining to the small (Kolmogorov) scales of the flow can be addressed. The PIV measurements are spatially filtered, and a correct estimate of the Kolmogorov scale cannot be obtained directly from the experimental setup. Direct numerical simulation does not have such a restriction since for an accurate simulation of turbulent flows the smallest scales have to be resolved, i.e., the grid has to be fine enough to capture the smallest scales. The knowledge of the behavior of the small scales is of interest since dissipation processes that directly affect the chemistry take place at these scales.

Third, the study is capable of focusing on the inlet flow conditions since these are more difficult to control and determine in experimental setups as opposed to DNS. The inlet flow conditions strongly affect the mixing characteristics behind the step, which in turn influence the volumetric heat release by combustion.

Another advantage of DNS is that it is an excellent source for a rich database of various statistics which can be used for model validation. The detailed information that one can obtain with DNS of the two-phase flow is virtually impossible to generate with experiment and will greatly help in understanding the physics involved in two-phase flows as well as validation of relevant models.

To elucidate the capabilities of DNS in capturing the details of the flow, the results from a two-dimensional simulation are presented. This investigation should provide a reasonable picture of the influence of countercurrent shear on two-dimensional coherent structures behind the step. Figure 3.3 shows the results for the flow with and without countercurrent shear. The strength of the countercurrent shear flow is set to 3%. The main effect of the suction is to destabilize the shear layer emanating from the step, causing it to burst further upstream than for the nonsuction case and leading to large vortical structures immediately after the step. This causes the reattachment length of the flow to decrease and the turbulence intensity right after the step to increase significantly. Another major characteristic of the suction case is the presence of a large vortex right after the step having clockwise rotation. This vortex grows in size until it sheds off quite abruptly. Then a new vortex forms/grows and the process repeats itself periodically.

The Stokes number of the dispersed phase was set to unity in these simulations. This results in a preferential distribution of the droplets as seen in Fig. 3.3 — the particles are preferentially concentrated around the periphery

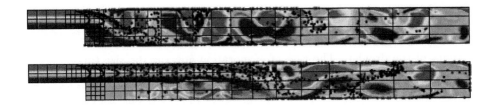

Figure 3.3 Two-phase flow over a backward-facing step with (top) and without (bottom) countercurrent shear. The carrier phase is visualized by vorticity contours.

of the vortices*. In the presence of the countercurrent shear, large vortical structures are observed behind the step, which cause the droplets to disperse after the step as opposed to the nonsuction case where the droplets disperse farther away from the step. In case of suction, the droplets are distributed around the large vortex behind the step and travel with the vortex when it sheds off.

3.3 PROBABILITY DENSITY FUNCTION MODELING

The PDF approach has its base in the study of kinetic theory of gases [5] and Brownian motion [6], and various existing PDF models have been reviewed recently by Minier and Peirano [7]. These models are categorized as one- and two-point PDF models. In the one-point PDF models, an equation governing the probability density of variables of interests of particle and fluid phase along the particle path is derived. These variables are known as the state vector, and various one-point PDF models differ in the selection of the variables of the state vector [8, 9]. The closed kinetic equations for isothermal flows were derived by Reeks [10], Hyland et al. [11], Pozorski and Minier [8], and Derevich [12]. The cases of nonisothermal flows were considered by Zaichik [13] and Pandya and Mashayek [14, 15]. In all these studies, particles were considered as nonevaporating. Derivation of the kinetic equation for nonisothermal flows with evaporating droplets is the subject of this chapter.

---

*Earlier, similar results were obtained by K. Kailasanath and E. Chang (1997. Numerical simulation of multiphase reacting flows. In: *Advanced computation and analysis of combustion*. Eds. G. Roy, S. Frolov, and P. Givi. Moscow, Russia: ENAS Publ. 122–35). (*Editor's remark.*)

Consider a two-phase nonisothermal turbulent flow in which droplets move under the influence of fluid drag force and their temperature, T_d, changes due to evaporation and the thermal interaction (driven by the temperature difference, $T - T_d$) with the carrier fluid. Here, T is the temperature of the fluid in the vicinity of the droplet. The rate of evaporation governs the size (diameter) of the droplets. A variety of equilibrium and nonequilibrium evaporation models available in the literature were recently evaluated by Miller et al. [16]. Here, the model which was used in the previous DNS work is selected [17]. The Lagrangian equations governing the time variation of the position $\mathbf{X}$, velocity $\mathbf{V}$, temperature T_d, and diameter d_d of the droplet at time t can be written as

$$\frac{dX_i}{dt} = V_i \tag{3.1}$$

$$\frac{dV_i}{dt} = \frac{f_1}{\tau_d}\left(U_i^* - V_i\right) \tag{3.2}$$

$$\frac{dT_d}{dt} = \frac{f_2}{3\mathrm{Pr}\sigma\tau_d}\left(T^* - T_d\right) - \frac{\rho^*\lambda f_3}{3\mathrm{Sc}\sigma\tau_d}\left(Y_s - Y^*\right) \tag{3.3}$$

$$\frac{dd_d}{dt} = -\left(\frac{18}{\rho_d}\right)^{0.5}\frac{2f_3\rho^*\tau_d^{0.5}}{\rho_d\mathrm{Re}_f^{1.5}\mathrm{Sc}d_d^2}\left(Y_s - Y^*\right) \tag{3.4}$$

with

$$\tau_d = \frac{\mathrm{Re}_f\rho_d d_d^2}{18}$$

$$Y_s = \exp\left[\frac{\gamma\lambda}{(\gamma-1)T_B}\left(1 - \frac{T_B}{T_d}\right)\right]$$

Here, all the variables are normalized by reference length, L_f, density, ρ_f, velocity, U_f, and temperature, T_f. Also, $\mathrm{Re}_f = \rho_f U_f L_f/\mu$, T_B is the boiling temperature of the droplet; $\sigma = C_l/C_p$ is the ratio of specific heats of liquid, C_l, and carrier-phase, C_p; Y_s is the vapor mass fraction at the surface of the droplet; Y^* is the vapor mass fraction in the vicinity of the droplet; Pr is the Prandtl number; Sc is the Schmidt number; $\lambda = L_v/C_p T_f$, L_v is the latent heat of evaporation; ρ is the carrier-phase density; ρ_d is the droplet density; γ is the ratio of the specific heats of the carrier gas; and the superscript $*$ indicates the value of a carrier-phase variable at the droplet location. Although the coefficients f_1, f_2, and f_3 appearing in Eqs. (3.2) and (3.3) are functions of $\mathrm{Re}_d = \mathrm{Re}_f\rho^* d_d|\mathbf{U}^* - \mathbf{V}|$, they are considered as constants ($f_1 = 1$, $f_2 = 2$, and $f_3 = 2$) during the derivation of the kinetic equation. This is justified when $\mathrm{Re}_d < 1$. The ensemble average of the Liouville equation for the phase-space density $W(\mathbf{x}, \mathbf{v}, \theta, d, t)$ can be written, using the Lagrangian equations, in the form:

$$\frac{\partial \langle W \rangle}{\partial t} + \frac{\partial}{\partial x_i}\left[v_i \langle W \rangle\right] - \frac{\partial}{\partial v_i}\left[\frac{(v_i - \langle u_i^* \rangle)\langle W \rangle}{\tau_d}\right]$$

$$- \frac{\partial}{\partial \theta}\left[\left(\frac{2(\theta - \langle T^* \rangle)}{3\mathrm{Pr}\sigma\tau_d} + \frac{2\rho^*\lambda(Y_s - \langle Y^* \rangle)}{3\mathrm{Sc}\sigma\tau_d}\right)\langle W \rangle\right]$$

$$- \frac{\partial}{\partial d}\left[\left(\frac{18}{\rho_d}\right)^{0.5}\frac{4\rho^*\tau_d^{0.5}}{\mathrm{Re}_f^{1.5}\mathrm{Sc}\rho_d d^2}(Y_s - \langle Y^* \rangle)\langle W \rangle\right]$$

$$= -\frac{\partial}{\partial v_i}\left[\frac{\langle u_i' W \rangle}{\tau_d}\right] - \frac{\partial}{\partial \theta}\left[\frac{2\langle t' W \rangle}{3\mathrm{Pr}\sigma\tau_d} + \frac{2\rho^*\lambda}{3\mathrm{Sc}\sigma\tau_d}\langle y' W \rangle\right]$$

$$- \frac{\partial}{\partial d}\left[\left(\frac{18}{\rho_d}\right)^{0.5}\frac{4\rho^*\tau_d^{0.5}}{\mathrm{Re}_f^{1.5}\mathrm{Sc}\rho_d d^2}\langle y' W \rangle\right] \qquad (3.5)$$

Here, $\mathbf{x}$, $\mathbf{v}$, θ, and d are phase-space variables corresponding to $\mathbf{X}$, $\mathbf{V}$, T_d, and d_d, respectively. This equation poses closure problems due to the presence of unknown correlations $\langle u_i' W \rangle$, $\langle t' W \rangle$, and $\langle y' W \rangle$. These types of problems appearing in the kinetic approach for the description of dispersed phase are accessible by methods such as Kraichnan's Lagrangian history direct interaction (LHDI), the functional formalism, and Van Kampen's method. Herein, the Furutsu–Novikov–Donsker formula in the functional framework [11, 14] is used to solve the present closure problems and to derive the closed expressions

$$\langle u_i' W \rangle = -\left[\frac{\partial}{\partial x_k}\lambda_{ki} + \frac{\partial}{\partial v_k}\mu_{ki} + \frac{\partial}{\partial \theta}\omega_i + \frac{\partial}{\partial d}\eta_i - \gamma_i\right]\langle W \rangle \qquad (3.6)$$

$$\langle t' W \rangle = -\left[\frac{\partial}{\partial x_k}\Lambda_k + \frac{\partial}{\partial v_k}\Pi_k + \frac{\partial}{\partial \theta}\Omega + \frac{\partial}{\partial d}\Upsilon - \Gamma\right]\langle W \rangle \qquad (3.7)$$

$$\langle y' W \rangle = -\left[\frac{\partial}{\partial x_k}\xi_k + \frac{\partial}{\partial v_k}\zeta_k + \frac{\partial}{\partial \theta}\varpi + \frac{\partial}{\partial d}\vartheta - \varsigma\right]\langle W \rangle \qquad (3.8)$$

Various tensors λ_{ki}, μ_{ki}, ω_i, η_i, γ_i, Λ_k, ... are obtained in terms of second-order correlations of u_i', t', and y' and various functional derivatives of V_k, X_k, T_d, and d_d with respect to u_i', t', and y'. These functional derivatives are obtained from the Lagrangian equations governing the particle trajectory, temperature, and diameter. Substituting Eqs. (3.6)–(3.8) into Eq. (3.5) and then taking the various moments give the macroscopic equations for droplets in inhomogeneous flows. These equations govern the statistical properties, such as mean number density, $\overline{N}$, droplet diameter, $\overline{d}$, droplet Reynolds stress, $\overline{v_i' v_j'}$, etc. These equations can be simplified in the case of homogeneous shear flows and computed for the purpose of assessments against the DNS data.

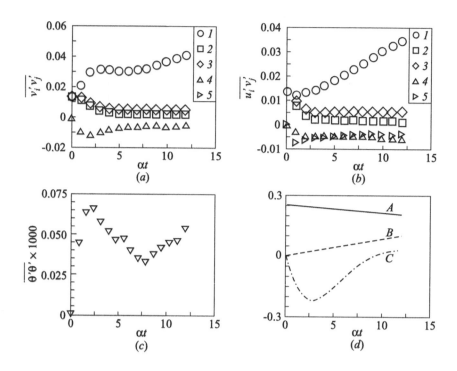

Figure 3.4 Temporal variations of statistical properties of evaporating droplets as obtained by DNS: (a) $\overline{v_i' v_j'}$, (b) $\overline{u_i' v_j'}$, (c) $\overline{\theta' \theta'}$, and (d): curve A — $0.25\overline{d}/d_0$, curve B — $10\sigma/d_0$, and curve C — skewness$/2$. The time axis has been normalized by the magnitude of the mean velocity gradient, α: 1 — $i = 1$, $j = 1$; 2 — $i = 2$, $j = 2$; 3 — $i = 3$, $j = 3$; 4 — $i = 1$, $j = 2$; and 5 — $i = 2$, $j = 1$.

For the purpose of assessment of macroscopic equations, DNS data are generated for evaporating droplets dispersed in homogeneous shear flows by extending the previous study [17]. In this particular case, some of the statistical properties of interests are: droplet Reynolds stresses, $\overline{v_i' v_j'}$; fluid-droplet velocity correlations, $\overline{u_i' v_j'}$; temperature correlation, $\overline{\theta' \theta'}$; and mean diameter, $\overline{d}$, of the droplets. The temporal variations of $\overline{v_i' v_j'}$, $\overline{u_i' v_j'}$, and $\overline{\theta' \theta'}$ are shown in Figs. 3.4a–3.4c, respectively. The varying evaporation rate for an individual droplet, which due to the stochastic nature of fluid flow field, results in polydispersed droplets even when the monodispersed droplets are injected in the flow. The variance σ in droplet diameter and skewness indicates the extent of polydispersity. The temporal variations of mean diameter $\overline{d}$, variance σ, and skewness are shown in Fig. 3.4d. The values of skewness suggest that the distribution of the droplet diameter is non-Gaussian in nature. These types of data can be used to assess macroscopic equations for the statistical properties.

3.4 CONCLUDING REMARKS

Our efforts on DNS and PDF modeling approaches for prediction of two-phase turbulent flows have been described in brief. The direct numerical simulation study has been focused on generating information useful in designing the experiment on the backward-facing step with countercurrent flow which closely represents the dump combustor flow situation. Direct numerical simulation allows one to quantify and visualize various effects arising due to different inlet flow conditions, Stokes number of the dispersed phase, and strength of the countercurrent flow. Also, DNS data have been generated for situations of evaporating droplets dispersed in homogeneous shear flows. The data are valuable in the development and assessment of predictive models. As an initial attempt, a PDF model has been obtained for a general situation of turbulent flow laden with evaporating droplets. This model is in the form of a closed equation which governs the transport of probability density of droplet position, velocity, temperature, and diameter. Performing DNS for inhomogeneous two-phase turbulent flow with evaporating droplets in backward-facing step geometry and assessment of the PDF model against the DNS data remains as important unfinished objectives of our ongoing research.

ACKNOWLEDGMENTS

This work was performed under ONR contract N00014-01-1-0122.

REFERENCES

1. Jacobs, G. B., D. A. Kopriva, and F. Mashayek. 2001. A particle tracking algorithm for the multidomain staggered-grid spectral method. AIAA Paper No. 2001-0630.

2. Jacobs, G. B., D. A. Kopriva, and F. Mashayek. 2002. Outflow boundary conditions for the multidomain staggered-grid spectral method. AIAA Paper No. 2002-0903.

3. Fischer, P. F. 1989. Spectral element solution of the Navier–Stokes equations on high performance distributed-memory parallel processors. Ph.D. Thesis. Massachusetts Institute of Technology, Massachusetts.

4. Forliti, D. J. 2001. Controlling dump combustor flows using countercurrent shear. Ph.D. Thesis. University of Minnesota, Minnnesota.

5. Boltzmann, L. 1964. Lectures on gas theory. Berkeley and Los Angeles: University of California Press. (Translated by S. G. Brush.)

6. Chandrasekhar, S. 1954. Stochastic problems in physics and astronomy. In: *Selected papers on noise and stochastic processes*. Ed. N. Wax. New York: Dover Publications.

7. Minier, J.-P., and E. Peirano. 2001. The pdf approach to turbulent polydispersed two-phase flows. *Physics Reports* 352:1.

8. Pozorski, J., and J. P. Minier. 1999. Probability density function modeling of dispersed two-phase turbulent flows. *Physical Review E* 59:855.

9. Mashayek, F., and R. V. R. Pandya. 2003. Analytical description of particle/droplet-laden turbulent flows. *Progress Energy Combustion Science* 29(4):329–78.

10. Reeks, M. W. 1992. On the continuum equations for dispersed particles in nonuniform flows. *Physics Fluids* 4:1290.

11. Hyland, K. E., S. McKee, and M. W. Reeks. 1999. Derivation of a pdf kinetic equation for the transport of particles in turbulent flows. *J. Physics A: Mathematical General* 32:6169.

12. Derevich, I. V. 2000. Statistical modeling of mass transfer in turbulent two-phase dispersed flows — 1. Model development. *Int. J. Heat Mass Transfer* 43:3709.

13. Zaichik, L. I. 1999. A statistical model of particle transport and heat transfer in turbulent shear flows. *Physics Fluids* 11:1521.

14. Pandya, R. V. R., and F. Mashayek. 2002. Kinetic equation for particle transport and heat transfer in nonisothermal turbulent flows. AIAA Paper No. 2002-0337.

15. Pandya, R. V. R., and F. Mashayek. 2002. Nonisothermal dispersed phase of particles in turbulent flow. *J. Fluid Mechanics* 475:205–45.

16. Miller, R. S., K. Harstad, and J. Bellan. 1998. Evaluation of equilibrium and nonequilibrium evaporation models for many-droplet gas–liquid flow simulations. *Int. J. Multiphase Flow* 24:1025.

17. Mashayek, F. 1998. Droplet–turbulence interactions in low-Mach-number homogeneous shear two-phase flows. *J. Fluid Mechanics* 367:163.

Chapter 4

LARGE-SCALE SIMULATIONS OF TURBULENT COMBUSTION AND PROPULSION SYSTEMS

A. Afshari and F. A. Jaberi

The basic objective of this chapter is to develop affordable high-fidelity models for numerical simulation of turbulent combustion systems. The models are based on the large-eddy simulation (LES) method and are being developed/tested for both single- and two-phase flows. Recently, LES of several methane jet flames was conducted in which the subgrid scalar correlations were modeled via the filtered mass density function (FMDF) methodology [1]. In some cases, the LES/FMDF was employed in conjunction with the equilibrium methane-oxidation model. This model is enacted via "flamelet" simulations, which consider a laminar counterflow flame configuration with a detailed kinetics mechanism. Comparison between experimental and numerical data on a piloted turbulent jet flame [2] indicates that the LES/FMDF/flamelet methodology is able to reproduce the experimental results with good accuracy while being very efficient. More recently, the LES method has been extended to two-phase flows, and some preliminary simulations of nonreacting particle dispersion in a turbulent jet have been conducted. Direct numerical simulation (DNS) of "selected" turbulent flames is also considered. The primary objective is to utilize the DNS-generated data for systematic assessment and validity appraisal of the LES closures.

4.1 INTRODUCTION

The development of predictive models for realistic combustion systems is a challenging task. It requires basic understanding of the relevant physics, systematic assessment of "submodels," and, of course, highly efficient and accurate numerical algorithms. The ability to produce the experimental data under a given set of conditions is necessary but not sufficient. Computational models should be able to predict the flow field for conditions that laboratory measurements are either not possible or very difficult to perform. Among numerical methods available for turbulent combustion, LES has the potential of being sufficiently accurate and affordable.

This work deals with the development of reliable LES models for turbulent reacting systems with and without particles/droplets. The work includes the following tasks: (i) development/application of LES to flows involving "simple" gas-phase fuels (methane and hydrogen) in various configurations with reduced and detailed kinetics models, (ii) development/application of LES to flows involving particle (droplet) dispersion and evaporation without combustion, and (iii) development/application of LES to flows involving particle dispersion, evaporation, and combustion with reduced kinetics models. In this chapter, the efforts on issues (i) and (ii) within the past year are briefly discussed. In addition to issues (i) to (iii), DNS of "simple" turbulent flames is also considered. The generated database is used for better understanding of the underlying processes and for systematic development and assessment of LES models. Ultimate validation is (and will be) done by comparing the LES results with experimental data. The tasks proposed originally have been modified in order to more effectively comply with the goals of the program.

4.2   THEORETICAL/COMPUTATIONAL APPROACH

In LES/FMDF calculations of single-phase flows, the resolved "hydrodynamic" field is obtained by solving the filtered form of the compressible Navier–Stokes equations [3]:

$$\frac{\partial \langle \rho \rangle_l}{\partial t} + \frac{\partial \langle \rho \rangle_l \langle u_i \rangle_L}{\partial x_i} = 0 \tag{4.1}$$

$$\frac{\partial \langle \rho \rangle_l \langle u_j \rangle_L}{\partial t} + \frac{\partial \langle \rho \rangle_l \langle u_i \rangle_L \langle u_j \rangle_L}{\partial x_i} = -\frac{\partial \langle p \rangle_l}{\partial x_j} + \frac{\partial \langle \tau_{ij} \rangle_l}{\partial x_i} - \frac{\partial \Theta_{ij}}{\partial x_i} + \langle \rho \rangle_l g_j \tag{4.2}$$

$$\langle p \rangle_l \approx \langle \rho \rangle_l R^0 \langle T \rangle_L \sum_1^{N_s} \frac{\langle \phi_\alpha \rangle_L}{W_\alpha} \tag{4.3}$$

$$\langle \tau_{ij} \rangle_l \approx \langle \mu \rangle_L \left(\frac{\partial \langle u_i \rangle_L}{\partial x_j} + \frac{\partial \langle u_j \rangle_L}{\partial x_i} - \frac{2}{3} \delta_{ij} \frac{\partial \langle u_k \rangle_L}{\partial x_k} \right)$$

$$\langle \mu \rangle_L = \Pr \langle k/c_p \rangle_L \tag{4.4}$$

where $\langle f(\mathbf{x}, t) \rangle_l$ and $\langle f(\mathbf{x}, t) \rangle_L = \langle \rho f \rangle_l / \langle \rho \rangle_l$ represent the filtered and the Favre-filtered values of the transport variable $f(\mathbf{x}, t)$ and ρ, u_i, p, and T are the fluid density, velocity, pressure, and temperature, respectively. In Eqs. (4.3) and (4.4), μ, k, c_p, and Pr are viscosity coefficient, thermal conductivity coefficient, specific heat, and Prandtl number, respectively; R^0 denotes the universal gas constant, and W_α is the molecular weight of species α. The hydrodynamic subgrid-scale (SGS) closure problem is associated with the SGS stress ($\Theta_{ij} = \langle\rho\rangle_l[\langle\langle u_i u_j\rangle\rangle_L - \langle u_i\rangle_L\langle u_j\rangle_L]$) which is modeled with a "standard" diffusivity closure. The scalar field (mass fractions and temperature) is obtained from the FMDF (as discussed below). In traditional LES methods, additional filtered equations are solved for the scalars. In these equations, the filtered chemical source/sink terms are not closed and need modeling. These terms are determined exactly with the knowledge of the FMDF.

The scalar FMDF is the joint probability density function of the scalars at the subgrid-level and is defined as [1]:

$$F_L(\boldsymbol{\psi}, \mathbf{x}, t) \equiv \int_{-\infty}^{+\infty} \rho(\mathbf{x}', t)\varsigma\left[\boldsymbol{\psi}, \boldsymbol{\phi}(\mathbf{x}', t)\right] H(\mathbf{x}' - \mathbf{x})\, d\mathbf{x}' \qquad (4.5)$$

where H denotes the filter function, and ς is the "fine-grained" density [4]. The scalar field, $\boldsymbol{\phi} \equiv \phi_\alpha$, $\alpha = 1, 2, \ldots, N_s + 1$, represents the mass fractions of the chemical species and the specific enthalpy and is obtained from the joint scalar FMDF [1]. The final form of the FMDF transport equation as derived from the original (unfiltered) scalar equations is

$$\frac{\partial F_L}{\partial t} + \frac{\partial[\langle u_i\rangle_L F_L]}{\partial x_i} = \frac{\partial}{\partial x_i}\left[\langle\rho\rangle_l\left(\langle D\rangle_L + D_t\right)\frac{\partial(F_L/\langle\rho\rangle_l)}{\partial x_i}\right]$$

$$+ \frac{\partial}{\partial \psi_\alpha}\left[\Omega_m(\psi_\alpha - \langle\phi_\alpha\rangle_L)F_L\right] - \frac{\partial[\widehat{S}_\alpha F_L]}{\partial \psi_\alpha} \qquad (4.6)$$

In Eqs. (4.5) and (4.6), S_α, $\boldsymbol{\psi}$, and Ω_m denote the production rate of species α, the "composition space" of scalar $\boldsymbol{\phi}$, and the SGS mixing frequency, respectively. The molecular diffusivity coefficient and the SGS diffusivity coefficient are denoted by D and D_t. The last term on the right-hand side (RHS) of Eq. (4.6) represents the effects of chemical reaction and is in a closed form. The second and the third terms on the RHS represent the effects of SGS mixing and SGS convection, respectively, and are modeled with closures similar to those used in Reynolds-Averaged Navier–Stokes/Probability Density Function (RANS/PDF) methods [4].

The most convenient means of solving Eq. (4.6) is via the "Lagrangian Monte Carlo" procedure [4]. With the Lagrangian procedure, the FMDF is represented by an ensemble of computational "stochastic elements" (or "particles") which are transported in the "physical space" by the combined actions of large-scale convection and diffusion (molecular and subgrid). In addition, transport in the "composition space" occurs due to chemical reaction and SGS mixing. All of these are implemented via a "stochastic process," described by the set of stochastic differential equations (SDEs) [1]. These SDEs are fully consistent with the original FMDF transport equation.

For gaseous flames, the LES/FMDF can be implemented via two combustion models: (1) a finite-rate, reduced-chemistry model for nonequilibrium flames and (2) a near-equilibrium model employing detailed kinetics. In (1), a system of nonlinear ordinary differential equations (ODEs) is solved together with the FMDF equation for all the scalars (mass fractions and enthalpy). Finite-rate chemistry effects are explicitly and "exactly" included in this procedure since the chemistry is closed in the formulation. In (2), the LES/FMDF is employed in conjunction with the equilibrium fuel-oxidation model. This model is enacted via "flamelet" simulations, which consider a laminar counterflow (opposed jet) flame configuration. At low strain rates, the flame is usually close to equilibrium. Thus, the thermochemical variables are determined completely by the "mixture fraction" variable. A flamelet library is coupled with the LES/FMDF solver in which transport of the mixture fraction is considered. It is useful to emphasize here that the PDF of the mixture fraction is not "assumed" a priori (as done in almost all other flamelet-based models), but is calculated explicitly via the FMDF. The LES/FMDF/flamelet solver is computationally less expensive than that described in (1); thus, it can be used for more complex flow configurations.

In DNS of single-phase flows, a complete set of compressible Navier–Stokes, energy, and scalar transport equations are calculated together with the equation of state and some constitutive relations [3]. In DNS of particle-laden flows, in addition to carrier-gas equations, the Lagrangian form of particle (droplet) equations are solved via "standard" difference schemes [5].

For LES of two-phase flows, the large-scale Eqs. (4.1)–(4.4) and the filtered energy equation are solved for the carrier-gas phase together with Mean Kinetic Energy Velocity (MKEV) [1] SGS stress and diffusivity closures. For the particle phase, a stochastic model is considered in which the SGS diffusivity (evaluated from large-scale quantities) is used to construct the residual or subgrid velocity of the carrier fluid at the particle location. The large-scale velocity field (which is explicitly calculated in LES) is interpolated to the particle location via a Lagrange interpolation scheme. The combination of large- and small-scale fluid velocity is used to move the particles (droplets) and to calculate the particle drag force.

The discretization procedure of the carrier fluid (in both LES and DNS) is based on the "compact parameter" scheme which yields up to sixth-order spatial accuracies. For DNS and LES of jet and mixing layer simulations, a finite difference method is used. For more complex configurations (i.e., dump combustor, bluff body), a finite-volume collocated grid method is used in which time differencing is based on a third-order, low-storage Runge–Kutta method. The preconditioning method has been employed to speed up the pseudotime step convergence for low-Mach-number flows in some of the simulations.

4.3   RESULTS AND DISCUSSION

As discussed above, LES/FMDF can be implemented with (1) nonequilibrium and (2) near-equilibrium combustion models. The former uses a "direct ODE solver" for the chemistry and can handle finite-rate chemistry effects. In the latter, a flamelet library is coupled with the LES/FMDF solver in which transport of the mixture fraction is considered. The latter approach has the advantage: it is computationally much less intensive and can be conducted with very complex chemical kinetics models. Below, some of the results recently obtained via Eq. (4.2) are presented. The flamelet library is generated with the full methane oxidation mechanism of the Gas Research Institute (GRI) [6] accounting for 53 species and about 300 elementary reactions.

With the near-equilibrium chemistry model, the results obtained by LES/FMDF/flamelet are compared with experimental data in a piloted turbulent jet

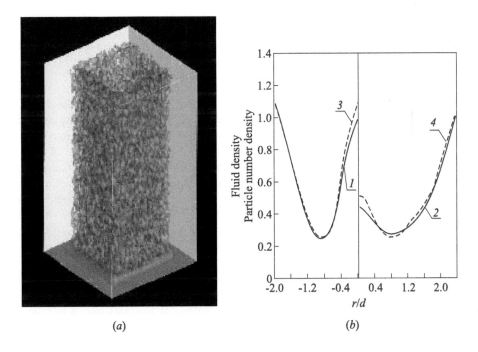

(a)   (b)

Figure 4.1 (*a*) Instantaneous number density of the Monte-Carlo particles as obtained by LES/FMDF in a near-equilibrium methane jet flame; and (*b*) radial (r/d) variations of the Reynolds-averaged values of the filtered fluid density as obtained by LES/FMDF and by finite difference simulation at two streamwise (x/d) locations (*1* — $x/d = 7.5$ (finite difference); *2* — $x/d = 15$ (finite difference); *3* — $x/d = 7.5$ (Monte-Carlo); and *4* — $x/d = 15$ (Monte-Carlo)).

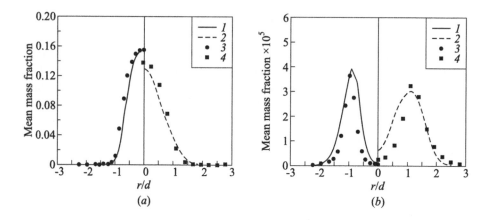

Figure 4.2 Radial (r/d) variations of the Reynolds-averaged values of the filtered mass fractions in a piloted methane jet flame (Sandia Flame D) at two streamwise (x/d) locations, as predicted by LES/FMDF and compared with experimental data obtained at Sandia: (*a*) mass fraction of CH_4 and (*b*) mass fraction of NO: *1* — $x/d = 7.5$ (LES/FMDF); *2* — $x/d = 15$ (LES/FMDF); *3* — $x/d = 7.5$ (experiment); and *4* — $x/d = 15$ (experiment).

flame obtained at the Combustion Research Facility of the Sandia National Laboratory [2]. This flame has been the subject of broad investigations by several other computational/modeling methodologies [2]. Three basic flames are considered in these experiments: flames D, E, and F. The configurations in all three flames are the same, but the jet inlet velocity is different. In flame D, the fuel jet velocity is the lowest, and the flame is close to equilibrium. The jet velocity increases from flames D to E to F, with noticeable nonequilibrium effects in the latter two. Herein, flame D with LES/FMDF/flamelet was simulated, and good agreement with experimental data was observed. Some sample results are presented in Figs. 4.1 and 4.2.

First, to demonstrate the operational procedure in LES/FMDF, the instantaneous number density of the Monte-Carlo elements or particles is shown in Fig. 4.1*a*. This figure shows the total number of particles within the computational domain. However, the weighted distribution of the particles is not, and should not be, uniform. Rather, it must be proportional to the flow density. This is demonstrated in Fig. 4.1*b* in which the ensemble-mean values of the weighted Monte-Carlo particle number density are compared with the fluid density as obtained from finite difference solution of the filtered density field. The very good correlation attained in this way verifies the consistency of the stochastic procedure and its Lagrangian Monte-Carlo solver.

Figure 4.2 shows the radial variations of the ensemble-averaged values of CH_4 and NO mass fractions. The good agreement as observed here (and also observed for other variables) indicates the predictive capability of the LES/ FMDF/flamelet methodology, at least for flames under equilibrium. Again, for the conditions that the flamelet assumption is not valid, the LES/FMDF procedure can be (and has been) implemented with a "direct ODE solver" for chemistry. This does not impose any limitation on the calculations other than an increase in computational time since chemical terms are closed in the formulation. Presently, LES/FMDF with a 12-step reduced methane–air kinetic mechanism [7] in conjunction with the ISAT routine [8] is considered.

For two-phase flows, so far, LES of nonreacting particle-laden dispersed turbulent flows in jet configuration is considered with a finite difference scheme similar to that used in the authors' single-phase LES.

Also, in collaborative efforts with Mashayek's group at the University of Illinois at Chicago, a generalized finite-volume LES code is under development which will be used for simulations of droplet dispersion, evaporation, and combustion in a dump combustor and swirl jet burner.

ACKNOWLEDGMENTS

The work was performed under ONR contract N00014-01-1-0843.

REFERENCES

1. Jaberi, F. A., P. J. Colucci, S. James, P. Givi, and S. B. Pope. 1999. Filtered mass density function for large eddy simulation of turbulent reacting flows. *J. Fluid Mechanics* 401:85–121.

2. Barlow, R. S. 2002. Sandia National Laboratories. Turbulent Diffusion Flame Workshop website //www.ca.sandia.gov/tdf/Workshop.html.

3. Poinsot, T., and D. Veynante. 2001. *Theoretical and numerical combustion.* Philadelphia, PA: R. T. Edwards Inc.

4. Pope, S. B. 1985. PDF methods for turbulent reactive flows. *Progress Energy Combustion Science* 11:119–92.

5. Mashayek, F., and F. A. Jaberi. 1999. Particle dispersion in forced isotropic low Mach number turbulence. *Int. J. Heat Mass Transfer* 42:2823–36.

6. Smith, G. P., D. M. Golden, M. Prenklach, N. W. Moriarty, B. Eiteneer, M. Goldenberg, C. T. Bowman, R. Hanson, S. Song, W. C. Gardiner, V. Lissianski, and Z. Qin. //www.me.berkeley.edu/gri-mech.

7. Sung, C. J., C. K. Law, and J.-Y. Chen. 1998. An augmented reduced mechanism for methane oxidation with comprehensive global parametric validation. *Combustion Institute Proceedings* 27:295–303.

8. Pope, S. B. 1997. Computationally efficient implementation of combustion chemistry using *in situ* adaptive tabulation. *Combustion Theory Modelling* 1:41–63.

Comments

Gore: Why is your agreement with NOx levels so good? Due to the low ppm levels, one could not expect the results to be this good. Results are very sensitive to temperature. Do you have any prediction of it? Also, did you consider radiation?

Jaberi: The results for major species and temperature are believed to be good to 10%. The NOx predictions also appear to be reasonably good. The radiation effect is estimated to be less than 5% in flame D.

Frolov: Your technique has been applied to unconfined constant-pressure flow conditions. What is really interesting for propulsion is the confined reactive flow with acoustic interactions, pressure gradients, etc. At your viewgraph regarding future work, I did not see any plan of using Monte-Carlo methodology to confined conditions. Can you use this technique for confined flow with pressure coupling?

Jaberi: The compressibility effects at large scales are already considered in our calculations, since we have used fully compressible LES equations and numerical methods. The compressibility effect on subgrid combustion is not presently modeled but will be included in the future. We can and we will use this technique for confined flows.

Lindstedt: How accurate are your CO_2 predictions downstream?

Jaberi: Within 10%.

Lindstedt: What GRI mechanism did you use in calculations?

Jaberi: GRI version 2.01.

Lindstedt: GRI 2.01 would underpredict NOx by 40%–45%. So you should not make too much of a point on data agreement for NOx.

Chapter 5

DIRECT SIMULATION OF PRIMARY ATOMIZATION

D. P. Schmidt

Within the scope of this work, the initial spray breakup process, providing information about the dense spray core, will be investigated. The formation of fuel drops will be simulated based on first-principles and will offer detailed insight into primary atomization. The three-dimensional, transient calculation will track the interface evolution through droplet formation and breakup. Because the results will be based on conservation laws, they will be extremely general. This will lead to better models that can be used with confidence in the engine design process.

5.1 INTRODUCTION

The fuel spray is a special challenge in the field of fluid mechanics. The spray is a complex multiphase flow that cannot be observed in full detail experimentally. The spray behavior is governed by a wide range of parameters, including the injection system, fuel properties, and surrounding flow. As a further complication, the physical mechanisms of atomization are not well understood. Researchers have attributed spray breakup to a long list of suspected phenomena: nozzle turbulence [1], the Kelvin–Helmholtz instability [2], the Rayleigh–Taylor instability [3], cavitation [4], and oscillations in the fuel injection system [5]. These differences in opinion indicate that there is great uncertainty in the physics of sprays [6]. Essential information is missing for the development of better spray models.

Current spray models in multidimensional computational fluid dynamics (CFD) codes achieve moderately good results under a limited range of conditions by speculating on the atomization process and including adjustable parameters. The user must estimate values for these parameters in order to match experimental results. Furthermore, the need to constantly vary these parameters may indicate that the model construction is not based on the correct assumptions. These unknown quantities create a significant uncertainty in the breakup behavior of the spray.

Mathematically, modeling these phenomena is also error-prone. Perturbation analyses have traditionally been used to study the behavior of instabilities. These mathematical analyses are valid for small disturbances on liquid sheets and jets. Some models further simplify the mathematics, using linearization and order-of-magnitude analyses. These techniques persist out of necessity, although Crapper *et al.* [7] showed that the linearized perturbation analysis does not always give the correct qualitative trends. Nonlinear analyses such as that of Mehring and Sirignano [2] represent progress, but are still limited to small-scale perturbations. The majority of the atomization process is still out of reach of current mathematical techniques.

Current spray models may not have the correct physics, may have unknown limits of applicability, and may contain empirical constants. In a recent test conducted by the author and United Technologies Research Center (UTRC), models of primary atomization, secondary atomization, droplet breakup, droplet collision, and turbulent dispersion were applied to an air blast spray. The predictions were compared to experimental data taken at UTRC. The predicted drop size was as much as 35% different from the measured values [8]. In contrast to the typical conference or journal publication, the models were not adjusted to make the agreement as close as possible. They were taken from the literature "as is." The conclusion is that physical models of high-speed spray behavior are still lacking, despite years of research in this area. Primary atomization, the beginning of the spray, is one area that is particularly poorly understood.

5.2 PAST WORK

Primary atomization, the formation of ligaments and drops by an atomizer, has already been a subject of study for over a century. The difficulty in experiments is that the numerous droplets reflect light, obscuring clear views of the atomization process. In addition, the high speed and small size of practical fuel injection means that the experimental images are often not clear. Dense sprays and nonspherical drops also make quantitative data difficult to obtain with laser-based diagnostics.

An alternative approach is to use direct numerical simulation (DNS). Numerical results can offer considerable detail and allow access to the complete fluid flow. For example, how big is the role of disturbances generated in the atomizer? Does the Rayleigh–Taylor instability matter in primary atomization? Do newly formed droplets immediately collide with ligaments and previously formed droplets? How useful are small-perturbation analyses for primary atomization? The disadvantage is that DNS is capable of simulating only a small part of the spray.

In contrast to traditional DNS, the simulation of two phases requires very advanced numerical techniques for handling the deforming free surface. With large density differences, stability can be a problem [9]. Also, interface tracking codes require several hundred percent more central processing unit (CPU) time than single-phase codes, making such simulations expensive.

There have been only a few major attempts to apply interface tracking to primary atomization [9, 10]. However, both of these codes treated all variables, excepting density, as continuous across the interface, which could have implications for their accuracy. For example, the discontinuity in viscosity causes the first derivative of velocity to change abruptly at an interface.

Other interface tracking schemes have been applied to single drops but have limitations that currently prevent them from being applied to sprays. Helenbrook has developed a very accurate spectral method, but this code is currently limited to two-dimensional, axisymmetric droplet simulations [11]. Cristini *et al.* have a three-dimensional multiphase flow code that is based on the boundary element method (BEM) [12]. However, the BEM will not predict separated flows or capture vorticity effects.

The proposed work will solve the complete Navier–Stokes equations. What distinguishes the proposed work from past efforts is the efficiency and accuracy of the code. The project, as described below, should be able to reveal physical details of the spray better than previous efforts.

The proposed approach will also represent an advance in multiphase fluid flow numerics. The numerical treatment of the interface will use the highly accurate approach of Nallapati [13]. The free surface will be a true interface, maintained coincidentally with cell faces on a deforming unstructured mesh. This approach allows the possibility of capturing the rapid velocity change near the interface that may play a pivotal role in spray behavior. This gradient controls the Kelvin–Helmholtz instability.

The Kelvin–Helmholtz instability is thought to be a main driver in the early steps of atomization. This instability is driven by the lift forces on the interface because of the relative velocity between the two phases. When a numerical method smears velocities across the interface with insufficient resolution, the consequence may be a retardation of the predicted instability growth. Since the Kelvin–Helmholtz instability is so important, the numerical method should be capable of capturing this effect accurately.

5.3   OBJECTIVES

In this work, direct interface tracking will be used to simulate the evolution of liquid jets and sheets into drops as shown in Fig. 5.1. The fundamental physics of spray flow without relying on assumptions about instabilities will be addressed.

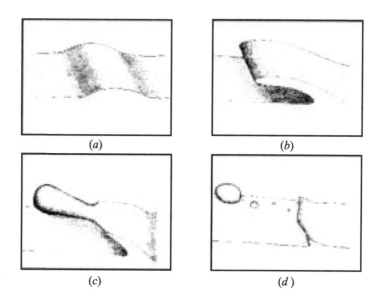

(a)   (b)

(c)   (d)

Figure 5.1 Anticipated results from the fully coupled simulation of gas–liquid flow. The pictures show the interface as it evolves, from image (<i>a</i>) to (<i>d</i>) during primary atomization. Based qualitatively on [9].

The effort will be based on new numerical techniques that permit good resolution of the interface condition. The code will use a deforming mesh that will keep the interface coincidental with cell boundaries, allowing the interface to be treated as a true boundary. Mesh adaptation will help provide fidelity at a reasonable cost. The results will reveal in new detail how high-speed atomization progresses.

It is anticipated that the results of calculations will show the governing mechanisms of primary atomization. They will indicate the relative importance of turbulence, the Kelvin–Helmholtz instability, the Rayleigh–Taylor instability, the initial perturbation level (attributable to cavitation or oscillations in fuel injection equipment), and other phenomena. The quantitative detail of the simulations will provide information and inspiration for the construction of a new generation of spray models. The proposed code can be used for other kinds of simulations, including wall impingement, liquid film flow, and impinging injections.

5.4   METHODOLOGY

The approach will use a moving, adaptive mesh in order to efficiently resolve the flow around the drops. The domain will be divided into tetrahedral cells, with cell

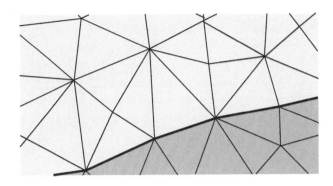

Figure 5.2 A view of the mesh near the interface. The heavy line shows the interface. The light color is the gas phase, and the dark color is the liquid phase. Note that the cell boundaries lie on the interface.

boundaries conforming to the gas–liquid interface. The conformal mesh allows the interface to be elegantly and simply represented as a boundary condition. A sketch is shown in Fig. 5.2. This approach stands in contrast to Volume of Fluid (VOF) methods where the surface tension must be represented as a force extracted from the density field.

As the fluid interface moves, the mesh will continuously move as well. The boundaries will continue to conform to the interfaces, while the interior mesh moves according to a smoothing algorithm. This approach requires the algorithm to solve the Reynolds transport theorem for a moving, deforming mesh. Like Arbitrary Lagrangian–Eulerian approaches, the mesh moves relative to the fluid (in the interior) or with the fluid (on the boundaries). However, the use of a tetrahedral, unstructured mesh allows adaptive refinement where necessary, without hanging nodes. The current algorithm does not globally remesh, which would degrade conservation properties, but gives the mesh a smoothly varying velocity. This velocity is determined by a smoothing algorithm that keeps the mesh quality adequate as the fluid volumes deform. This approach requires cells to occasionally locally change connectivity, which is called "mesh flipping."

The most tangible benefit will be the ability to model a large domain while capturing the formation of small drops. In contrast, fixed-mesh schemes require an extremely fine mesh in order to resolve the smallest drops that can form from atomization. This makes the simultaneous resolution of large-scale ligaments and small-scale drops very computationally expensive. The proposed work will efficiently adapt the mesh to be faster and more accurate and will not be subject to these limitations.

The process of mesh adaptation, though challenging, has a significant speed and accuracy advantage over VOF methods that use a fixed underlying mesh

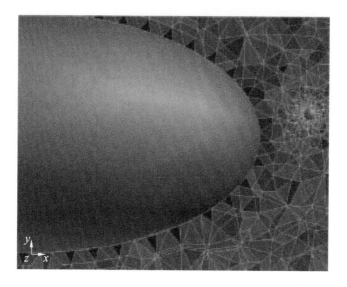

Figure 5.3 A sketch of a mesh around a large and small drop that is adaptively refined.

to resolve the interface. In order to properly resolve atomization, the mesh of a VOF scheme must be small enough to refine the smallest drops. The mesh must be fine wherever the smallest drops may form or may travel, which would enormously drive up the cost. This difficulty cannot be easily avoided, because sprays generate a large range of drop sizes. The proposed method will have a fine mesh only in the vicinity of the small drops or features that move with the droplets, as shown in Fig. 5.3. This "smart meshing" technique provides resolution only where it is needed.

The advantage of this scheme cannot be understated. A quick hand calculation shows the importance of this adapting mesh. In a high-speed atomizing jet, one might expect small droplets to be on the order of 2 μm. The largest drops and the ligaments that form droplets will be on the order of 100 μm. So the overall domain size should be much larger than 100 μm, yet the mesh resolution must be much smaller than 2 μm. For a cubic millimeter computational domain with 0.25-micrometer mesh spacing, the number of cells required for uniform meshing would be 64 billion cells. The current work avoids this problem by adaptive meshing. This efficiency translates into the ability to simulate larger domains with the formation of numerous ligaments and drops.

The differencing scheme is second-order accurate in space, when applied to a regular mesh. For distorted meshes, the scheme has a lower formal order of accuracy. This behavior is analogous to the behavior of second-order central differencing on rectangular meshes. No artificial viscosity will be required, because of the moderate Reynolds numbers of the flow (with Reynolds number based on relative velocity). The mesh size will be fine enough to resolve the smallest eddies, because the mesh size is constrained by the droplet sizes. A very conservative estimate of the Kolmogorov scale is no smaller than 10 μm, which is on the order of the drop sizes.

The code uses a staggered mesh, with contravariant velocities located at the center of mesh faces. Pressure is eliminated by applying a discrete curl operator to the discrete momentum equations. This gives a set of equations for a three-dimensional stream function. The resulting matrix is symmetric, positive definite and is solved using a conjugate gradient method. This scheme conserves kinetic energy, as shown by Perot [14]. The technique also happens to be quite fast.

To further bring down the computational cost of the simulations, the code will be parallelized. The code will run on a network of desktop computers. Each computer will perform part of the calculation and communicate its results to the others. This approach allows a collection of cheaper computers to be used instead of an expensive supercomputer. The Fluid Dynamics Group at the University of Massachusetts shares a 180 CPU Beowulf cluster for large, parallel computations. These computer resources will be able to simulate the formation of tens or hundreds of drops. Though this will only represent a fraction of the entire spray, it will certainly be sufficient to reveal the interesting physics behind spray breakup.

5.5 TASKS

To date, the code is capable of simulating a single, deforming drop, as shown in Figs. 5.4*a* and 5.4*b*. The code will be expanded to provide the required capabilities for this project. The work will proceed through the following phases:

1. Adaptive mesh development;

2. Gas–liquid coupling;

3. Changes in topology;

4. Simulation of spray physics; and

5. Documentation of the spray results.

The phases will be executed over the next years. Increasingly complex and realistic simulations will be performed as the code development progresses.

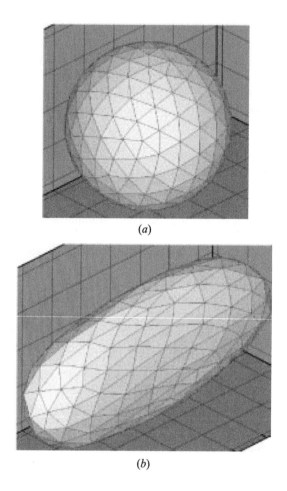

(a)

(b)

Figure 5.4 Preliminary results with the proposed numerical method: (a) an initially spherical drop is given a velocity perturbation; and (b) a deforming droplet is given in the absence of a surrounding gas.

ACKNOWLEDGMENTS

The work was performed under ONR contract N00014-02-1-0507.

REFERENCES

1. Huh, K., and A. D. Gosman. 1990. Atomization mechanism of fuel injection. *ILASS-90-Americas.*

2. Mehring, C., and W. A. Sirignano. 2000. Planar-liquid-stream distortion from Kelvin–Helmholtz and capillary effects. *8th ICLASS Meeting Proceedings.* Pasadena, CA.

3. Beale, J. C., and R. D. Reitz. 1999. Modeling spray atomization with the Kelvin–Helmholtz/Rayleigh–Taylor hybrid model. *Atomization Sprays* 9(6):623–50.

4. Karasawa, T., M. Tanaka, K. Abe, S. Shiga, and T. Kurabayashi. 1992. Effect of nozzle configuration on the atomization of steady spray. *Atomization Sprays* 2:411–26.

5. Chaves, H., and F. Obermeier. 1996. Modeling the effect of modulations of the injection velocity on the structure of diesel sprays. SAE Paper No. 961126.

6. Smallwood, G. J., and Ö. L. Gülder. 2000. Views on the structure of transient diesel sprays. *Atomization Sprays* 10:355–86.

7. Crapper, G. D., N. Dombrowski, W. P. Jepson, and G. A. D. Pyott. 1973. A note on the growth of Kelvin–Helmholtz waves on thin liquid sheets. *J. Fluid Mechanics* 57:671–72.

8. Chiappetta, L. M. 2001. Tests of a comprehensive model for airblast atomizers including the effects of spray initialization (primary atomization), secondary breakup, collision and coalescence. UTRC Report.

9. Tauber, W., and G. Trygvasson. 2000. Primary atomization of a jet. *ASME FED Summer Meeting Proceedings.* Boston.

10. Lafaurie, B., T. Mantel, and S. Zaleski. 1998. Direct numerical simulation of liquid jet atomization. *3rd Conference (International) on Multiphase Flow.* Lyon, France.

11. Helenbrook, B. T. 2001. Numerical studies of droplet deformation and break-up. *ILASS Americas Conference Proceedings.* Dearborn, MI. 123–28.

12. Cristini, V., J. Blawzdziewicz, and M. Lowenberg. 2001. An adaptive mesh algorithm for evolving surfaces: Simulations of drop breakup and coalescence. *J. Computational Physics* 168:445–63.

13. Nallapati, R. 2001. Numerical simulation of free-surface flows using a moving unstructured staggered grid method. MS Thesis. University of Massachusetts, Amherst.

14. Perot, J. B. 2000. Conservation properties of unstructured staggered mesh schemes. *J. Computational Physics* 159:58–59.

Comments

Mashayek: There would appear to be resolution problems with high Reynolds numbers. What Reynolds numbers of droplets do you expect?

Schmidt: Pretty low, about 150.

Smirnov: You are using Navier–Stokes equations. What will be the boundary conditions?

Schmidt: Classic conditions with constant surface tension.

Smirnov: For validation, which model problem do you plan to solve?

Schmidt: There is not much data available. Perhaps we will use Kelvin–Helmholtz instability.

Smirnov: The droplet ejection problem will seriously challenge the model.

Frolov: In your preliminary calculations, what kind of internal motion in the deformed droplet did you observe? Is that resembling Hill vortex, or is it something else?

Schmidt: In the example presented, the drop was externally driven without the effect of gas phase. Therefore, no vortical motions were observed.

Chapter 6

EXTINCTION AND RELIGHT
IN OPPOSED PREMIXED FLAMES

E. Korusoy and J. H. Whitelaw

Extinction and extinction times of turbulent unforced and forced symmetric opposed methane–air flames with lean equivalence ratios were examined qualitatively with photographs and chemiluminescence images. Thermocouples and a laser-Doppler-velocimeter quantified the temperature and velocity in the stagnation plane. Thus, local and complete extinction were related to high local mean strain rates deduced from the velocity measurements. The photographs of unforced flames showed that their axial position was unstable with nozzle separations of 2.0 exit diameters and greater, while experimental access was restricted with separations less than 0.2 diameters. This defined a range of nozzle separations over which the arrangement could be used to examine extinction, and the photographs showed further that it occurred only with single-brush flames. The effects of equivalence ratio, bulk velocity, and nozzle separation were examined with reference to the condition $0.7 : 2.0$ m/s $: 0.4D$ (equivalence ratio : bulk velocity : nozzle separation distance). Twin-brush flames were observed at richer equivalence ratios, smaller bulk velocities, or larger nozzle separations, and these merged into a single-brush flame that eventually extinguished with leaner mixtures, higher velocities, or smaller separations. Nozzle separations of $1.0D$ or greater had near-uniform profiles of strain rate in the stagnation plane, while smaller values led to an increase from the axis to a peak at the edge of the impingement region. The amplitude of the peak increased as the separation was reduced. Thus, the flames quenched from large to small radii at the smaller separations. Complete extinction occurred when the local mean strain rate at the nominal stagnation point exceeded a critical value, quantified here as a function of equivalence ratio. The strain rates at the stagnation plane were at least 40% greater with combustion than with isothermal flow and caused incomplete combustion, lower temperatures, and nonlinear dependence of local strain rate on bulk velocity and nozzle separation. Measurements of instantaneous and fluctuating quantities in unforced flows provided evidence of local intermittent extinction and relight as a consequence of instantaneous strain rates above and below the critical value. When the amplitude of the fluctuations was further increased by oscillations, imposed with acoustic drivers, twin-brush flames merged to a single brush and the critical mean strain rate decreased. A reduction in the frequency of oscillations had the same effect, and extinction times were greatly reduced by modest increases in bulk velocity or a reduction in nozzle separation.

6.1   INTRODUCTION

This experimental investigation was motivated by the requirements of lean-premixed methane–air flames in modern gas-turbine combustors and the periodic extinction and relight observed close to the lean limit [1]. The first involves low equivalence ratios with possible dynamic effects, and the second involves a strain rate mechanism that may imply oscillations in bluff-body stabilized flames at all equivalence ratios. Opposed flames are used here to examine the nature of extinction, and to a lesser extent ignition; to quantify extinction velocities and times; and to determine limitations of this comparatively simple arrangement. The same arrangement was used in investigations of the corresponding isothermal flow [2].

The following section describes the flow configurations and instrumentation and provides an assessment of the uncertainties associated with the results of Section 6.3. The results are presented in two parts, with new information in unforced and forced flows, respectively.

6.2   EXPERIMENTAL SETUP

The two identical nozzles and their dimensions described previously [2] had contractions of area ratio 9.0 and followed a fifth-order polynomial [3] to a diameter of 25 mm, Fig. 6.1. The nozzle separation was varied between 0.2 and 2.0 exit diameters with bulk velocities from 1.49 to 7.00 m/s, and, since these velocities corresponded to Reynolds numbers of 2,000 and 10,000, a perforated plate was located at the end of the contraction with 4-millimeter diameter holes and 50% solidity. A subsequent straight pipe, two exit diameters in length, allowed the wakes to diminish and the small-scale turbulence to develop [4]. The two jets were mounted on a frame that allowed the separation to be varied while maintaining the same geometric axis. The compressed air and gas supply of natural methane was filtered, and the flow was measured with calibrated rotameters to accuracy better than 3%, while the centerline velocities were matched within 0.1 m/s.

A surrounding coaxial nitrogen flow is sometimes necessary in opposed flame experiments to prevent stabilization of a flame in the shear layer between the

jet and the surroundings. The lean
premixed flames of this investigation did not stabilize in the shear
layer, and a coaxial nitrogen flow
was not necessary. It is expected,
however, that the results presented
in the next section will be quantitatively similar to those reported
by previous authors where a surrounding co-flow was present. Previous experiments [5] showed that
a co-flow of nitrogen, of bulk velocity 0.25 times the inner jet, had
no effect on the exit velocity profile,
and it was reported [6] that a general coaxial flow of any density with
a specific momentum flux less than
that of the inner jet did not affect

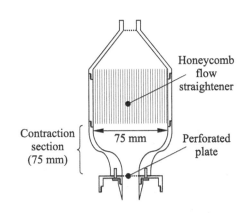

Figure 6.1 Sketch of the general shape of
the two identical nozzles (not to scale).

the exit flow. This suggests that it did not affect the profile of strain rate at the
stagnation plane. Moreover, streamline visualizations [5, 7, 8] showed that the
mixing layer at the boundary between the inner and outer jets did not reach the
stagnation plane within $1.5R$ of the axis so the co-flow could not affect the flame
in this region.

Photographs were obtained with a camera (Kodak Digital Science DC210
Zoom Camera) and interfaced to a computer (INTEL 586, 200 MHz, 32 MB
RAM) using commercially available software (Windows 95, Kodak DS Picture
Easy, Adobe PhotoDeluxe 2.0) and are presented in true color. Chemiluminescence images were acquired with a charge-coupled device (CCD) camera (Prox-
itronic fast motion HF1) in conjunction with a filter (431 nm for the CH-radical
emission line). The integration time was 0.5 ms per instantaneous image, and
the camera was triggered with a gated oscilloscope interfaced to an imaging
computer (INTEL 586, 200 MHz, 512 MB RAM) with 750 instantaneous images per set, acquired over 75 s. The mean chemiluminescence intensity was
acquired from the image set using custom software and was presented in false
color, and the uncertainties associated with estimating absolute intensity render
chemiluminescence imaging a qualitative tool, which produces maps indicative
of relative heat-release rates. The CCD camera was also used for instantaneous
images of ignition.

Profiles of mean and root-mean-squared (rms) temperatures were measured
at 1-millimeter intervals along the stagnation plane. The latter was identified
by the zero mean axial velocity as described below, of the opposed flames with
thermocouple junctions comprising butt-welded bare platinum with platinum–
rhodium (13%) wires of 50-micrometer diameter, with an aspect ratio of approx-

imately 100, and negligibly small conductive heat loss from the beads [9], while the thermocouple wire was oriented parallel to the flame in order to minimize temperature gradients [8]. The thermocouple was supported by wires of the same material of 150-micrometer diameter clad in a twin-bore mineral holder and housed in a stainless steel protective tube. The output of the thermocouple was amplified with a custom built low-noise differential amplifier and interfaced to a computer (INTEL 586, 200 MHz, 32 MB RAM), using an analog to digital converter card (16-bit ANALOGIC HSDAS) in conjunction with custom software. The uncertainty in the axial positioning of the probe was estimated to be ± 0.1 mm and ± 1 mm in the radial position, while the uncertainty in temperature from digitization noise was estimated to be less than ± 2 K [9]. The reproducibility of the measurements was determined by repeating a point measurement 30 times with a standard deviation of 20 K, equivalent to 1% of the mean.

6.3   RESULTS

Unforced Flames

The photographs in Figs. 6.2, 6.3, and 6.4 show the range of nozzle separations over which the opposed methane–air flame arrangement is likely to be useful for quantification of extinction strain rates. A combination of visual observation and photographic and chemiluminescence imaging determines qualitatively the position of locally quenched regions and the pattern of quenching leading to complete extinction, while mean temperature measurements allow comparison with laminar flows and quantify quenching. The effects of combustion on the flow field are quantified in terms of velocity measurements, and mean strain rates explain the pattern of quenching. Real-time temperature traces identify regions of intermittency and relight, which are linked to fluctuating strain rates inferred from measurements of velocity fluctuations.

Figure 6.2 corresponds to photographs of unstable and stable twin-brush flames, the former with a separation of $2.0D$ and similar to previous observations with propane–air mixtures [10]. Visual observation showed that the axial position of the unstable methane–air flames changed intermittently by up to $0.5D$ but the distance between the flames remained constant, and the instability could not be removed by changing the equivalence ratio or the bulk velocity. Thus, the range of separations likely to be useful for measurements of the extinction strain rate is limited to less than $2.0D$, while restricted experimental access prevented investigations with separations less than $0.2D$. Separations between 0.2 and 2.0 diameters yielded stable twin- and single-brush flames, as in Figs. 6.2 and 6.3, with the equivalence ratios, bulk velocities, and separations shown in the captions. The stable single-brush flame at $\Phi = 0.7$, 2.00 m/s, and $0.4D$ is used here as a reference against which to compare other flames.

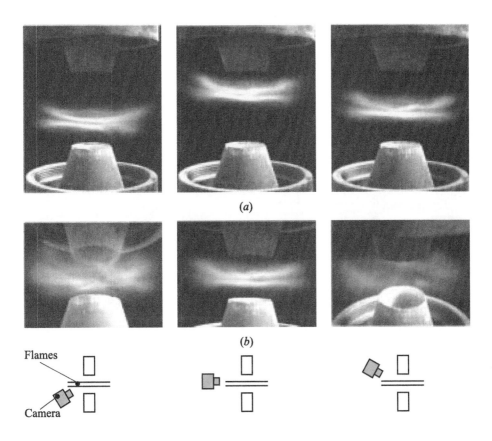

Figure 6.2 Unstable and stable twin-brush flames: (a) $\Phi = 0.7$, $U_b = 2$ m/s, and $H/D = 2.0$ (unstable, at three times); and (b) $\Phi = 0.7$, $U_b = 2$ m/s, and $H/D = 1.0$ (stable, at three angles shown below). (Refer color plate, p. II.)

Twin-brush flames did not extinguish but merged into single-brush flames at the stagnation plane with a reduction in equivalence ratio or higher strain rates, and extinction occurred only in the single-brush flames. Reduction in the equivalence ratio, from 0.7 of the reference condition to 0.65, extinguished the single-brush flame, while an increase to 0.8 resulted in two brushes, one on each side of the stagnation plane, which moved apart with further increase to stoichiometry. In contrast to the effects of equivalence ratio, an increase in bulk velocity from 2.00 to 2.48 m/s reduced the thickness of the single brush with extinction at higher values, while a reduction led to two brushes at 1.73 m/s, which moved further apart as the bulk velocity was reduced to 1.49 m/s. The twin-brush flames of Fig. 6.2 moved together from a flame-separation of 5 mm

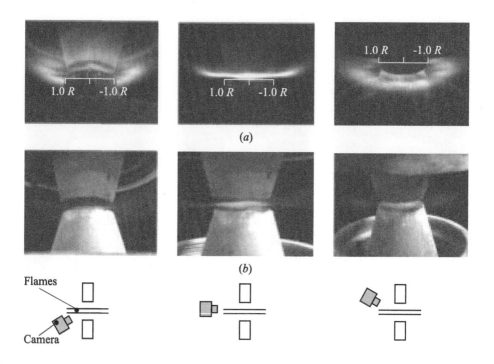

Figure 6.3 Stable single-brush flames and their extinction: (*a*) $\Phi = 0.7$, $U_b = 2$ m/s, and $H/D = 0.2$ (stable); and (*b*) $\Phi = 0.7$, $U_b = 2.16$ m/s, and $H/D = 0.2$ (partial extinction). Flames imaged at three different times and three angles shown below. (Refer color plate, p. III.)

as the nozzle separation was reduced from $1.0D$ until they merged to a single brush at the stagnation plane at the separation of $0.4D$.

The effects on flame-brush separation of equivalence ratio, bulk velocity, and nozzle separation were in qualitative agreement with the trends noted in propane–air flames [10]. The observation that extinction occurred only in the single-brush flame restricted the range of the three parameters useful for examining extinction to leaner mixtures, higher velocities, and smaller separations than the reference condition. Thus, single-brush flames and their extinction are the main topic of the remaining investigations, although selected measurements in twin-brush flames are presented for comparison with past temperature trends in laminar twin-flames and to contrast trends in the single-brush flames.

The angled images from above and below the stable single-brush flame in Fig. 6.3 indicate partial quenching in the form of a dark ring at a radial distance of $1.0R$ from the axis, while integration over the line of sight implied that it could

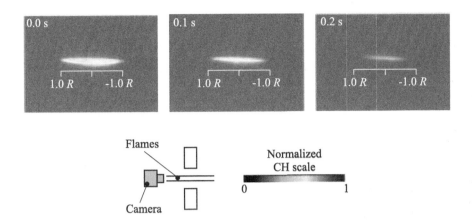

Figure 6.4 Instantaneous chemiluminescence images of an extinguishing single-brush flame: $\Phi = 0.7$, $U_b = 2.26$ m/s, $H/D = 0.2$, and image intervals $= 0.1$ s. (Refer color plate, p. III.)

not be seen when viewed from the side, as in the center image. The dark region also occurred, to a lesser extent, at $0.4D$ and was absent at larger separations. The quenched region at the smallest separation grew outwards to leave only the center part of the flame as the bulk velocity increased from 2.00 to 2.16 m/s, and temporal variation in the radial position of the edge of the flame disk was apparent. Extinction occurred when the bulk velocity was increased to 2.26 m/s, and the instantaneous chemiluminescence images of Fig. 6.4 demonstrate that the extinction process required a time of 0.3 s as the quenched region grew inwards from large to small radii. Observation confirmed that the pattern of extinction was the same at stoichiometry and likely to be related only to the distribution of strain rate on the stagnation plane.

The images in Fig. 6.3 show that the single-brush flame at the separation of $0.2D$ was weakest in the dark region at $1.0R$, and a minimum in the local temperature was likely to result from a maximum in the local strain rate at this radius, while the maximum strain rate was expected to reduce with an increase in separation as the dark region became less pronounced. As previously stated, there was no coaxial nitrogen flow, and previous visualizations showed that the mixing layer between the jet and the surroundings did not reach the stagnation plane within $1.5R$ of the axis and could not have been responsible for the observations at $1.0R$. The following extends the photographic information with quantitative measurements of mean temperature in the stagnation plane.

Figure 6.5 presents measured profiles of the mean and rms temperature in the stagnation plane as functions of bulk velocity and separation. The single-brush

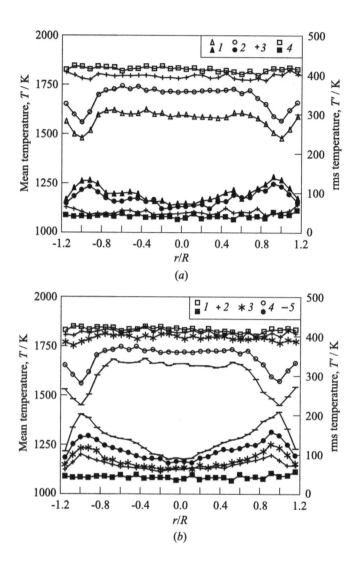

Figure 6.5 Profiles of mean (open symbols) and rms (filled symbols) temperature in the stagnation plane. (*a*) Effect of bulk velocity, $\Phi = 0.7$, $H/D = 0.4$: *1* — 2.48 m/s, *2* — 2.00, *3* — 1.73, and *4* — 1.49 m/s; *1* and *2* refer to single brush, *3* and *4* refer to twin brush. (*b*) Effect of normalized nozzle separation, $\Phi = 0.7$, $U_b = 2.00$ m/s: *1* — 1.00, *2* — 0.8, *3* — 0.6, *4* — 0.4, and *5* — 0.2; *1*, *2*, and *3* refer to twin brush, *4* and *5* refer to single brush.

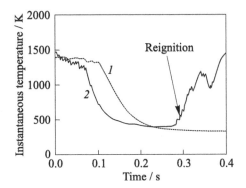

Figure 6.6 Temporal variation of temperature to extinction (*1*: $\Phi = 0.65$, $U_b = 2$ m/s, and $H/D = 0.4$) and with intermittent quenching (*2*: $\Phi = 0.7$, $U_b = 2$ m/s, and $H/D = 0.2$).

flames, considered here, occurred with bulk velocities of 2.00 m/s or greater and with nozzle separations of 0.4 or less, while the twin-brush flames, discussed in a later subsection, occurred at smaller velocities and larger separations.

The mean temperature at the axis of the single-brush flame at the reference condition above was 1717 K, and it reduced with radius to a minimum of 1565 K at $1.0R$, the location of the dark region in Fig. 6.3. Both values were lower than the unstrained adiabatic flame temperature, estimated at 1850 K, as expected in a turbulent flow and from the previous findings of incomplete combustion in single-flames [11]. Selected measurements confirmed that a reduction in equivalence ratio from 0.7 caused a decrease in the temperature at the axis and showed that the flame extinguished at 0.65. It is also apparent from Fig. 6.5 that an increase in bulk velocity from 2.00 to 2.48 m/s reduced the temperature at the axis from 1717 to 1600 K and in the dark region from 1565 to 1480 K, probably due to increasingly incomplete reaction, which led to extinction at higher velocities. The figure shows that the temperature at the axis also dropped from 1717 to 1657 K as the separation was reduced from $0.4D$ to $0.2D$, and the large decrease from 1657 to 1444 K at $1.0R$ explains the increasingly pronounced dark region in the images of Fig. 6.3.

The real-time temperature trace at the axis, Fig. 6.6, shows the variations in temperature until extinction at 1230 K, similar to reported values of 1250 K [12], after which the thermocouple measured an exponential temperature decay slowed by its finite thermal inertia. The temperatures were close to a previously calculated value of 1200 K and provide support for the chemical extinction mechanism that was proposed [13]. It is apparent from Fig. 6.6 that the amplitude of high-frequency temperature fluctuations was small, in part a consequence of the finite thermal inertia, while extinction and relight occurred in the dark region at $1.0R$ from the axis and are discussed later in the text.

Figure 6.5 shows that large temperature fluctuations occurred in the regions of low mean temperature close to extinction with rms of temperature fluctuations in the single-brush flames up to 100 K. The rms increased with radius to a maximum of 207 K in the quenched dark region and the minimum of mean temperature. The real-time temperature trace in the dark region at the separation of $0.2D$, Fig. 6.6, shows that the flame quenched locally at 0.08 s with relight at 0.28 s, and the intermittent quenching explains the large temperature fluctuations and low mean temperatures close to extinction. The probability distribution function of temperature in the dark region $1.0R$ from the axis in the single-brush flame with $0.4D$ separation was bimodal with increasingly separated peaks as the bulk velocity increased from 2.00 to 2.48 m/s, and this suggests that the flame was quenched for longer periods before relight as the velocity increased.

The mean axial velocity was zero in the stagnation plane, and profiles of the radial component included the single-brush flame at the reference condition at the equivalence ratio of 0.7 and the effects of the three main variables. The radial velocity increased along the stagnation plane from the axis as the hot flame products expanded; and the flow was accelerated by the favorable pressure gradient, up to a maximum of $4.5U_b$ at $1.4R$, after which visualizations [5, 7, 8] showed that the mixing layer between the radial jet and the surroundings reached the stagnation plane. It was also apparent that the qualitative shape and symmetry of the profiles were preserved with an increase in equivalence ratio. The mean radial velocity within $0.5R$ of the axis of the single-brush flame increased almost linearly with bulk velocity from 2.00 to 2.48 m/s and was nonlinear further from the axis where the temperature measurements of Fig. 6.5 suggested incomplete combustion, and linear extrapolation increasingly overestimated the radial velocity by up to 16% in the dark region. A reduction in nozzle separation from $0.4D$ to $0.2D$ also increased the amplitude of the peak in the profile of radial velocity from $4.8U_b$ to $8.6U_b$, probably due to a larger pressure gradient along the axis, but, in contrast to the effects of equivalence ratio and bulk velocity, the width of the peak increased and the changes away from the axis were disproportionately larger with qualitative alterations to the shape of the profile.

Mean radial strain rates were calculated from the velocities, and their gradients and the profiles showed that the strain rate in the linear region close to the axis of the single-brush flame at the reference condition was constant at 800 s^{-1} and the maximum value of 1316 s^{-1} was at $1.0R$, where Figs. 6.3 and 6.6 showed a dark region and a minimum in temperature. An increase in equivalence ratio resulted in larger strain rates at the stagnation plane but without extinction, as a consequence of higher temperatures and greater density changes during combustion. The position of the peaks was unaffected, and their amplitude increased proportionally with increases close to the axis so that the profiles appeared to scale with mixture fraction and probably with the heat-release rate. The strain rate within $0.5R$ of the axis increased linearly from 540 to 920 s^{-1}, with bulk velocities from 2.00 and 2.48 m/s, with a smaller non-linear increase in the value at the peak of the profile, from 1315 to 1401 s^{-1}, probably as a consequence of local reduction in the heat-release rate due to the high strain rates, suggesting that the strain rates did not scale with bulk velocity close to extinction. A reduction in separation from $0.4D$ to $0.2D$ resulted in a relatively large increase in the amplitude of the peak in the strain rate distribution, from 1300 to 3000 s^{-1}, where the dark region became increasingly pronounced and despite obvious reduction in the local heat-release rate. The increase in strain rate with nozzle separation in the linear region close to the axis, from 807 to 940 s^{-1}, was disproportionately smaller than the effects away from the axis as the flame approached extinction, and this confirms that reduction in separation resulted in changes in the shape of the profile of strain rate.

The flame quenched locally at the peak of the strain rate distribution when the value exceeded 1000 s^{-1} and formed the dark region of Fig. 6.3. The amplitude of the peak in the partially extinguished flame of Fig. 6.3, where only the mid-part of the flame was left, was 3000 s^{-1}. The region with local strain rates greater than the critical value of 1000 s^{-1} moved towards the axis with an increase in bulk velocity and explained the pattern of extinction from large to small radii. Complete extinction occurred when the local mean strain rate at all points along the stagnation plane was greater than the critical value, and the shape of the profiles showed that this criterion was satisfied when the strain rate at the nominal stagnation point exceeded the critical value.

Forced Flames

Chemiluminescence images of unquenched and partially extinguished single-brush flames, with bulk velocities of 2.00 and 2.16 m/s and separations of $0.4D$ and $0.2D$, are shown in Fig. 6.7 together with images of flames forced at 100 Hz and triggered at 90-degree intervals in phase angle. The forcing caused the rms of the axial velocity fluctuations at the nominal stagnation point to increase from $0.40U_b$ to $0.70U_b$ and from $0.44U_b$ to $0.74U_b$ with separations of $0.4D$ and $0.2D$, respectively. The figure shows that the average luminosity of the forced flames was a function of phase angle, with a minimum and a maximum at 90° and 270°, respectively, and comparison of the forced and unforced images shows that the minimum and maximum luminosities were less and greater than the average luminosity of the unforced flame. The previous observations

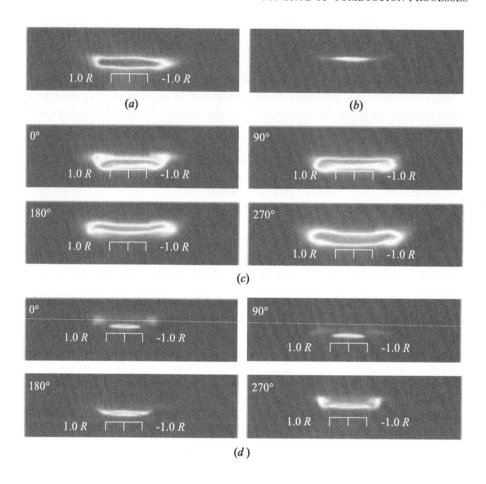

Figure 6.7 Chemiluminescence images of a single-brush flame, unforced: (a) and (b); and forced at 100 Hz: (c) and (d); (a) and (c): $\Phi = 0.7$, $U_b = 2$ m/s, and $H/D = 0.4$; (b) and (d): partially quenched, $\Phi = 0.7$, $U_b = 2.16$ m/s, and $H/D = 0.2$. (Refer color plate, p. IV.)

of a negative correlation between luminosity and instantaneous strain rate [14] at the larger separation of $0.8D$, and the findings in unforced flows that a drop in luminosity and temperature resulted from an increase in strain rates, make it likely that the minimum and maximum luminosity corresponded to minimum and maximum instantaneous strain rates. Comparison of the forced and partially quenched flame disks at the separation of $0.2D$ demonstrates that the forced flame quenched and reignited periodically at the edge of the flame disk.

The images of the single-brush flame were complemented by profiles of temperature, velocity, and calculated strain rates at the stagnation plane (not shown here), and the effects of forcing amplitude and frequency on the reduction in the mean extinction strain rate close to the axis were quantified while previous results [15] were extended to include the effects of bulk velocity and separation on extinction times. The amplitude of imposed oscillations was quoted in terms of the rms of the axial velocity fluctuations at the nominal stagnation point normalized by the bulk velocity [14].

The mean temperature of the single-brush flame close to the axis was reduced from 1717 to 1514 K as the forcing amplitude was increased to $1.0U_b$, and from 1560 to 772 K at the edge of the flame disk, in the region of maximum mean strain rate in unforced flows, probably as a result of periodic quenching. The mean radial velocity within $0.2R$ of the axis, where the reduction in mean temperature was smaller, decreased by less than 10% as the forcing amplitude increased and decreased by up to 25% in the region of largest temperature drop. The mean radial strain rate decreased by 16% with forcing, from 807 to 677 s^{-1} close to the axis, and the peak at $1.0R$ was in the region of the largest temperature drop with magnitude reduced by 25%, from 1315 to 985 s^{-1}, as expected from the reduction in mean velocity. Thus, the increase in forcing amplitude resulted in a reduction in the mean temperature of the flame, particularly at the peak of the mean strain rate distribution where it was weakest, and this led to a reduction in the mean velocity. Extinction was observed at large amplitudes due to large instantaneous strain rates from which the flame was unable to relight.

Complete extinction occurred when the forcing amplitude was greater than $1.0U_b$, and the critical mean radial strain rate close to the axis was 730 s^{-1}, smaller than the critical value of 1000 s^{-1} in the unforced flow. It is apparent that instantaneous strain rates did not vary from 0 to 2000 s^{-1}, despite the large axial rms of $1.0U_b$, and it is likely that coherent flapping was responsible in part for the fluctuations. Previous experiments [14] showed that forced flames withstood instantaneous maximum strain rates 15% larger than the critical unforced values, and this, together with the critical unforced value of 1000 s^{-1} observed here, suggests that the instantaneous values were greater than 1150 s^{-1}. In addition, the critical forced value of 730 s^{-1}, some 270 s^{-1} less than that without forcing, implies that instantaneous maximum values may have been as great as 1270 s^{-1}.

6.4   CONCLUDING REMARKS

The range of useful nozzle separations was limited to less than $2.0D$ to prevent flapping and to $0.2D$ or greater due to restricted experimental access. Thus, the effects of equivalence ratio, bulk velocity, and nozzle separation were determined with respect to a reference condition of 0.7, 2.0 m/s, and $0.4D$, which gave rise to a single-brush flame. Extinction occurred with leaner mixtures, higher bulk velocities, or smaller separations, namely, as the equivalence ratio was reduced to 0.65, the bulk velocity increased to greater than 2.48 m/s, or the nozzle separation reduced below $0.2D$. Twin-brush flames were observed at richer equivalence ratios, lower bulk velocities, or larger separations and were not of interest here since they did not extinguish.

Local quenching by local strain rates in excess of 1000 s^{-1} was evident from low local temperatures of the order of 1444 K, at $1.0R$ from the axis of symmetry. Photographs showed an increasingly pronounced dark region at this radius as the nozzle separation was reduced from $0.4D$ to $0.2D$. Relight at the lower strain rates at greater radii was prevented when the local strain rate at $1.0R$ reached 3000 s^{-1}, leaving a central flame at 2.16 m/s, $0.2D$. Chemiluminescence images showed that a further increase in bulk velocity led to complete extinction at 2.26 m/s in 0.3 s.

During the extinction process, the edge of the central flame disk moved inwards towards the axis as the local strain rate reached 1000 s^{-1}. The critical mean strain rate at the axis increased, from 640 to 1840 s^{-1}, with equivalence ratios from 0.6 to stoichiometry. The values were independent of separation, to within 10%, despite increasing nonuniformities in the profiles of the local mean strain rate from $1.0D$ to $0.2D$. It is likely that the critical local values of the strain rate, quantified in the combusting flow, were not geometry specific and could be applied to determine conditions of local quenching in more complex geometries. This was not possible with the bulk strain rate, which was a strong function of nozzle separation in addition to equivalence ratio.

Maximum and minimum instantaneous local strain rates that were above and below the critical value were responsible for intermittent local quenching and relight, leading to high levels of temperature fluctuations that lowered the mean temperature. Imposed oscillations led to local extinction and relight at the forcing frequency supporting the strain rate mechanism proposed in ducted self-oscillating flames [1].

The amplitude of acoustic forcing was measured in terms of the increase in the rms of the velocity fluctuations at the stagnation plane normalized by the bulk velocity [14], and amplitudes of up to $1.02U_b$ reduced the overall temperature of the flame from 1717 to 1514 K on the axis and from 1558 to 772 K at $1.0R$. This, in turn, led to a reduction in the mean radial strain rate close to the axis and with larger reductions further from it. The critical mean radial strain rate at the stagnation point decreased with forcing amplitude, and a sharp drop towards zero occurred at amplitudes of over $1.0U_b$. The effects of forcing were greater at lower frequencies with which the high strain rate part of the forcing cycle was applied for a longer time. The gradual weakening of the flame during forced extinction, despite low mean strain rates, shows that it was caused by peak instantaneous values from which the flame was unable to relight completely. The measurements of extinction times extended previous results [14] to show that, in addition to the increase with equivalence ratio, a 10% increase in bulk velocity reduced the extinction times by a factor of 5 and a reduction in separation from $1.0D$ to $0.4D$ had the same effect.

ACKNOWLEDGMENTS

The work was performed under ONR contract N00014-99-1-0832.

REFERENCES

1. De Zilwa, S. R. N., I. Emiris, J. H. Uhm, and J. H. Whitelaw. 2001. Combustion of premixed methane and air in ducts. *Proc. Royal Society London* 457:1915–49.

2. Korusoy, E., and J. H. Whitelaw. 2001. Opposed jets with small separations and their implications for the extinction of opposed flames. *Experiments Fluids* 31:111–17.

3. Bell, J. H., and R. D. Mehta. 1988. Contraction design for small wind tunnels. NASA Contractor Report No. 177488.

4. Villermaux, E., Y. Gagne, and E. J. Hopfinger. 1993. Self-sustained oscillations and collective behaviours in a lattice of jets. *Applied Scientific Research* 51:243–48.

5. Rolon, J. C., D. Veynante, J. P. Martin, and F. Durst. 1991. Counter-jet stagnation flows. *Experiments Fluids* 11:313–24.

6. Favre-Marinet, M., E. B. Camano, and J. Sarboch. 1999. Near-field of coaxial jets with large density differences. *Experiments Fluids* 26:97–106.

7. Kostiuk, L. W. 1991. Premixed turbulent combustion in counterflowing streams. Ph.D. Thesis No. 16831. Churchill College, University of Cambridge, UK.

8. Mastorakos, E. 1993. Turbulent combustion in opposed jet flows. Ph.D. Thesis. Imperial College, London, UK.

9. Heitor, M. 1985. Experiments in turbulent reacting flows. Ph.D. Thesis. Imperial College, University of London, UK.

10. Kostiuk, L. W., K. N. C. Bray, and R. K. Cheng. 1993. Experimental study of premixed turbulent combustion in opposed streams. Part II — Reacting flow field and extinction. *Combustion Flame* 92:396–409.

11. Yamaoka, I., and H. Tsuji. 1991. The effect of back diffusion of intermediate hydrogen on methane–air and propane–air flames diluted with nitrogen in a stagnating flow. *Combustion Flame* 86:135–46.

12. Sato, J. 1982. Effect of Lewis number on extinction behaviour of premixed flames in a stagnation flow. *19th Symposium (International) on Combustion Proceedings.* Pittsburg, PA: The Combustion Institute. 1541–48.

13. Law, C. K., D. L. Zhu, and G. Yu. 1986. Propagation and extinction of stretched premixed flames. *21st Symposium (International) on Combustion Proceedings.* Pittsburg, PA: The Combustion Institute. 1419–26.

14. Sardi, E., A. M. K. P. Taylor, and J. H. Whitelaw. 2000. Extinction of turbulent counterflow flames under periodic strain. *Combustion Flame* 120:265–84.

15. Sardi, E., and J. H. Whitelaw. 1999. Extinction timescale of periodically strained, lean counterflow flames. *Experiments Fluids* 27:199–209.

Chapter 7

INFLUENCE OF MARKSTEIN NUMBER ON THE PARAMETRIC ACOUSTIC INSTABILITY

N. J. Killingsworth and R. C. Aldredge

Laminar, premixed methane–air flames propagating through the annulus of a Taylor–Couette (TC) combustor were studied experimentally. Flame speeds measured during a stage of quasiplanar propagation near the middle of the combustor attained through action of the primary acoustic instability were found to agree well with laminar-flame speeds reported in the literature. Axial and circumferential velocity fluctuations of the flow field were measured during flame propagation using Laser Doppler Velocimetry (LDV), and critical acoustic-velocity amplitudes for the onset of the secondary, parametric acoustic instability were determined. The influence of the equivalence ratio on the critical acoustic-velocity amplitude determined from the LDV measurements is shown to agree well qualitatively with theoretical predictions. Finally, the stability of sufficiently rich methane–air flames found in the reported experiments, in contrast to unstable flame propagation found by earlier investigators for all rich propane–air mixtures, is explained through consideration of the variation of a Markstein number with the equivalence ratio for methane–air mixtures, which differs qualitatively from that for propane–air mixtures.

7.1 INTRODUCTION

Background

Premixed flames may be influenced by the Darrieus–Landau hydrodynamic instability [1, 2] when the chemical heat release is sufficiently large. Hydrodynamically unstable premixed flames are not always observed, however, because of stabilizing influences of buoyancy and thermal diffusion. The long-wavelength flame-surface wrinkles are attenuated by buoyancy for downward propagating flames, and thermal-diffusive effects stabilize small-wavelength wrinkles when the

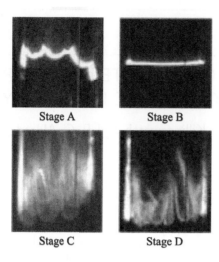

Stage A   Stage B

Stage C   Stage D

Figure   7.1   Four stages of propagation found in the TC burner: Stage A — wrinkled flame, due to the Darrieus–Landau instability; Stage B — flat flame, due to the primary pyroacoustic instability; Stage C — pulsating cellular flame, due to the secondary pyroacoustic instability; and Stage D — turbulent flame.

Lewis number based on the limiting reactant is sufficiently large [3]. Therefore, the Darrieus–Landau instability is expected to result in enhanced flame-surface wrinkling only when it is able to overcome stabilizing influences of buoyancy and thermal diffusion. Additionally, a flame under confinement may be influenced by acoustic waves, and coupling between the flame and acoustic-wave dynamics may result in unstable flame propagation as well.

Experimental studies of downward flame propagation in tubes of circular and annular cross-section ([4] and [5], respectively) away from the open end and toward the closed end have revealed four distinct stages of propagation as the flame traverses the length of the tube. Upon ignition, the flame surface is initially wrinkled due to the Darrieus–Landau instability, and, if the amplitude of the wrinkles is sufficiently large as the flame enters the bottom half of the tube, sound is generated at the fundamental tube frequency as a result of a primary pyroacoustic instability [4–7]. As the amplitude of the axial acoustic waves associated with the primary pyroacoustic instability steadily increases, the flame becomes flat. Once the amplitude of the axial acoustic waves reaches a critical value, a secondary pyroacoustic instability develops that is associated with the onset and growth of transverse acoustic waves and cells on the flame surface that pulsate with a period twice that of the acoustic waves [4, 8, 9]. For nearly stoichiometric mixtures of either propane or methane and air, continued rapid growth of both axial and transverse acoustic-wave amplitudes was found to lead to a transfer of energy from acoustic fluctuations to random, turbulent fluctuations and flame acceleration [4, 5]. Figure 7.1 shows examples of annular flame fronts found for each of the four stages of propagation in the TC burner used in the present study.

The mechanisms that govern pyroacoustic instability must satisfy the Rayleigh criterion [10], which states that over the cycle of an acoustic wave the flame must add more energy during the positive-pressure phase than during the negative-pressure phase. If this is accomplished then an acoustic wave will be amplified, otherwise it will die out as a result of viscous damping. The overall heat release by the flame must therefore periodically change in a specific manner so as to feed the acoustic waves. One method in which this may be achieved is through coupling between acoustic-velocity fluctuations and flame-surface-area variations. Variations in surface area along the flame front are associated with variations in local flame curvature, the effect of which on the local heat release of the flame may be characterized through the Markstein number (Ma) [11]. A positive (negative) Markstein number would indicate that positive curvature (i.e., a flame concave towards the burnt gases) would decrease (increase) the local reaction-zone temperature and thus have a stabilizing (destabilizing) effect on flame propagation. Therefore, influences associated with the value of the Markstein number are integral to the stability of a premixed flame subject to periodic variation in flame curvature induced by an acoustic field.

Focus of Present Work

In an earlier study of downward flame propagation in a tube open at the top and closed at the bottom using propane–air mixtures [4], it was found that flame flattening occurred in lean but not in rich mixtures when the flame reached the middle of the tube. Consistent with these findings, it is hypothesized that flames in rich propane–air mixtures are not flattened by the primary acoustic instability [9] because the Markstein number for propane–air mixtures decreases strongly with increasing equivalence ratio and can therefore be significantly less for rich mixtures than for lean mixtures [11]. On the other hand, methane–air mixtures have a Markstein number that increases with an increasing equivalence ratio, and so it might be possible for flames in rich methane–air mixtures to be flattened by the primary acoustic instability. To evaluate this hypothesis, flame propagation in lean and rich methane–air mixtures, having different Markstein numbers but similar laminar-flame speeds, can be compared to determine the influence of the Markstein number on the ability of the primary acoustic instability to flatten the flame. An evaluation of this hypothesis is one aim of the present work.

Another aim is an examination of the influence of the Markstein number on the effect of the secondary pyroacoustic instability, associated with the development of transverse acoustic waves and pulsating cells on the flame [4, 5]. In addition, laminar-flame speeds measured during the stage of a nearly flat surface (Stage B in Fig. 7.1) are compared with those reported in the literature.

7.2 EXPERIMENTAL PROCEDURE

The experimental configuration is shown in Fig. 7.2. The TC combustor features an anodized aluminum inner cylinder with an outer radius of 7.9 cm and a length of 60 cm. The outer cylinder has an inner radius of 9 cm and a length of 62 cm and is constructed of Pyrex to allow optical accesses into the annulus. The annulus has a width of 1.1 cm and is open to atmosphere at the top and closed at the bottom. The two cylinders can be rotated independently in either direction up to 3450 revolutions per minute to generate turbulence within the annulus, but for the current study the cylinders are held stationary.

The reactants are metered using two digital mass-flow controllers and then seeded with 2-micrometer siliconcarbide particles. The seeded gas enters the annulus via 900 holes 2 mm in diameter around the base of the inner cylinder. The upward mean-velocity of the reactant mixture is maintained at 10 cm/s; five annulus volumes of the reactant mixture are allowed to flow through the combustor before ignition. The reactants are ignited at the top of the apparatus by two spark plugs, with the two sets of electrodes 180° apart. Additionally, a loose paper lid is placed over the top of the annulus because it was found to provide more consistent ignition (it is thought to reduce dilution of the mixture with ambient air near the electrodes), but not found to have any other substantial effect on the experiments. A one-component LDV system was used to measure the instantaneous flow velocity at one point, employing a backscatter configuration with the probe perpendicular to the centerline of the cylinders. When measuring circumferential velocities, the two beams lie in a horizontal plane and are symmetric about the centerline of the cylinders, such that as the beams pass through the curved glass any beam deflection is symmetric about the centerline of the fiberoptic probe. However, the fringe spacing is slightly altered due to refocusing of the beams as they pass through the curved glass. This results in a 0.03% error in the velocity of the circumferential component. More details about the exper-

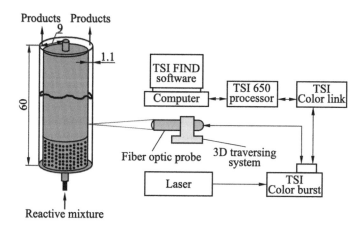

Figure 7.2 A schematic of the TC burner and velocity measurement setup. Dimensions are in cm.

imental setup and LDV system used can be found in [12]. Flame propagation is recorded at 30 frames per second using a Sony CCD-TR3300 video camera.

7.3 RESULTS

Planar-Flame Speeds

While the flame is planar, under the influence of the pyroacoustic instability, its speed is measured by recording the downward movement of the flame front using a video camera, taking into account the upward mean flow of reactants. Each calculated flame speed plotted in Fig. 7.3 represents the ensemble average of measurements from five experiments containing quasi-planar completely annular flames. The error bars show the experimental uncertainty of these measurements. The planar-flame speeds from this study are compared to laminar-flame speeds from experimental studies of weakly stretched

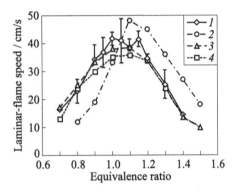

Figure 7.3 Methane–air planar-flame speeds measured in the TC burner (1) compared with those obtained in earlier studies: 2 — [15], 3 — [13], and 4 — [14].

flames, in which the laminar burning velocity is the extrapolated value at zero stretch [13, 14]. These studies use two different experimental techniques: the counterflow twin-flame technique [13] and spherically expanding flames [14]. Flame speeds obtained in an earlier study [15] using the same burner are also shown in this plot.

The planar-flame speeds from the present study agree well with those of [13] and [14] for slower flames, well within experimental uncertainty. However, for faster flames the agreement is not as good; this is because the quasi-planar flames in the TC combustor are somewhat wrinkled, resulting in an increase in the mass burning rate. In addition, as can be seen in Fig. 7.3, the results of this study do not agree with those of [15]. This is attributed to improvement of the gas-flow control system since the earlier study. The previous system resulted in equivalence ratios with uncertainties as high as 7% (95% confidence). Therefore, the equivalence ratio at 1.1 is known only to within ±0.075 and could have any value between 1.025 and 1.175. The newer system incorporates digital mass-flow controllers and has a maximum equivalence ratio uncertainty of 1.5% (99.7% confidence). Since the flame speed of a methane–air mixture is very sensitive to the proportions of air and fuel, uncertainty in the equivalence ratio directly translates to uncertainty in the flame speed for a known equivalence ratio. In addition, there exists experimental uncertainty directly associated with determining the flame speed because of the variation in the surface area of the nearly-planar flame from experiment to experiment for flames close to stoichiometric. Uncertainty bars are shown for the measurements of each equivalence ratio in Fig. 7.3. Note that if the curve from [15] is shifted to the left in Fig. 7.3, it agrees reasonably well with that of the present study, which suggests that the deviation between this work and [15] is explained by the improvement in experimental accuracy of the equivalence ratio measurements.

Parametric Instability

Searby and Rochwerger [9] developed a model describing the effect of an acoustic field on the stability of a laminar, premixed flame, treated as a thin interface between two fluids of different densities and under the influence of a periodic gravitational field. Their model is an extension of the work by Markstein [8] and is consistent with the more recent flame theory of Clavin and Garcia-Ybarra [16]. Bychkov [17] later solved the problem analytically, presenting the following linear equation for the perturbation amplitude, f, of a flame under the influence of an acoustic field [17]:

$$A\frac{d^2 f}{dt^2} + BU_f k\frac{df}{dt} - CU_f^2 k^2 f + C_1 gkf = 0 \qquad (7.1)$$

where U_f is the laminar-flame speed, k is the wave number of the wrinkled flame, and g is the effective acceleration, which in general form can take into account a time-dependent acceleration. The dimensionless coefficients A, B, C, and C_1 are given in [17] and depend on the wave number k and on flame parameters such as the gas expansion ratio Θ, flame thickness L_f, Markstein number Ma, and Prandtl number Pr. Equation (7.1) has one solution under which the flame is unstable due to the Darrieus–Landau instability and another for which the flame is unstable due to the parametric instability. For all other solutions, the acoustic field stabilizes the flame.

Stability plots for two methane–air flames using the analytical solution of Bychkov [17] with parameters for equivalence ratios of 0.8 and 1.3 are shown in Fig. 7.4. The y-axis corresponds to the acoustic velocity normalized by the laminar-flame speed, and the x-axis is the wave number scaled by the flame thickness. However, within this model the Markstein parameter (Eq. (8) in [17]) was modified to incorporate an effective Lewis number to allow use of the model for mixtures at and near stoichiometric composition [11]. A frequency of 920 radians per second is selected, which corresponds to the quarter-wavelength frequency of the TC burner during the parametric instability, and the activation

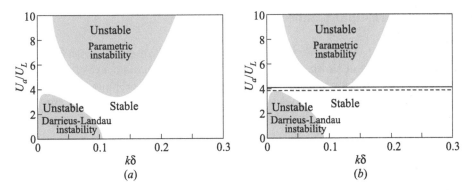

Figure 7.4 Stability plots for laminar flames in methane–air mixtures with equivalence ratios of 0.8 (*a*) and 1.3 (*b*).

energy ($E/R = 23{,}873$ K) is from [18]. Other parameters were obtained from a web interface [19] that computes chemical equilibrium and transport properties using CHEMKIN [20–22] and STANJAN [23].

The model predicts that methane–air flames with an equivalence ratio of 0.8 propagating downward are always unstable as a result of either the Darrieus–Landau instability or the secondary pyroacoustic instability, also known as the parametric instability, yet flames with an equivalence ratio of 1.3 will be stabilized by a small band of normalized acoustic-velocity amplitudes between 3.7 and 4. Although it was found experimentally that a methane–air flame with an equivalence ratio of 0.8 can be stabilized, the results of the model agree qualitatively with the experimental findings, specifically that a methane–air flame with an equivalence ratio of 1.3 is stable over a larger range of acoustic amplitudes than one with an equivalence ratio of 0.8.

For flames that exhibit the parametric instability, the velocity at which the exponential growth of velocity fluctuations started for each experiment was noted. These critical velocities are shown in Fig. 7.5, normalized by the laminar-flame speeds reported in [13]. All points shown on this plot represent the ensemble average of measurements from five experiments, and the error bars indicate the standard deviation about the mean value. The other curve on this plot was calculated using the analytical model of a premixed flame under the influence of an oscillating gravitational field by Bychkov [17], as described above. Each point represents the smallest normalized acoustic velocity at the most unstable reduced wave number that resulted in the parametric instability. The experimental results show the same trend as the theoretical model mixtures with an equivalence ratio of 0.9, which require the smallest normalized acoustic velocity to trigger the parametric instability while flames on either side require larger values.

Effect of Markstein Number

It is interesting to compare the behavior of methane–air flames with different equivalence ratios but similar flame speeds to elucidate the influence of the Markstein number. Markstein numbers for methane–air and propane–air mixtures over a range of equivalence ratios are shown in Fig. 7.6. The Markstein numbers were calculated using the theory of Bechtold and Matalon [11] and using the same parameters previously used to find the critical velocities of the parametric instability. Flames with methane–air equivalence ratios of 0.8 and 1.3 have very similar laminar-flame speeds of 25 and 24 cm/s, respectively, yet their growth rates of the parametric instability are somewhat different, 28.3 and 14.3 s^{-1}. Similarly, when comparing mixtures with equivalence ratios of 0.9 and 1.2, with laminar-flame speeds of 33 and 34 cm/s, the same trend exists; they have corresponding growth rates of 27.7 and 19.6 s^{-1}, respectively. The methane–air mixture with the lower Markstein number has a larger growth rate of the parametric instability. Likewise, the peak axial velocity fluctuations are higher in the case of the mixture with the lower Markstein number. This behavior was not addressed in earlier studies, which employed only lean mixtures.

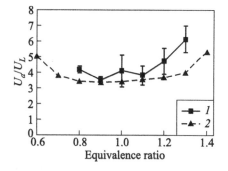
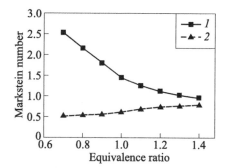

Figure 7.5 Normalized critical axial acoustic velocity for onset of the parametric instability plotted vs. equivalence ratio: *1* — experiment and *2* — analytical [17].

Figure 7.6 Markstein number variation with equivalence ratio for propane–air (*1*) and methane–air (*2*) mixtures.

7.4 CONCLUDING REMARKS

In this work, the velocity field ahead of the flame propagating in a TC combustor was measured using LDV for methane–air premixed laminar flames with equivalence ratios ranging from 0.7 to 1.4. The results for rich methane–air flames help clarify important parameters that influence the occurrence of the four stages of propagation. For example, this work shows that mixtures with similar flame speeds can have very different behaviors due to their level of susceptibility to stretch effects as characterized through the Markstein number. In the case of methane–air flames, lean mixtures have a smaller Markstein number and are therefore more prone to instabilities for a given flame speed than rich mixtures with the same flame speed. The opposite trend was found for propane–air flames; rich propane–air flames are more unstable than lean flames. This result is in qualitative agreement with recent theory [9, 24] describing a premixed laminar flame in an acoustic field. It was also found that the strength of the parametric instability was much greater for lean methane–air flames. In summary, the flame speed and the Markstein number are critical in determining how a mixture will be affected by acoustic waves. Rich methane–air flames were found to have regions of stability due to the periodic acceleration of the flame by acoustic waves, whereas Searby and Rochwerger [9] found that all rich propane–air flames are unstable. This difference is attributed to the opposite trends of the Markstein number with an equivalence ratio for these two mixtures.

ACKNOWLEDGMENTS

The work was performed under ONR contract N00014-96-1-0419.

REFERENCES

1. Darrieus, G. 1938. *Propagation d'un front de flame*. Paris: La Technique Moderne.
2. Landau, L. D. 1944. On the theory of slow combustion. *Acta Physicochimica* 19:77.
3. Williams, F. A. 1985. *Combustion theory*. Reading, MA: Addison-Wesley.
4. Searby, G. 1992. Acoustic instability in premixed flames. *Combustion Science Technology* 81:221.
5. Vaezi, V., and R. C. Aldredge. 2000. Laminar-flame instabilities in a Taylor–Couette combustor. *Combustion Flame* 121:356.
6. Pelce, P., and D. Rochwerger. 1992. Vibratory instability of cellular flames propagating in tubes. *J. Fluid Mechanics* 239:293.
7. Clanet, C., G. Searby, and P. Clavin. 1999. Primary acoustic instability of flames propagating in tubes: Cases of spray and premixed gas combustion. *J. Fluid Mechanics* 385:157.
8. Markstein, G. H. 1964. In: *Nonsteady flame propagation*. Ed. G. H. Markstein. New York: The MacMillan Comp.
9. Searby, G., and D. Rochwerger. 1991. A parametric acoustic instability in premixed flames. *J. Fluid Mechanics* 231:529.
10. Rayleigh, J. W. S. 1878. The explanation of certain acoustical phenomena. *Nature* 18:319.

11. Bechtold, J. K., and M. Matalon. 2001. The dependence of the Markstein length on stoichiometry. *Combustion Flame* 127:1906.

12. Vaezi, V., E. S. Oh, and R. C. Aldredge. 1997. High-intensity turbulence measurements in a Taylor–Couette flow reactor. *J. Experimental Thermal Fluid Science* 15:424.

13. Vagelopoulos, C. M., F. N. Egolfopoulos, and C. K. Law. 1994. Further considerations on the determination of laminar flame speeds with the counterflow twin-flame technique. *22nd Symposium (International) on Combustion Proceedings*. Pittsburgh, PA: The Combustion Institute. 1341.

14. Hassan, M. I., K. T. Aung, and G. M. Faeth. 1998. Measured and predicted properties of laminar premixed methane/air flames at various pressures. *Combustion Flame* 115:539.

15. Aldredge, R. C., V. Vaezi, and P. D. Ronney. 1998. Premixed-flame propagation in turbulent Taylor–Couette flow. *Combustion Flame* 115:395.

16. Clavin, P., and P. Garcia-Ybarra. 1983. The influence of the temperature dependence of diffusivities on the dynamics of flame fronts. *J. de Mecanique Appliquee* 2:245.

17. Bychkov, V. 1999. Analytical scalings for flame interaction with sound waves. *Physics Fluids* 11:3168.

18. Muller, U. C., M. Bollig, and N. Peters. 1997. Approximations for burning velocities and Markstein numbers for lean hydrocarbon and methanol flames. *Combustion Flame* 108:349.

19. Dandy, D. S. 2000. http://grashof.engr.colostate.edu/tools.

20. Kee, R. J., G. Dixon-Lewis, J. Warnatz, M. E. Coltrin, and J. A. Miller. 1986. A FORTRAN computer code package for the evaluation of gas-phase, multicomponent transport properties. Livermore, CA: Sandia National Lab.

21. Kee, R. J., F. M. Rupley, and J. A. Miller. 1990. The CHEMKIN thermodynamic database. Livermore, CA: Sandia National Lab.

22. Kee, R. J., F. M. Rupley, and J. A. Miller. 1991. CHEMKIN II: A Fortran chemical kinetics package for the analysis of gas phase chemical kinetics. Livermore, CA: Sandia National Lab.

23. Reynolds, W. C. 1986. The element potential method for chemical equilibrium analysis: Implementation in the interactive program STANJAN. Stanford University, Stanford, CA.

24. Bychkov, V. V., and M. A. Liberman. 2000. Dynamics and stability of premixed flames. *Physics Reports* 325:115.

Comments

Lindstedt: Have your results on Markstein number been compared to work done earlier in this area?

Aldredge: No.

Ghoniem: Were fluctuating pressure measurements made?

Aldredge: We did not measure pressure fluctuations. Only velocity was measured.

Chapter 8

PREVAPORIZED JP-10 COMBUSTION AND THE ENHANCED PRODUCTION OF TURBULENCE USING COUNTERCURRENT SHEAR

D. J. Forliti, A. A. Behrens, B. A. Tang, and P. J. Strykowski

Experiments are described using premixed prevaporized JP-10 in a dump combustor modified to accommodate counterflow control at the dump plane. The primary benefit of operation under prevaporized conditions is the careful regulation of the JP-10/air stoichiometry, thereby providing independent assessment of flame speed modification using counterflow technology. Studies were also conducted to evaluate the fundamental nature of a planar nonreacting turbulent countercurrent shear layer. An abrupt transition in flow behavior at a counterflow level of $U_2/U_1 \approx -0.13$ provides the first evidence of global instability in a self-similar turbulent shear layer.

8.1 INTRODUCTION

Considerable insight into the dynamics of countercurrent shear layers has been developed through the Navy research programs [1, 2]. Countercurrent shear has been used successfully to control nonreacting flows ranging from low-subsonic to over Mach 2, and at temperatures between ambient and 1650 °C (3000 °F). Recent studies have documented the advantages of countercurrent shear over other active and passive control strategies, for unconfined shear flows as well as the confined flows of critical importance to Navy combustion systems. The present work examines the application of counterflow control to the combustion of prevaporized JP-10 in a dump combustor operating near stoichiometric conditions at a heat-release rate in the range of 100–200 kW, as well as the ability of counterflow to enhance the production of turbulence in a self-similar nonreacting shear layer.

8.2   PREVAPORIZED JP-10 COMBUSTION

JP-10 was selected for the reacting flow experiments conducted in the countercurrent dump combustor, since JP-10 is presently the fuel of choice for Navy weapon systems. Furthermore, prevaporization of the fuel was desired since much of the dump combustor research in the literature is for operation in gas phase, thus prevaporization would be the best way to judge the performance of the counterflow technology in a dump configuration. Additionally, prevaporization is advantageous in general for controlling NO_x production through controlling combustion product temperatures.

JP-10 is a stable fuel having a fairly low vapor pressure. The flashpoint for JP-10 is approximately 55 °C (130 °F), which suggests that insufficient fuel concentrations are generated through evaporation near ambient temperatures. In order to achieve gas-phase mixtures near stoichiometric equivalence ratios, preheating of the primary air is required. At a particular equivalence ratio, there is a minimum temperature that must be maintained to avoid incomplete evaporation and/or condensation. This temperature varies between approximately 45 °C (110 °F) at $\phi = 0.7$ and 60 °C (140 °F) at $\phi = 1.5$, for vaporized JP-10 in air. Although it is difficult to make a rigorous connection between the lower limit of combustion and the flashpoint, the relatively high flashpoint for JP-10 along with these high minimum temperatures would support the idea of JP-10 having a relatively high lower limit for self-sustained combustion.

8.3   COMBUSTION FACILITIES

The combustor is a modified dump geometry, which allows for variable counterflow through the application of suction to a cavity below the trailing edge, as shown in the facility sketch in Fig. 8.1. Premixed/prevaporized JP-10 is introduced and forms the primary stream; a splitter plate separates the primary forward flow from the vacuum cavity residing below, which is used to establish counterflow in the neighborhood of the trailing edge of the step. To the right of the splitter plate, the flow enters the test section, which can be visualized through a quartz window. The evaporation process takes place in a heated chamber upstream of the combustor. The mass flow rate through the facility provides primary stream velocities between approximately 10–20 m/s, corresponding to Reynolds numbers based on combustor step height from 9600 to 19,200.

Two Chromalox KSEF-430M air heaters are placed in series in the chamber to heat the main airflow to approximately 95 °C (200 °F). The JP-10 is sprayed into the vaporization chamber downstream of the heaters using a BEX JPL26B spray nozzle. The vaporization chamber is designed to allow for a sufficient

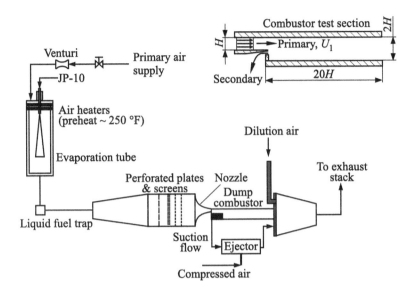

Figure 8.1 JP-10 prevaporization chamber and side view of the test section.

residence time to ensure complete vaporization of the fuel. A liquid trap is mounted at the base of the vaporization chamber to monitor the evaporation process and maintain accurate equivalence ratios.

The fuel and air mixture leaves the vaporizer and is fed to the combustor assembly, which consists of a diffuser, plenum with flow conditioning, nozzle, and combustor test section. The diffuser transits the flow from a 50.8-millimeter circular duct to a 160×120 mm rectangular-shaped plenum. The flow conditioning consists of two perforated plates, which enhance the flow uniformity and serve as effective flame arresters. The two-dimensional nozzle is constructed from sectors of a large-radius pipe and has an area ratio of 6:1. Pitot probe surveys showed that this setup produced a uniform top-hat velocity distribution across the area of the nozzle exit [2].

The height of the channel upstream of the dump plane is 20 mm; the step height is also fixed at 20 mm. The trailing-edge thickness was chosen to be 1 mm, which is expected to impact the flame anchoring process in the presence of counterflow. The suction gap height is 3 mm and was chosen based on previous studies of nonreacting flow in the chamber [2]. The spanwise dimension is 160 mm (providing an aspect ratio of 8:1), and the streamwise dimension from the step to the outlet is 400 mm. These dimensions are representative of other dump combustors studied in the literature, to which present results were compared [3, 4].

8.4   RESULTS AND DISCUSSION:
COMBUSTION STUDIES

The vaporized JP-10 mixed with air is fed to the dump combustor, where the reaction is initiated using a continuous spark igniter. The igniter is inserted into the combustor such that the spark is located in the natural low-velocity recirculation bubble. To avoid condensation of JP-10 on surfaces within the flow circuit, the facility is operated with heated air for approximately 1 hour to achieve a thermal steady state before igniting the combustor. Fiberglass insulation is used to reduce heat loss and the potential for fuel condensation. For the results discussed below, a total mass flow rate of 0.041 kg/s was used, and the reactants entered the combustor plenum at a temperature of approximately 95 °C.

The objective of this research is to study the impact of counterflow on the combustion within the dump combustor under both stable and unstable burning conditions. In order to explore the burning characteristics of JP-10 in the dump combustor, a very lean mixture was fed to the combustor and flame ignition was attempted. If ignition was not possible, the fuel flow rate was slightly increased and ignition was again attempted until a flame was successfully ignited. Ignition was ultimately achieved at an equivalence ratio of $\phi = 0.9$, resulting in a bright blue flame within the combustor. Visual observations of the flame at this equivalence ratio suggested that unsteady partial flashback was occurring, as the flame was propagating slightly upstream of the trailing edge, although the flashback process appeared to be periodic and unsteady. Operation at slightly higher equivalence ratios resulted in complete flashback, and the flame was stabilized near the perforated plate located in the combustor plenum.

Visualization of the combustion process was recorded by capturing short-time-exposure images of the chemiluminescence of the flame. A series of images is shown in Fig. 8.2. The time delay between the pictures is large; thus, temporal evolution of the flame is not captured. These images are representative of the entire sample of photographs, which suggest a periodic process is occurring within the combustor. The images have been organized from top to bottom to reflect the observed combustion instabilities found in the literature for facilities of this geometry operated near stoichiometric conditions [4], as illustrated to the right of each image.

The schematics shown in Fig. 8.2 illustrate the combustion instability that is believed to be present in the dump combustor and has been observed in similar dump combustors by Keller et al. [3] and Smith and Zukoski [4]. The cycle begins with the formation of a strong vortex at the trailing edge, which manipulates the flame. The vortex grows and moves downstream, eventually interacting with the lower wall. The flame propagates from the recirculation region toward the upper wall of the combustor. Eventually, the burning slows down

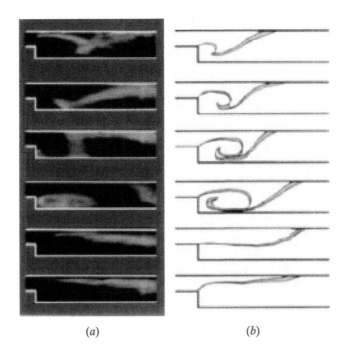

(a) (b)

Figure 8.2 Chemiluminescence images for JP-10 at $\phi = 0.9$ (a) and corresponding instability modes in the dump combustor (b).

as nearly all reactants present within the combustor become consumed. The pressure within the combustor begins to drop as the combustion products are exhausted. The decreasing pressure creates an impulse of flow into the combustor, which creates the strong spanwise coherent vortex, thus completing the cycle. Although this is not a rigorous proof of the presence of a thermoacoustic instability in the present facility, there are strong similarities that give circumstantial evidence which supports the idea that a thermoacoustic instability is present.

Although detailed measurements of the dump combustor at this condition have yet to be made to substantiate and quantify the presence of a thermoacoustic instability, it is not an unexpected result since many studies have reported thermoacoustic instabilities when the equivalence ratio is near unity [3–5]. Application of counterflow at this operating condition is the subject of ongoing experiments. It is expected based on cold-flow studies that suction should enhance turbulent energy production and three-dimensionality which should create a compact yet intense burning zone with reduced spanwise coherence, thereby aiding in the control of the thermoacoustic instability.

8.5   ENHANCED PRODUCTION OF TURBULENCE

Despite recent evidence of the benefits of countercurrent shear for the production of turbulence in confined geometries [1, 2], a systematic examination of the unconfined planar turbulent countercurrent shear layer has not been undertaken. The motivation for this study was in part to provide a fundamental framework on which to advance compact JP-10 combustion for Navy systems, but also to further improve the understanding of separated flows in general, to which local counterflow is ubiquitous. Initial attempts to examine this flow were undertaken by Humphrey and Li [6], using a novel facility designed explicitly for this purpose and shown in Fig. 8.3a. Their facility consisted of opposing momentum-driven streams separated by splitter plates and bounded by the walls of a closed-circuit wind tunnel. While the purpose of the study was to create a mixing region between the opposing streams, the flow sets up a global stagnation zone at a scale commensurate with the test section itself. Essentially, the primary and secondary streams came to rest, causing each to turn and exhaust 180 degrees from the direction in which they entered the facility.

A more recent investigation by Alvi *et al.* [7] examined countercurrent shear for the purpose of developing high convective Mach numbers, as illustrated in Fig. 8.3b. A highly disturbed shear layer was observed displaying rapid spatial growth, but turbulent flow statistics were not reported. Khemakhem [8] attempted to isolate planar countercurrent shear in the unconfined environment

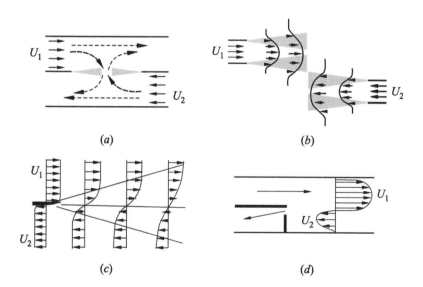

Figure 8.3 Confined and unconfined countercurrent geometries.

of Fig. 8.3c, but the scale of the facility prevented the attainment of the far-field boundary conditions. The geometry, shown in Fig. 8.3d, illustrates the confined domain considered for the reacting flow work described earlier. While the isothermal flow characteristics were documented in the 2001 ONR Annual Report [2], the results are facility dependent and cannot be readily extrapolated to other geometries. These limitations provided the motivation for the studies described below of a fully developed turbulent planar shear layer of the type illustrated in Fig. 8.3c.

8.6   SHEAR LAYER FACILITY

The experiments designed to create the countercurrent shear layer were conducted in a modified Eiffel-type wind tunnel having cross-sectional dimensions of 45×45 cm. The primary stream velocity was held invariant throughout the studies at $U_1 = 32$ m/s, a value selected to achieve conditions of self-similarity within the streamwise and cross-stream dimensions of the tunnel, while allowing sufficient spatial resolution using Particle Image Velocimetry (PIV). The quality of the primary flow was evaluated using hot wires when the lower tunnel was removed to allow free entrainment of ambient fluid into the shear layer. Power spectra indicated weak discrete peaks associated with the fan blade passage and overall free-stream turbulent levels consistently below 0.15% of U_1 (signals filtered between 1 Hz and 5 kHz). Two co-planar Nd:YAG laser light sheets (up to 230 mJ/pulse) were introduced into the test section to illuminate the forward and reverse streams, made possible by the introduction of atomized olive oil droplets (a Laskin nozzle was used, creating droplets generally on the order of 0.35 μm [9]). Multiple overlapping domains (between 9 and 12) of typically 5×5 cm were needed to capture the full extent of the shear layer while maintaining sufficient spatial resolution to map turbulent quantities. All images were captured by a Kodak MegaPlus ES:1.0 camera, containing two 1008×1018 pixel CCD arrays and processed in cross-correlation mode; see Tang [10] for details.

   Significant effort was made to establish the two-dimensionality of the base flow. The upper and lower facility walls were adjusted to minimize pressure and free-stream velocity variations within the test section and, in particular, over the region where self-similarity was established. Comprehensive studies were conducted of the flow at velocity ratios of U_2/U_1 equal to 0, -0.08, -0.13, -0.19, -0.24, and -0.30. At each velocity ratio, data from neighboring domains were patched and averaged over 500 image pairs. Collapse of the mean flow quantities occurs rather quickly; however, the higher order turbulent statistics do not settle down until larger streamwise distances are reached. Hence, the onset of self-similarity was determined by data convergence in the Reynolds stress $\overline{u'v'}$ to avoid the premature prediction of the onset of local flow equilibrium.

8.7   RESULTS AND DISCUSSION: SHEAR LAYER STUDIES

The results in Fig. 8.4 summarize the mean shear layer development at all velocity ratios. Note that y_{50} and x_0 are the cross-stream locations of the mean shear layer velocity and the virtual origin, respectively. The normalized time-averaged profiles have been spatially averaged over the entire self-similar domain to produce representative composite profile shapes at each value of U_2/U_1. The collapse is excellent and indicates that countercurrent shear layers do not display shape dependence with velocity ratio, in contrast to findings for coflowing layers by Mills [11] and Yule [12]. The reader is reminded that while the normalized profiles collapsed, the dimensional profiles indicated increased shear with increasing magnitude of U_2/U_1, as U_1 is held invariant in these studies. Furthermore, the spatial growth rate of the shear layer increases with the magnitude of the velocity ratio, collapse made possible by employing the stretching parameter σ in the abscissa.

Perhaps the most important kinematic marker of shear layer spatial growth is the development of the transverse fluctuating velocity v'. This is shown with composite profiles in Fig. 8.5a. Measurements taken for a single-stream shear layer, i.e., for $U_2/U_1 = 0$, and for modest levels of counterflow (e.g., at $U_2/U_1 = -0.08$ and -0.13) indicate shape invariance consistent with the behavior observed in the mean quantities. At these velocity ratios, the peak magnitude of v' (rms) is approximately $0.11\Delta U$, consistent with the literature for single-stream and coflowing shear layers [13]. What is most remarkable, however, is an abrupt transition observed in the profiles at counterflow levels above $\sim 0.13U_1$. The transition leads to transverse turbulent kinetic energy increasing by nearly

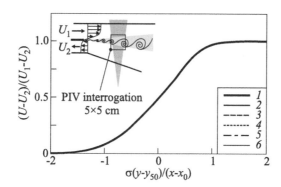

Figure 8.4 Self-similar U-distributions showing insensitivity to velocity ratio: 1 — $U_2/U_1 = 0$; 2 — -0.08; 3 — -0.13; 4 — -0.19; 5 — -0.24; and 6 — -0.30 (actually, all curves almost coincide).

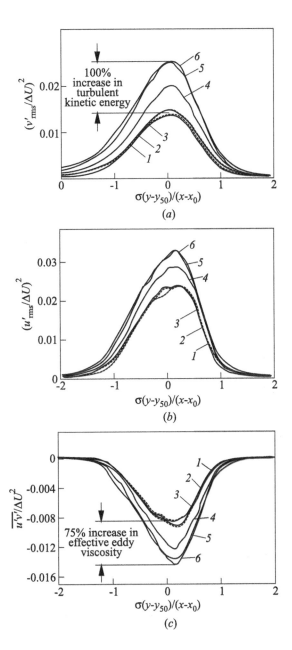

Figure 8.5 Self-similar fluctuations: (a) transverse; (b) streamwise; and (c) $\overline{u'v'}$-distributions. $1 - U_2/U_1 = 0$; $2 - -0.08$; $3 - -0.13$; $4 - -0.19$; $5 - -0.24$; and $6 - -0.30$.

a factor of two between 13% and 30% counterflow, with v' ultimately achieving a value of nearly $0.16\Delta U$. Comparable changes are observed in the streamwise velocity fluctuations u' as seen in Fig. 8.5b, where only small discrepancies are seen in the profiles at $|U_2| < 0.13U_1$. Initial indications of the nature of the turbulence caused by the transition can be evaluated by comparing the isotropy ratio $(v'/u')_{max}$, which increases from a nearly constant value of 0.75 for velocity ratios less than 13% counterflow to 0.88 for velocity ratios above 13%.

The Reynolds stress distributions in Fig. 8.5c indicate that turbulent momentum transport is also modified at high levels of counterflow. Since the mean velocity profiles (shown in Fig. 8.4) display independence of U_2/U_1, but the Reynolds stress experiences enhanced transport, the overall turbulent production of the layer is considerably increased above 13% counterflow. In fact, a comparison of the self-similar stress profiles in Fig. 8.5 indicates that a common state is achieved for $|U_2| < 0.13U_1$, and a second common self-similar state is achieved for $|U_2| > 0.24U_1$. From 0% to 13% counterflow, the turbulent profiles collapse, indicating a mechanism for generating the turbulence that scales with the growth rate parameter σ and velocity difference ΔU. Above 13% counterflow, there is an increase in turbulence level across the entire cross-stream extent of the layer. This increase seems to be dependent on velocity ratio, but not on the parameter σ. Since the mean profiles display similar shape, there is likely an additional mechanism for turbulence production when $|U_2|$ is greater than approximately $0.13U_1$.

Modifications in the turbulent statistics ultimately impact the spatial development of the shear layer. The growth rate for coflowing shear layers and those with modest levels of counterflow generally follow the trends as predicted by the model of Brown and Roshko, based on mixing-length assumptions [14]. However, the data reveal a deviation from the model predictions at $|U_2/U_1|$ greater than approximately 13% and illustrate one of the primary benefits of counterflow, namely, rapid spatial growth conducive to compact combustion.

The transition is striking, both in terms of both turbulent quantities and mean shear layer growth rates, and warrants further examination. While comparisons to linear stability theory would seem indefensible given the self-similar turbulent state of the flow, there appears to be some evidence of such a connection. Huerre and Monkewitz [15] predicted a transition from convective to absolute instability at a velocity ratio near -0.136 for planar shear layers. Studies by Strykowski and Niccum [16, 17] illustrated the global flow implications of this transition in spatially developing laminar shear layers, observing the onset of discrete-frequency oscillations and the insensitivity to external forcing. However, spectral records of the turbulent shear layer with and without counterflow [10] do not reveal the onset of discrete excitation in the supercritical regime. The spectra do show that the turbulent transport is dominated by large scales over a considerable range at lower frequencies. Hence, while a direct connection to linear stability concepts is unwarranted, the evidence of the flow transition described above near $U_2/U_1 = -0.13$ is quite compelling.

8.8 CONCLUDING REMARKS

Prevaporized JP-10 has been successfully burned in a dump combustor at nearly stoichiometric equivalence ratios. The flammability of the JP-10/air mixture seems to be limited to fairly high equivalence ratios, which has led to what is believed to be a thermoacoustic instability, which flashes back into the combustor plenum if the equivalence ratio is slightly above 0.9. Although JP-10 will continue to be used in the dump combustor studies for future work, the potential limited range of equivalence ratio operation may motivate the use of another fuel in conjunction with JP-10, which can be used over a wider range. This would allow for a more thorough examination of the countercurrent dump combustor performance.

ACKNOWLEDGMENTS

This work was performed under ONR contract N00014-01-1-0644.

REFERENCES

1. Forliti, D. J., and P. J. Strykowski. 2000. Examining the application of counterflow in dump combustors. *13th ONR Propulsion Conference Proceedings*. Minneapolis, MN. 57–62.

2. Forliti, D. J., and P. J. Strykowski. 2001. Performance and control of dump combustor flows using countercurrent shear. *14th ONR Propulsion Conference Proceedings*. Chicago, IL. 48–53.

3. Keller, J. O., L. Vaneveld, D. Korschelt, G. L. Hubbard, A. F. Ghoniem, J. W. Daily, and A. K. Oppenheim. 1982. Mechanism of instability in turbulent combustion leading to flashback. *AIAA J.* 20:254–62.

4. Smith, D. A., and E. E. Zukoski. 1985. Combustion instability sustained by unsteady vortex combustion. AIAA Paper No. 85-1248.

5. Keller, J. O., J. L. Ellzey, R. W. Pitz, I. G. Shepherd, and J. W. Daily. 1988. The structure and dynamics of reacting plane mixing layers. *Experiments Fluids* 6:33–43.

6. Humphrey, J. A. C., and S. Li. 1981. Tilting, stretching, pairing and collapse of vortex structures in confined counter-current flows. *J. Fluids Engineering* 103:466–70.

7. Alvi, F. S., A. Krothapalli, and D. Washington. 1996. Experimental study of a compressible counter-current turbulent shear layer. *AIAA J.* 34(4):728–35.

8. Khemakhem, A. S. D. 1997. An experimental study of turbulent countercurrent shear layers. Ph.D. Thesis, University of Minnesota.

9. Gerbig, F. T., and P. B. Keady. 1985. Size distributions of test aerosols from a Laskin nozzle. *Microcontamination* 3:56–61.

10. Tang, B. A. 2002. An experimental investigation of planar countercurrent turbulent shear layers. M.S. Thesis, University of Minnesota.

11. Mills, R. D. 1968. Numerical and experimental investigations of the shear layer between two parallel streams. *J. Fluid Mechanics* 33:591–616.

12. Yule, A. J. 1971. Two-dimensional self-preserving turbulent mixing layers at different free stream velocity ratios. Reports & Memoranda No. 3683.

13. Birch, S. F., and J. M. Eggers. 1972. Critical review of the experimental data for developed free turbulent shear layers. NASA Special Publication SP-321. 11–40.

14. Brown, G. L., and A. Roshko. 1974. On density effects and large structure in turbulent mixing layers. *J. Fluid Mechanics* 64:775–816.

15. Huerre, P., and P. A. Monkewitz. 1985. Absolute and convective instabilities in free shear layers. *J. Fluid Mechanics* 159:151–68.

16. Strykowski, P. J., and D. L. Niccum. 1991. The stability of countercurrent mixing layers in circular jets. *J. Fluid Mechanics* 227:309–43.

17. Strykowski, P. J., and D. L. Niccum. 1992. The influence of velocity and density ratio on the dynamics of spatially developing mixing layers. *Physics Fluids* 4:770–81.

Comments

Grinstein: I sensed that you tried avoiding the use of the words "global instability." Why?

Strykowski: Because the evidence is only circumstantial.

Chapter 9

MIXING CONTROL FOR JET FLOWS

M. Krstić

The research presented opens a new area for flow control in combustion systems — mixing by active feedback control. Considered herein is a two-dimensional (2D) jet, and simple control strategies are proposed that, with small control effort, generate a large increase in turbulent kinetic energy of the jet flow. The control is applied by microjets or microflaps at the lip of the jet and requires only the measurement of pressure at the jet lip. It is demonstrated that the controller enhances mixing of massless particles, particles with mass, and a passive scalar.

9.1 INTRODUCTION

The destabilization (mixing) of a jet flow is desirable in many applications, including combustion, noise suppression at jet engine exhaust, infrared signature reduction of engine exhaust for platform stealth, and employment of a lighter and cheaper material for the lift flap such as that used in C-17 aircraft.

Previous Work

Several passive and open-loop controls have been proposed previously.

Tabs: Grinstein *et al.* [1] and Mi and Nathan [2] investigated experimentally and numerically jet nozzles with *delta* or *mushroom* tabs.

Noncircular nozzle: Gutmark and Grinstein [3] reviewed experimental and numerical work of noncircular nozzle jets including rectangular and elliptic shapes. These nozzles cause a vortex evolution mechanism called axis switching which is responsible for mixing enhancement.

Secondary jet: Gutmark *et al.* [4] examined the effect of streamwise co-flow at the jet exit with a constant feeding frequency for soot reduction in a combustion chamber. Parekh *et al.* [5] performed experimentally and Freund and Moin [6] examined computationally a transverse forcing jet at the main nozzle exit. The amplitude and forcing frequency were constant. Lardeau *et al.* [7] numerically examined two auxiliary jets with or without pulsating and at an angle to the main jet.

Objective and Results of This Project

The objective is to demonstrate that mixing in jet flows can be generated by feedback and to develop systematic tools for generating specific mixing profiles. The control inputs are applied only at the lip of the jet via velocity actuation. This actuation can be applied as blowing and suction using microjets or by redirecting the flow using microflaps. The measurements are also taken only at the lip of the jet via pressure sensing. Hence, no invasive or unrealistic down-stream measurements or actuation are employed. In the study presented here it is demonstrated that the pressure measurement at the lip of the jet, with an appropriate time delay, is equivalent to pressure measurement downstream.

To assess the effectiveness of the controllers designed, several visualization techniques are employed. First, massless dye particles are injected in the flow; their motion traces the "streaklines" of the flow. Second, heavier particles sim-ulating droplets in combustion applications are injected in the flow, and their motion is followed. Third, the distribution of a passive scalar is calculated to analyze diffusive mixing.

9.2   JET FLOW MODEL AND SIMULATION TECHNIQUES

Governing Equations

The governing equations are the normalized Navier–Stokes and continuity equa-tions for incompressible viscous flows:

$$\frac{\partial u_i}{\partial t} + \frac{\partial u_i u_j}{\partial x_j} = -\frac{\partial p}{\partial x_i} + \frac{1}{\mathrm{Re}_D}\frac{\partial^2 u_i}{\partial x_j^2}$$

$$\frac{\partial u_i}{\partial x_i} = 0$$

where u_i are the velocities, p is the pressure, $\mathrm{Re}_D = \rho U_0 D/\mu$ is the Reynolds number, U_0 is the maximum inlet velocity, D is the jet pipe diameter, ρ is density, and μ is constant viscosity. All variables are nondimensionalized by U_0 and D.

The numerical method used in this research was developed by Akselvoll and Moin [8] for backward facing step and was modified by Bewley *et al.* [9] for plane

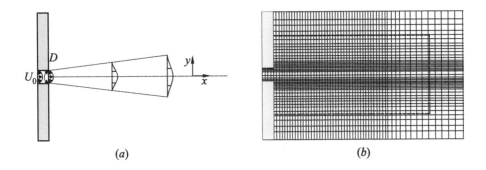

Figure 9.1 Computational domain (a) and grid configuration (b).

channel. A modification was made to adapt the geometry for this study. The simulation was performed on a computational domain consisting of a channel of $l_x = 2$ and $D = 1$ and an open field of $L_x = 50$ and $L_y = 40$ with a buffer zone $L_b = 10$ on each open boundary. The computation domain is shown in Fig. 9.1a.

Boundary Conditions

The convection boundary condition for the velocity field suggested by Sani and Gresho [10] is used for the open boundaries:

$$\frac{\partial \widehat{u}_i}{\partial t} + U_c \frac{\partial \widehat{u}_i}{\partial n} = 0$$

where U_c is a convection velocity and is chosen to be half of U_0 in the streamwise direction and a quarter of U_0 in the normal direction, and n is the direction normal to the boundary. This boundary condition is computed at full step by a forward-in-time backward-in-space (FTBS) scheme and interpolated to each Runge–Kutta (RK) substep.

In the buffer zone, i.e., between the *physical* domain and the open boundary, the damping terms are added to the right-hand side (RHS) of Navier–Stokes equations:

$$\frac{\partial u_i}{\partial t} + \frac{\partial u_i u_j}{\partial x_j} = -\frac{\partial p}{\partial x_i} + \frac{1}{\mathrm{Re}_D} \frac{\partial^2 u_i}{\partial x_j^2} - \sigma(x_j)[u_i - \widehat{u}_i(x_j)]$$

where $\widehat{u}_i$ is the boundary value computed by the convection boundary condition, and the damping coefficient is calculated by an exponential function, $\sigma(x_j) = \alpha(x_j - x_{j_b})^\beta / L_b^\beta$. The x_{j_b} is the coordinate of the interface between the physical domain and the buffer zone; $\alpha = 1$ and $\beta = 3$ were chosen for this study.

The overlap of streamwise- and normal-direction buffer zone is divided by a 45-degree split. The Neumann boundary condition $\nabla p \cdot \mathbf{n} = 0$ is applied along all the boundaries.

Spatial Discretization

The origin of the coordinate system is at the center of jet exit. A one-dimensional (1D) stretching formula used by Colonius *et al.* [11] is adapted for grid distribution in the streamwise direction; in the normal direction, the grid location is calculated by a hyperbolic tangent function inside the jet core and the 1D stretch function outside the jet core. The grid number in the open field is 211×175 in the streamwise and normal directions, respectively. Figure 9.1*b* shows a sample grid configuration. The *physical* domain is inside the thick dashed line.

The second-order accurate central difference scheme is used in interior nodes, and the second-order accurate inward biased scheme is used on boundary nodes. Thus, the overall accuracy is kept second-order accurate in space. A staggered grid is used to compute the pressure at the cell center and velocities at surrounding grid lines.

Temporal Discretization

Fine resolution in the normal direction is necessary around the shear layers, and it gives severe limitation on the time step for numerical stability. Thus, it is preferred to compute the derivatives in the normal direction implicitly, while the derivatives in the streamwise direction are treated explicitly. This leads to a hybrid time-integration scheme with a low-storage third-order RK (RK3) scheme for explicitly treated terms and a second-order Crank–Nicholson scheme for implicitly treated terms. The overall accuracy is thus second order in time. The discretized Navier–Stokes equations have the forms:

$$\frac{u_i^k - u_i^{k-1}}{\Delta t} = \beta_k \left[B\left(\mathbf{u}^k\right) - B\left(\mathbf{u}^{k-1}\right) \right] + \gamma_k A\left(\mathbf{u}^{k-1}\right) + \zeta_k A\left(\mathbf{u}^{k-2}\right) - 2\beta_k \frac{\delta p^k}{\delta x_i}$$

$$A(u_i) = \nu \frac{\delta^2 u_i}{\delta x_1^2} - \frac{\delta u_1 u_i}{\delta x_1}$$

$$B(u_i) = \nu \frac{\delta^2 u_i}{\delta x_2^2} - \frac{\delta u_2 u_i}{\delta x_2}$$

where β_k, γ_k, and ζ_k are the coefficients of the RK3 scheme; $A(u_i)$ and $B(u_i)$ are the explicit and implicit operators, respectively; and $\nu = 1/\mathrm{Re}_D$ is the normalized kinematic viscosity.

Poisson Solver

The fractional step method is used to advance each RK substep, and a Poisson equation needs to be solved for pressure update. The current geometry leads to a nine-diagonal matrix for the Poisson solver. An LU-decomposition solver was made to solve this nine-diagonal matrix. The singularity of the matrix is removed by prescribing the reference value of zero at the last cell (N_{x_1}, N_{x_2}) in the open field.

9.3   SIMULATION OF OPEN-LOOP JET FLOW

Reynolds numbers in the preliminary results, 100 and 150, are above the critical Reynolds number 30 stated in [12]. Figure 9.2 shows the vorticity plot at time step 10,000.

The jet flow is inherently unstable. This is verified theoretically [13] and experimentally [14]. Numerical simulations show that the dominant unstable mode at a low Reynolds number (< 200) is the low-frequency sinuous mode. This is also observed in the present study.

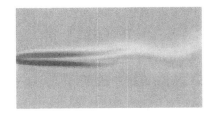

Figure 9.2 Vorticity plot — uncontrolled.

9.4   DESTABILIZATION AND MIXING OF MASSLESS PARTICLES

Two mixing enhancement controllers with a similar feedback law were simulated. The controller consists of a pair of actuators acting in the streamwise direction at the jet exit in an antisymmetric fashion and a pair of sensors measuring the pressure difference across the original jet diameter. The control laws are:

– Pressure-difference sensor downstream:

$$U_1(t) = -U_2(t) = K_u p_y(x_1, t)$$

– Pressure-difference sensor at jet exit with delay:

$$U_1(t) = -U_2(t) = K_u p_y(0, t - \tau)$$

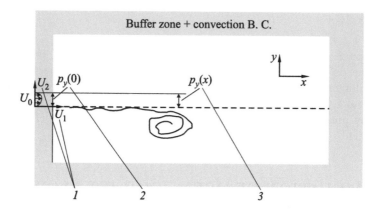

Figure 9.3 The mixing control law: *1* — actuators, *2* — sensor at exit with delay, and *3* — sensor downstream.

In the above expressions, U_1 and U_2 are the actuators, K_u is the proportional gain, $p_y(x_1, t)$ is the transverse pressure difference across the jet diameter $[p_{y=+0.5} - p_{y=-0.5}]$ at downstream location x_1 and time t, and τ is the delay. A saturation is set to the actuator for practical purposes and to ensure numerical stability. The two different control laws are illustrated in Fig. 9.3.

The vorticity plots (Fig. 9.4) and streakline plots (Fig. 9.5) of both control laws show promising results. These plots are taken at the time step 10,000. The two different controls generate similar flow patterns. The *time scale* provided by the time delay is comparable to the *length scale* provided by the downstream sensor. Therefore, further investigation will focus on the control law with time delay. The flow is sensitive to the choice of gain and time/space delay. One can achieve a great variety of mixing patterns with the simple feedbacks alone, presented here. One can expand the degrees of freedom by combining multiple feedback terms with different delay values.

9.5 MIXING OF PARTICLES WITH MASS

The Basset–Boussinesq–Oseen (BBO) equation can be simplified for gas–particle flows with a very small density ratio between the carrier phase and the discrete phase ($\sim 10^{-3}$) and with the assumption of one-way coupling such that only the carrier phase has influence on the particle but not vice versa. The particles are governed by these nondimensional motion equations:

$$\frac{d\mathbf{v}}{dt} = \frac{f}{\mathrm{St}}(\mathbf{u} - \mathbf{v}); \qquad \frac{d\mathbf{x}}{dt} = \mathbf{v}$$

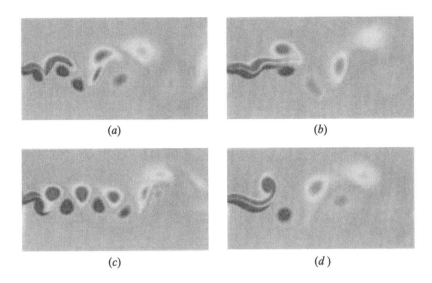

Figure 9.4 Vorticity plots of mixing controllers with different parameters: (a) downstream sensor at $x = 5$; (b) downstream sensor at $x = 10$; (c) jet exit sensor with delay $= 200$; and (d) jet exit sensor with delay $= 500$.

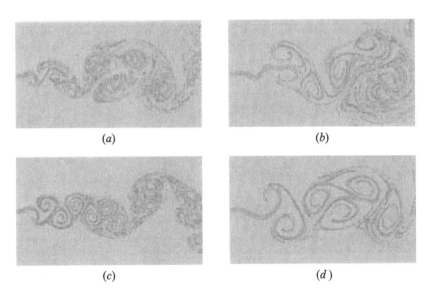

Figure 9.5 Streakline plots of mixing controllers with different parameters: (a) downstream sensor at $x = 5$; (b) downstream sensor at $x = 10$; (c) jet exit sensor with delay $= 200$; and (d) jet exit sensor with delay $= 500$.

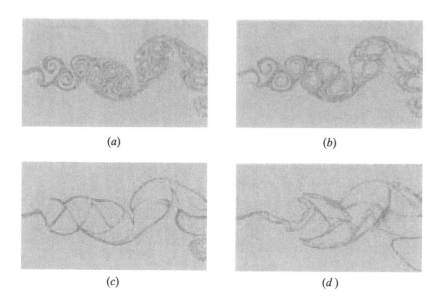

Figure 9.6 The dispersion of particles with different Stokes numbers: (*a*) St = 0, streaklines; (*b*) St = 0.1; (*c*) St = 1; and (*d*) St = 10.

where $\mathbf{v}$ is the particle velocity, $\mathbf{x}$ is the particle coordinate, $\mathbf{u}$ is the carrier-phase velocity at the same location, f is a correction factor, and St is the Stokes number of the particle. The correction factor $f = 1 + 0.15\mathrm{Re}_p^{0.687}$ is a function of the particle's Reynolds number Re_p, which is defined as $D_p|\mathbf{u} - \mathbf{v}|/\nu$ with the particle diameter D_p. The Stokes number is the ratio of the particle's momentum response time to the flow-field time scale:

$$\mathrm{St} = \frac{\rho_p D_p^2/18\mu}{D_0/U_0}$$

By definition, a larger Stokes number represents a larger or heavier particle, and a smaller or lighter particle has a smaller Stokes number.

Three different Stokes numbers, 0.1, 1, and 10, were simulated with the same density, $\rho_p = 10^3$, in this research. The particles are fed into the flow at the jet exit every time step and are advanced in time by the explicit RK3 scheme used for the carrier phase. The simulation results are shown in Fig. 9.6.

(1) For St $\ll$ 1, the particles will mostly follow the fluid motion acting like a tracer;

(2) For St $\sim$ 1, the particles centrifuge out of the vortex cores and concentrate on the vortex peripheries; and

(3) For St $\gg$ 1, the carrier fluid has very limited influence on the particle motion.

9.6   MIXING OF PASSIVE SCALAR

The passive scalar evolution is simulated to analyze diffusive properties in the presence of the controller. The *passive* stands for the one-way coupling assumption. The scalar field has the highest value $S = 1$ representing the full density at the jet exit and the lowest value $S = 0$ representing a zero density at the ambient field originally. The normalized scalar evolution equation is

$$\frac{\partial S}{\partial t} + \frac{\partial u_j S}{\partial x_j} = \frac{1}{\mathrm{Re}_D} \frac{1}{\mathrm{Sc}} \frac{\partial^2 S}{\partial x_j^2}$$

where Sc is the Schmidt number defined as the ratio of fluid viscosity to the scalar's molecular diffusivity D_s, $\mathrm{Sc} = \nu / D_s$. The Schmidt number was chosen to have unity value in this study.

The results in Fig. 9.7 show that a more uniform scalar field is obtained by generating larger vortex rings, while the smaller eddies cause high-concentration spots of scalar. This observation suggests a longer time delay for better mixing enhancement, which is in contradiction with the observation one would make based solely on streaklines. The smaller eddies have stronger rotation, which can attract more fluid particles into the core but also restrict the molecular diffusion inside the vortices; on the other hand, the larger eddies have weaker structures which tend to spread and allow diffusion to take place.

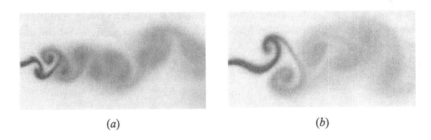

(a)   (b)

Figure 9.7 Scalar field plots of mixing enhancement controllers: (*a*) jet exit sensor with delay = 200 and (*b*) jet exit sensor with delay = 500.

ACKNOWLEDGMENTS

This work was performed under ONR Grant N00014-98-1-0700.

REFERENCES

1. Grinstein, F. F., E. J. Gutmark, T. P. Parr, D. M. Hanson-Parr, and U. Obeysekare. 1996. Streamwise and spanwise vortex interaction in an axisymmetric jet. A Computation and Experimental Study. *Physics Fluids* 8(6):1515–24.

2. Mi, J., and G. J. Nathan. 1999. Effect of small vortex-generators on scalar mixing in the developing region of a turbulent jet. *Int. J. Heat Mass Transfer* 42:3919–26.

3. Gutmark, E. J., and F. F. Grinstein. 1999. Flow control with noncircular jets. *Annual Review Fluid Mechanics* 31:239–72.

4. Gutmark, E. J., T. P. Parr, K. J. Wilson, and K. C. Schadow. 1995. Active control in combustion systems with vortices. *4th IEEE Conference on Control Applications Proceedings*. IEEE. 679–84.

5. Parekh, D. E., V. Kibens, A. Glezer, J. M. Wiltse, and D. M. Smith. 1996. Innovative jet flow control: Mixing enhancement experiments. AIAA Paper No. 96-0308.

6. Freund, J. B., and P. Moin. 1998. Mixing enhancement in jet exhaust using fluidic actuators: Direct numerical simulations. ASME. FEDSM98-5235.

7. Lardeau, S., E. Lamballais, and J.-P. Bonnet. 2002. Direct numerical simulation of a jet controlled by fluid injection. *J. Turbulence* 3:002.

8. Akselvoll, K., and P. Moin. 1995. Large eddy simulation of turbulence confined coannular jets and turbulent flow over a backward facing step. Report TF-63. Thermosciences Division, Dept. of Mech. Eng., Stanford University.

9. Bewley, T. R., P. Moin, and R. Teman. 2001. DNS-based predictive control of turbulence: An optimal benchmark for feedback algorithms. *J. Fluid Mechanics* 447:179–225.

10. Sani, R. L., and P. M. Gresho. 1994. Resume and remarks on the open boundary condition minisymposium. *Int. J. Numerical Methods Fluids* 18:983–1008.

11. Colonius, T., S. K. Lele, and P. Moin. 1993. Boundary conditions for direct computation of aerodynamic sound generation. *AIAA J* 31(9):1574–82.

12. Schlichting, H., and K. Gersten. 2000. *Boundary layer theory*. 8th ed. Berlin: Springer-Verlag.

13. Michalke, A. 1994. Survey on jet instability theory. *Progress Aerospace Sciences* 21:159–99.

14. Rockwell, D. 1992. External excitation of planar jets. *J. Applied Mechanics* 34:883–90.

Comments

Ghoniem: Are you trying to use distributed sensors in filament-based control?

Kristic: No, I am not suggesting that.

Edwards: If you are looking for an application for stable jet, you may want to consider cutting jets.

Chapter 10

CHARACTERISTICS AND CONTROL OF A MULTISWIRL SPRAY COMBUSTOR

E. J. Gutmark, G. Li, and S. Abraham

The present research program deals with the control of an industrial gas-turbine spray combustor with multiple swirlers and distributed fuel injection for rapid mixing and stabilization. It focuses on investigating the mixing patterns and flame structure in a triple-swirl stabilized combustor and the development of control strategies for improved performance of modern industrial combustors utilizing this technology. It is performed in collaboration with GE Aircraft Engines and Goodrich Corp. The experimental data, taken by Laser Doppler Velocimetry (LDV) and Particle Image Velocimetry (PIV), depict the characteristics of the complex flow field downstream of different triple-swirler combinations for non-reacting flow. These measurements provide information for the time-averaged or steady flow field and instantaneous or unsteady flow field to benchmark Large-Eddy Simulations that are performed by the Naval Research Laboratory. Similar experimental approaches will be applied to the reacting flow test rig and will reveal the combustion characteristics and mixing patterns that are related to the flow features of different swirler geometry combinations.

10.1 INTRODUCTION

The present research deals with the control of an industrial gas-turbine spray combustor with multiple swirlers and distributed fuel injection for rapid mixing and stabilization [1]. The research focuses on investigating the mixing patterns and flame structure in a multiple-swirl stabilized combustor and develops control strategies for improved performance [2]. It is performed in collaboration with Delavan Gas Turbine Products, a division of Goodrich Aerospace. The research is applicable in both propulsion and power generation systems, such as land-based gas turbines [3].

In a complex geometry, such as the present Triple Annular Research Swirler (TARS) in which several co-swirling or counter-swirling flows interact, the vortical evolution and breakdown are complex. Detailed velocity measurements are necessary to resolve the physical processes that involve mixing and interaction of three distinct air streams [4]. As the first phase of this project, nonreacting velocity flow fields downstream of TARS were studied by LDV and PIV. Basic flow structures are discerned by the experimental data, and comparison of flow structure caused by different swirler combinations is made.

10.2 EXPERIMENTAL SETUP

The research was conducted in a cold-flow combustor test rig simulating the exact geometry of the hot combustion rig. This test rig (Fig. 10.1) is set up vertically. The air is introduced at the bottom and is conditioned through an air conditioning section which is composed of perforated cone, screens, and honeycomb; is settled in 3.8-inch diameter circular chamber; is fed into TARS; and then exits to the atmosphere. The pressure drop across TARS is 4%. The TARS features three separate airflow passages and independent liquid fuel supply lines that can be controlled separately (Fig. 10.2).

The time-averaged velocity field downstream of the TARS, including axial and tangential mean and turbulent velocity components, was measured using a Phase Doppler Particle Analyzer (PDPA) system.

By installing the test rig on a frame, stereo PIV measurements were taken for different swirler combinations with a confined tube under similar operating conditions.

10.3 RESULTS AND DISCUSSIONS

Time-averaged mean and turbulent velocity data without a confined combustion chamber were used to compare the basic characteristics of the flow field for the different multiple swirler combinations listed in Table 10.1.

The different swirlers in the table are labeled by the different vane angles of inner, intermediate, and outer air passages. The flow was ducted through the TARS nozzle and exited to ambient atmosphere. The outlet of the TARS was located at the inlet dump plane of the combustion chamber. This location was defined as $z/R = 0$, where R is the radius of the TARS outlet.

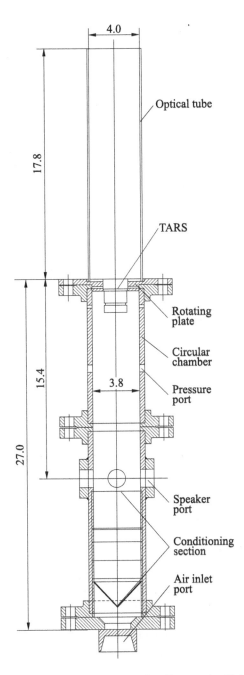

Figure 10.1 Schematic drawing of nonreacting flow at the University of Cincinnati (units are in inches).

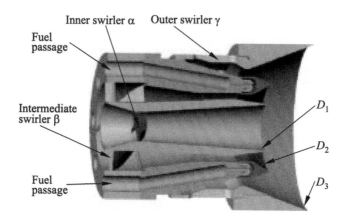

Figure 10.2 Cross-section drawing of TARS.

Table 10.1 Combinations of triple annular research swirlers included in the study; "c" in the swirler label denotes counter-clockwise rotation relative to the incoming flow.

	Swirler 304545	Swirler 3045c45	Swirler 304545c	Swirler 504545
γ, °	30	30	30	50
Number of vanes	8	8	8	8
β, °	45	45c	45	45
Number of vanes	4	4	4	4
α, °	45	45	45c	45
Number of vanes	4	4	4	4
Swirl No.	0.39	0.22	0.36	0.40
M, kg/s	0.040	0.046	0.041	0.046

Velocity Mapping of Swirler 504545

Mapping of the axial and tangential velocity components was performed in a plane perpendicular to the axis at $z/R = 0.1$ and in the streamwise plane which included the axis of the nozzle. The mean and root-mean-squared (rms) axial and tangential velocity components mapping at $z/R = 0.1$ for Swirler 504545 are shown in Fig. 10.3. The following flow structures are observed in this figure:

(1) The axial velocity mapping shows a central axisymmetric recirculation zone that was formed around the axis with a diameter of about $1R$ at this plane; and

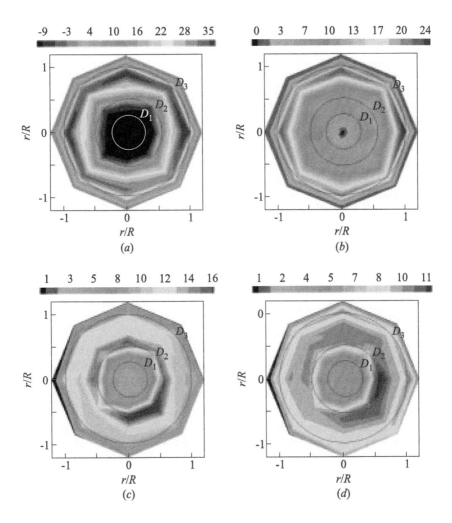

Figure 10.3 Velocity mapping of Swirler 504545 at $z/R = 0.1$ cross-sectional plane: (a) mean axial velocity (m/s); (b) mean tangential velocity (m/s); (c) rms axial velocity (m/s); and (d) rms tangential velocity (m/s). (Refer color plate, p. V.)

(2) Close observation revealed that at $z/R = 0.1$, the flow region outside of the Center Toroidal Recirculation Zone (CTRZ) is not fully axisymmetric and several peaks can be discerned near $r/R = 0.9$ in the mean and $r/R = 0.6$ in rms axial velocity mapping and in the tangential velocity mapping.

These peaks may be correlated to the number of vanes in the inner, intermediate, and outer swirlers, which were 4, 4, and 8, respectively. At the location of

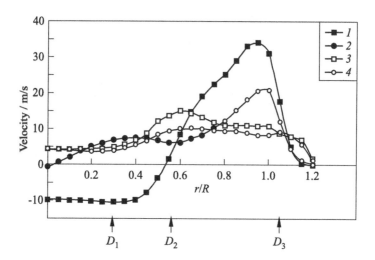

Figure 10.4 Circumferential-averaged velocity profile at $z/R = 0.1$ for Swirler 504545: *1* — mean axial; *2* — mean tangential; *3* — rms axial; and *4* — rms tangential.

$z/R = 0.1$, the nonaxisymmetric flow induced by the swirler vanes was not fully mixed to produce uniform flow, and some locations (corresponding to the vane geometry) have larger velocity components than others.

The contour plots of Fig. 10.3 were averaged circumferentially to produce radial profiles of the axial and tangential velocity components as shown in Fig. 10.4 for Swirler 504545 at $z/R = 0.1$. The locations of the peak velocity of the mean and tangential components are at $r/R = 0.9$ and $r/R = 0.95$. The peaks of the rms velocity of the two components are at nearly the same locations at $r/R = 0.6$, where the shear layer of the annular jet is located. The CTRZ has a radius of $r/R = 0.5$, which is primarily contributed by the inner swirler and partially by the intermediate swirler.

The mean and rms axial and tangential velocities in the streamwise plane for this swirler are shown in Fig. 10.5. This plane extends from $z/R = 0.1$ to $z/R = 4.7$. The negative region of mean axial velocity near the centerline shows that the region of the center recirculation zone extends to about $z/R = 5$ (extrapolated from the data). The momentum of the strong airflow in the annular jet is gradually reduced as the jet merges into the main stream at about $z/R = 2$. High magnitude indicates that tangential velocity is confined to a small region near the nozzle exit. A secondary region of increased tangential component level is centered at about $z/R = 1.5$. This amplified tangential velocity may be related to a circumferential flow instability mode. The rms mapping shows that the multiple swirling flows mix well before reaching the $z/R = 2$ plane.

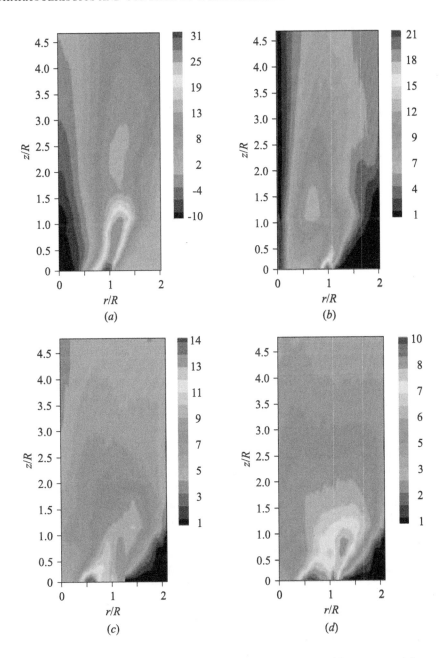

Figure 10.5 Velocity mapping for half streamwise plane of Swirler 504545: (a) mean axial; (b) mean tangential; (c) rms axial; and (d) rms tangential. (Refer color plate, p. VI.)

Comparison of Different Swirler Combinations

The circumferentially averaged mean and turbulent axial and tangential velocity components were calculated at an axial distance of $z/R = 0.1$ for the different swirler combinations listed in Table 10.1. The important flow features that can be discerned in these comparisons are:

(1) The counterrotating action of the intermediate swirler (Swirler 3045c45) causes significant reduction in the diameter of the recirculation zone from $r/R = 0.55$ to $r/R = 0.35$, while reducing the peak axial velocity and spreading the peak over a wider range of radial distance. It seems that the intermediate counter-clockwise rotating swirling flow pushes the inner flow inside and mixes with the outer flow in the radial direction better than the other swirlers. The intermediate flow has a significant effect on the inner flow field, as can be seen from the rms of both axial and tangential components that increase significantly in the inner flow region. This enhanced mixing suggests that a counterrotating swirler will be a good choice to improve the inner region fuel atomization but may reduce the flame stabilization.

(2) The swirling action of the inner swirler has a negligible effect on the flow field because the inner flow is mostly fully developed. This can be seen from the nearly identical profiles of Swirlers 304545 and 304545c.

(3) The larger swirling angle of the outer Swirler (504545) compared to Swirler 304545 leads to higher magnitude of both the axial and the tangential velocity components. There was no apparent effect of this increased swirl on the axial or tangential rms levels.

(4) For Swirlers 3045c45 and 304545c, the flow region between inner and intermediate swirlers is subject to the same counterrotating shear layer, but the resulting flows are significantly different. One possible reason is that the outer shear layer, which is formed by the flows through the outer and intermediate air passages, affects the inner shear layer region.

Comparison of Data With/Without a Confined Tube

Data for Swirler 504545 with and without a confining tube that simulates the combustion chamber were taken under the same operating condition for the streamwise plane (Fig. 10.6). Figures 10.6a and 10.6b show that the jet and CTRZ are significantly enlarged by the tube both in radial and in axial directions. There is an external recirculation zone, which is located at the corner of the expansion tube. Another interesting observation is that the maximum axial velocity area, which is spread between $r/R = 1 \sim 1.2$ and $z/R = 0 \sim 1.2$ in

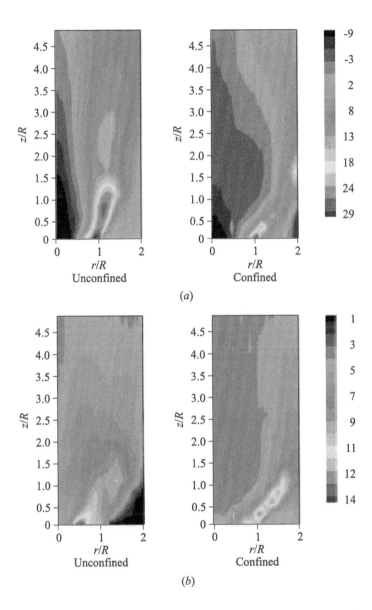

Figure 10.6 Comparison of axial velocity and rms axial velocity for Swirler 504545 with and without a confined tube: mean (a) and rms (b) axial velocities (m/s) on half streamwise plane. (Refer color plate, p. VII.)

the unconfined case, tends to be damped immediately after exiting the confined nozzle. The turbulent axial velocity component (Fig. 10.6*b*) is reduced and is more uniformly distributed in the center region of the jet. This suggests that the confined tube acts to damp the turbulence by creating a larger uniform CTRZ. These cold-flow observations should have a significant effect on the combustion characteristics in reacting flow studies.

10.4 PARTICLE IMAGE VELOCIMETRY RESULTS

Initial PIV results of Swirler 304545 are presented to show the three-dimensional (3D) vector field and vorticity downstream of the nozzle with a 4-inch confining tube. These results were obtained by averaging 100 images, as the instantaneous images could not be used to discern the dominant structure of the flow field. The mean axial, tangential, and radial velocity components; vorticity; and 3D vector are shown in Fig. 10.7.

In this swirler, due to the fact that all three swirlers were installed in the same swirl direction, the main flow structure is dominated by a helical rolling around the centerline of the fuel nozzle, as can be seen from the 3D vector field (Fig. 10.7*e*). A single large vortex that is located in the center of this rolling structure splits it into two sections. Axial velocity near the nozzle recirculates towards the nozzle in a conical pattern (Fig. 10.7*b*). The flow further downstream near the center moves away from the nozzle. The axial vortex determines the stagnation point location of the recirculation region. The plot of the axial velocity (Fig. 10.7*a*) shows a conical structure that produces strong shear along its widespread edges. This shear action could improve the fuel and air mixing and assist in anchoring the flame to stabilize combustion. Future reacting flow tests will further investigate these features.

The vorticity mapping (Fig. 10.7*d*) suggests that the intense vortical region develops at 1.5–2 nozzle diameters downstream of the nozzle. This strong vortex could contribute to the stabilization of the flame.

10.5 CONCLUDING REMARKS

Laser Doppler Velocimetry measurements were performed on a TARS in a cold-flow test rig that simulates an identical hot combustion rig. Axial and tangential velocity mapping of the flow in different cross-sectional planes and along a centerline streamwise plane revealed several important flow structures in the triple annular swirling flow with co-swirling and counter-swirling cases. An axisymmetric recirculation zone was formed in the center of the flow, but the outer

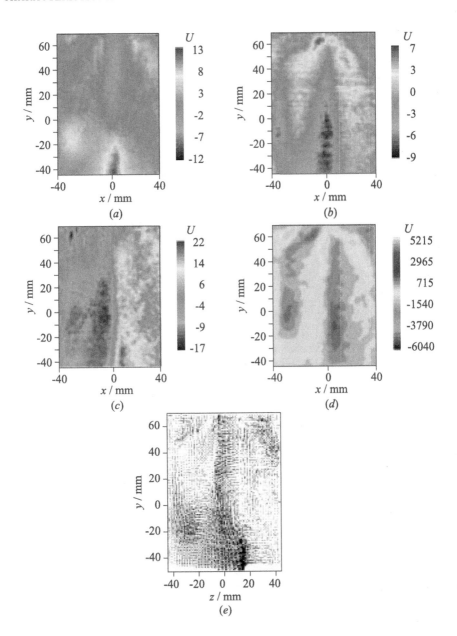

Figure 10.7 Particle Image Velocimetry results on a streamwise plane downstream of Swirler 304545 at 68 scfm, 2-inch diameter exhaust nozzle, and $L_{mt} = 0''$: (a) mean axial velocity component; (b) mean tangential velocity component; (c) mean radial velocity component; (d) vorticity; and (e) 3D vector. (Refer color plate, p. VIII.)

regions showed some nonaxisymmetric features due to incomplete internal mixing. The strong rms magnitude induced by the counter-clockwise intermediate swirler suggests that this configuration could be beneficial for enhancing fuel and air mixing. However, the same configuration resulted in a reduced central recirculation zone, which may lead to flame destabilization.

The comparison between confined and unconfined results for Swirler 504545 shows the significant effect of the confining tube on the swirling flow in terms of jet expansion and decay rate, enlargement of the CTRZ, and reduced turbulent intensity which becomes more uniform in the center jet region. This fact emphasizes the importance of the expansion ratio at the combustor inlet.

Compared with LDV results, the PIV shows a different recirculation zone structure. The underlying reason needs to be further investigated. The strong helical structure observed emphasizes the need to investigate the reacting flow in order to reveal the influence of dominant flow structures on the combustion characteristics.

Reacting Flow Test Rig

The reacting flow test rig was constructed at the Gas Dynamics and Propulsion Laboratory at the University of Cincinnati. The 3D geometry is shown in Fig. 10.8. For better optical observation and laser measurement, the combustion chamber has an octagonal shape with four flat quartz windows. Four

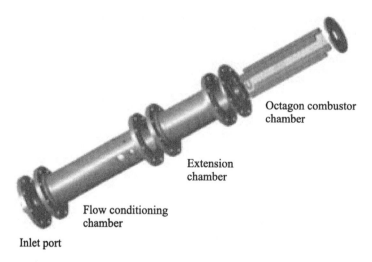

Figure 10.8 Solid model of the reacting flow test rig.

75-watt loudspeakers are installed upstream to force flow at certain instability modes.

ACKNOWLEDGMENTS

The work was performed under ONR contract N00014-02-1-0756.

REFERENCES

1. So, R., S. Ahmed, and H. Mongia. 1985. Jet characteristics in confined swirling flow. *Experiments Fluids* 3:221–30.
2. Paschereit, C., E. Gutmark, and W. Weisenstein. 1999. Flow–acoustic interactions as a driving mechanism for thermoacoustic instability. *Physics Fluids* 11(9): 2667–78.
3. Mongia, H., R. Santoro, S. Menon, E. Gutmark, and J. Gore. 2001. Combustion research needs for helping development of next-generation. AIAA Paper No. 2001-3853.
4. Sarpkaya, T. 1971. Vortex breakdown in conical flows. *AIAA J.* 9:1792–99.

Comments

Frolov: I wonder how you are going to apply cold homogeneous flow measurements to understand the dynamics of reactive heterogeneous flow in realistic combustors? I anticipate this will be not a simple task as the flows I mentioned differ considerably.

Gutmark: We will compare cold- and hot-flow measurements to gain understanding. The cold-flow and hot-flow facilities are identical and will be run in parallel. We can always learn from cold-flow studies. Next, we will study reacting flow fuel mixedness, flow field, what creates stabilities and their effects, etc.

Smirnov: Your approach of using water droplets in cold flow to mimic the droplet dynamics in reacting flow will not be accurate.

Gutmark: I think it should work fine. On campus, we use a cold-flow facility with water but can change the vaporization rate. Off campus, we can perform any combination of variables.

Whitelaw: I come back to the question of my Russian colleagues. This cold-flow research will take too long and generate a lot of not-so-useful information, which can't be applied to reacting flows. Why don't you burn first and then reevaluate candidate geometries in detail?

Gutmark: To understand the mechanisms, we must examine the cold flow first. Earlier work was practical but not fundamental. We try to understand the mechanisms and gain additional understanding about control.

Ghoniem: As you increase the number of swirl vanes, you increase the number of places where the flow separates. Thus, pressure drop is affected. How will you handle this?

Gutmark: We document every configuration we test. Some have a higher pressure drop penalty than others. We consider this using active/passive combustion control.

Chapter 11

SWIRLING JET SYSTEMS FOR COMBUSTION CONTROL

F. F. Grinstein and T. R. Young

A hybrid simulation approach is used to investigate the flow patterns in an axisymmetric swirl combustor configuration. Effective inlet boundary conditions are based on velocity data from solving Reynolds-Averaged Navier–Stokes (RANS) equations or actual laboratory measurements at the outlet of a fuel-injector nozzle, and Large-Eddy Simulation (LES) is used to study the unsteady nonreactive swirl flow dynamics downstream. Case studies ranging from single-swirler to more complex triple-swirler nozzles are presented to emphasize the importance of initial inlet conditions on the behavior of the swirling flow entering a sudden expansion area, including swirl and radial numbers, inlet length, and characteristic velocity profiles. Swirl of sufficient strength produces an adverse pressure gradient which can promote flow reversal or vortex breakdown, and the coupling between swirl and sudden expansion instabilities depends on the relative length of the inlet. The flow is found to be very sensitive to the detailed nature of the radial velocity profiles. The critical challenge of specification of suitable inlet boundary conditions to emulate the turbulent conditions in the laboratory experiments is raised in this context.

11.1 INTRODUCTION

The present studies are devoted to the axisymmetric swirl combustor configuration shown schematically in Fig. 11.1. It involves a primary fuel nozzle, within which air is passed through a swirler arrangement to mix and atomize the fuel. Coupling swirling flow motion with sudden expansion to the full combustor diameter provides an effective way of enhancing the fuel–air mixing and stabilizing combustion. Because of performance requirements on the design of gas-turbine

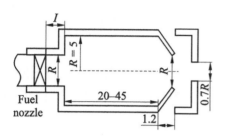

Figure 11.1 Axisymmetric swirl combustor (outlets used are indicated on the right). Dimensions are in cm.

engines, there is considerable interest in identifying optimal swirl and geometrical conditions to achieve specific practical goals in actual flight regimes, such as reduced emissions, improved efficiency, and stability.

Numerical simulations of compressible flows developing in both space and time with precise control of initial and boundary conditions are ideally suited in the quest to recognize and understand the local and global nature of the flow instabilities driving the combustor performance — which are the main focuses of this work. Numerical experiments can be used to isolate suspected fundamental mechanisms from others which might confuse issues. The extensive space/time diagnostics available based on the simulation database can be exploited to develop analytical and conceptual bases for improved modeling of the turbulent flame.

In simulations for engineering problems involving turbulent combustion, a RANS description of the flow [1] and simplistic combustion models (e.g., [2]) are typically combined. This involves simulating only the mean flow-field features and modeling the effects of the entire range of turbulent scales. The restricted information provided by this approach, regarding the fluid dynamics, combustion, and their different interactions, precludes adequate prediction of the important phenomena required to achieve effective control of the combustion processes, such as combustion-induced flow instabilities, cycle-to-cycle variations, and combustion oscillations associated with unsteady vortex dynamics.

Large-Eddy Simulations provide a cost-effective alternative between RANS and (prohibitively expensive) direct numerical simulations in full-scale, three-dimensional combustor configurations (e.g., [3, 4]). However, although LES is capable of simulating flow features which cannot be handled with RANS, such as significant flow unsteadiness and strong vortex–acoustic couplings, the added advantage comes at the expense of computational cost, since LES is typically an order of magnitude more expensive than RANS. As a consequence, hybrid simulation approaches restricting the use of LES to flow regions where it is crucially needed, and using RANS for the remaining regions, can be used for practical flow configurations (e.g., [5]).

The hybrid simulation approach used here for the swirl combustor configuration in Fig. 11.1 involves effective boundary conditions emulating the fuel nozzle and LES to study the flow within the combustor. Case studies ranging from single-swirler to more complex triple-swirler nozzles were investigated. The inlet boundary conditions used to initialize the combustor flow involve velocity and turbulent intensities based on data at the outlet of the fuel-injector nozzle from RANS or actual laboratory measurements. A more detailed report of the authors' joint studies with the University of Cincinnati (UC) has been presented elsewhere [6]; the authors' ongoing research addressing the interaction between combustion and flow dynamics is reported separately [7].

11.2 NUMERICAL SIMULATION MODEL

Simulation of turbulent reacting flows encompasses dealing with a broad range of length and time scales. The largest scales of turbulent flows are related to the specific geometry and regime considered, and the smallest scales are associated with the dissipation of turbulent energy through viscosity. In conventional LES [8], the governing equations are low-pass filtered to remove the dynamics of the smallest eddies, the effects of which are represented by explicit subgrid scale (SGS) closure models. A promising LES approach is Monotonically Integrated LES (MILES) [4, 9–11], which involves solving the unfiltered Navier–Stokes equations using high-resolution monotone algorithms; in this approach, implicit tensorial (anisotropic) SGS models [10], provided by intrinsic nonlinear high-frequency filters built into the convection discretization, are coupled naturally to the resolvable scales of the flow. The MILES approach provides an attractive alternative when seeking improved LES for inhomogeneous (inherently anisotropic) high-Reynolds-number turbulent flows such as those studied here.

The three-dimensional MILES model used in the present work involves structured grids and solves the time-dependent compressible flow conservation equations for total mass, energy, momentum, and species concentrations with appropriate boundary conditions and an ideal gas equation of state. The explicit finite-difference numerical method [10] is based on splitting integrations for convection and other local processes (e.g., molecular viscosity and thermal conduction) and coupling them using a timestep splitting approach. Convection is based on the use of the NRL FAST3D code, implementing direction-splitting, fourth-order Flux-Corrected Transport algorithm, second-order predictor–corrector integration, and Virtual Cell Embedding to handle the complex geometrical features [12].

Inlet swirl inflow conditions are discussed below. The outflow boundary conditions at the combustor outlet involve advection of all flow and species variables with U_c, where the instantaneous mean streamwise outlet boundary velocity U_c is periodically renormalized to ensure that the time-averaged mass flux coincides with that at the inlet; these convective boundary conditions are enforced in conjuction with soft relaxation of the outflow pressure to its ambient value. Two types of outlets were considered (Fig. 11.1). Viscous wall regions in the combustor cannot be practically resolved for the moderately-high Reynolds

numbers involved here. Near-wall boundary condition models are used in conjuction with adiabatic free-slip wall conditions; the local wall conditions involve
imposing the impermeability condition for the velocity component normal to the
wall and implicit boundary conditions on the tangential velocity component W;
the latter conditions are implemented through specification of the surface shear
stress in the spirit of a rough-wall model (e.g., [13]). Additional inflow/outflow
numerical boundary conditions required for closure of the discretized equations
are chosen based on characteristic analysis (CA) as in previous jet simulation
studies (e.g., [4]). Resolution tests in selected cases involved additional runs on
$126 \times 307 \times 126$ and $56 \times 136 \times 56$ grids.

11.3   SWIRL INITIAL CONDITIONS

Numerous efforts have been devoted to study swirling flows in various combustion systems such as gas-turbine engines and diesel engines. The swirling flows
are used to improve and control the mixing process between fuel and oxidant
streams and in order to achieve flame stabilization and an enhanced heat-release
rate [14, 15]. Velocity field characteristics of swirling flow combustors have been
investigated extensively but were limited to specific geometry and were primarily focused on swirling jet flows [16] or on the interaction of two co-swirling or
counter-swirling streams [17]. One of the main features that are usually observed in swirling flows is the formation of a Center Toroidal Recirculation Zone
(CTRZ) around the axis of the jet. This CTRZ is used to stabilize the combustion process in a compact region within the combustion chamber. Experimental
results showed that the CTRZ is a quasi-axisymmetric bubble developed by vortex breakdown which is associated with swirling flow exceeding a certain swirling
strength [14, 16]. Swirling flow is introduced in practical combustor configurations by appropriately forcing tangential or azimuthal velocity components (e.g.,
introduced through guide vanes, tangential entry swirlers, or a rotating honeycomb). Multiple factors, including inlet conditions and geometry [18], tangential
velocity profile, and axial velocity [19], are known to affect the process of vortex
evolution and the breakdown process.

Swirl conditions of various degrees of complexity were considered to initialize
the simulations at the inlet. This included:

(1) idealized inflow boundary conditions involving a top-hat profile for the axial
velocity $U(r)$, the zero radial velocity V, and a tangential velocity profile
$W(r)$ from RANS of swirling turbulent pipe flows [20];

(2) top-hat velocity profiles based on experimental data from a practical coannular swirl/counter-swirl (GEAE LM-6000) combustor configuration [3];
and

(3) more complex velocity profiles based on RANS or laboratory studies of the flow within a triple-swirler fuel injector.

The combustor flows investigated here were characterized by peak inlet free-stream Mach numbers between 0.05 and 0.3 and by standard temperature and pressure (STP) conditions. Swirl (S) and radial (R) numbers, defined in terms of circumferentially-averaged velocity data by

$$
S = \frac{\int\limits_0^{R_0} \rho U W r^2 \, dr}{R_0 \int\limits_0^{R_0} \rho U^2 r \, dr} \, ,
$$

$$
R = \frac{\int\limits_0^{R_0} \rho U V r \, dr}{\int\limits_0^{R_0} \rho U^2 r \, dr}
$$

where the inlet radius R_0 is chosen to be half of the combustor radius R (Fig. 11.1), were considered. Here, S and R typically varied between 0 and 0.75 and between 0 and 0.5, respectively. Other than passive excitation due to the swirl, the flow was unforced and was allowed to naturally develop its unsteadiness. Typical Reynolds numbers involved, based on the peak mean inlet axial velocity and diameter, were Re > 70,000.

11.4   RESULTS AND DISCUSSION

LM-6000 Inlet Conditions

The LM-6000 device involves a coaxial dual-swirl (swirl/counter-swirl) configuration; it is being developed as operational hardware by General Electric Aircraft Engines (GEAE) for gas-turbine applications and used to test computational modeling capabilities in lean premixed turbulent combustion regimes [3]. The laboratory-measured, nozzle-outlet swirl velocity profiles used in [3] are compared in Fig. 11.2a for $S = 0.56$ with those of the turbulent pipe-flow case. The latter case involved the simplest swirl initial conditions and was reported elsewhere [6]. Inflow boundary conditions at the inlet used a top-hat profile for the axial velocity, zero radial velocity, a CA-based floating pressure condition [21], and a swirl tangential velocity profile selected as typical of those found in RANS studies of swirling turbulent pipe flows [20]; profiles associated with various body force distributions were reported in [20]; the profile corresponding to a constant body force was used in the present simulations. Radial velocities are finite but

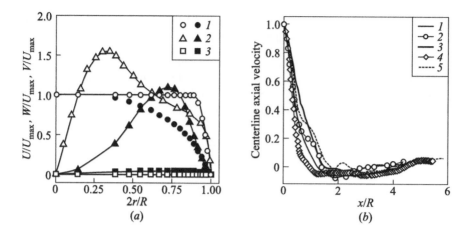

Figure 11.2 (*a*) Turbulent pipe-flow (open symbols) and GEAE LM-6000 (filled symbols) radial profiles of inlet velocity components (normalized by the peak mean inlet axial velocity) for $S = 0.56$: *1* — axial U; *2* — tangential W; and *3* — radial V. (*b*) Sensitivity of LM-6000 normalized centerline velocity to choice of floating inflow boundary condition; streamwise variable is scaled with inlet diameter R: previous LES [3]: *1* — fine grid; *2* — coarse grid (fix T and V); present work: *3* — MILES (fix ρ and V); *4* — MILES (fix P_0 and float V); and ·5 — OEEVM (fix ρ and V).

very small (i.e., $R = 0.012$) for the LM-6000 case, in contrast with identically zero prescribed radial velocities for the pipe-flow case.

For model testing purposes, the actual conditions in the previous nonreactive LES in the LM-6000 configuration [3] were first emulated with present (and other) simulation models; extensive comparative model studies will be reported separately [7] and are only illustrated here. Model testing studies included both the particular geometry and the operating conditions ($P_0 = 6$ atm, $T = 660$ K, premixed methane–air mixture, equivalence ratio $\phi = 0.56$, and inlet diameter $D_0 = 3.4$ cm). At the combustor inlet, Dirichlet conditions can be used in principle for primary flow variables other than the velocities. However, because subsonic inflow is involved, at least one physical quantity must be allowed to float at the inlet, and there is no indication from laboratory experiments on whether any particular (floating) inlet variable should be preferred. The approach in [3] was to use a condition derived using CA based on assuming fixed temperature and velocity components at the inlet boundary. Two other inlet boundary condition approaches were also tested based on:

(1) fixing the inlet mass density at the inflow and allowing pressure (and temperature) to float through a CA-based condition [21], and

(2) allowing the inlet radial velocity to float.

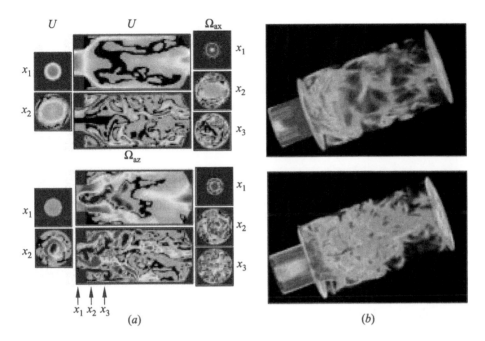

Figure 11.3 Instantaneous flow visualizations (*a*) and volume renderings of the vorticity magnitude Ω (*b*) for $S = 0.56$; LM-6000 (top), turbulent pipe flow (bottom). Instantaneous visualizations involve axial velocity levels between $-0.33U_0$ (dark blue) and U_0 (dark red) and vorticity values between $-R/(2U_0)$ (blue) and $+R/(2U_0)$ (red); more details are in [6]. (Refer color plate, p. IX.)

Comparison of time-averaged centerline velocity from nonreactive LES in [3] with present simulations using MILES and a One-Equation Eddy-Viscosity Model (OEEVM) [7] indicates that the near-inlet flow is captured fairly well with LES but can be somewhat sensitive to the actual choice of the specific inlet floating condition (Fig. 11.2*b*).

The sensitivity of the axisymmetric combustor flow dynamics to the actual choice of inlet velocity conditions was also examined. Figure 11.3 compares the results of initializing the simulations with the turbulent-pipe or LM-6000 swirling conditions and otherwise identical initial conditions ($S = 0.56$, $U_0 = 100$ m/s, STP). The flow visualizations depict the significant effects on the combustor vortex dynamics of changing the specifics of the velocity profiles used to initialize the LES, with noticeably more-axisymmetric features observed in the flow features for the LM-6000 case. The LM-6000 initial velocity conditions (Fig. 11.2*a*) involve a peak tangential velocity component located farther away from the axis and a more moderate radial gradient of the axial velocity. A clear consequence of these initial condition specifics, apparent in Fig. 11.3*a*, is that the LM-6000

swirl flow is dominated by the axisymmetric zone at the sudden expansion and
the adverse pressure gradient region necessary for occurrence of vortex breakdown phenomenon is not present. The associated unsteady flow patterns can
be observed in terms of the corresponding volume visualizations of the vorticity
magnitude in Fig. 11.3b, depicting characteristically more vortical and helically-
dominated modes in the swirling pipe-flow case.

Triple Annular Research Swirler

Based on a design by GEAE and Goodrich Aerospace typically used in industrial dry low-emissions combustors (e.g., [22]), this model gas-turbine fuel injector (Fig. 11.4) features multiple independent fuel supply lines for efficient fuel
distribution and multiple air inlets to obtain co- and counterrotating swirling
air streams. The entire combustion air is supplied through the mixer/fuel injector which is located at the combustor dome, denoted Triple Annular Research
Swirler (TARS)*.

The mixer includes three air passages equipped with swirlers leading into
the combustion chamber. The two central coaxial passages feature axial swirlers,
while the external air passage has radial swirling vanes. Air blast fuel atomizers

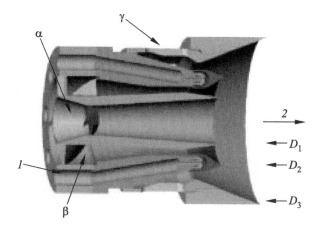

Figure 11.4 Multiswirl fuel injector (TARS): *1* — injected fuel; α — injected air
inner swirler; β — injected air intermediate swirler; γ — injected air outer swirler; and
2 — air–fuel mixture.

---

*For more details see Gutmark, E. J., G. Li, and S. Abraham. Characteristics and control
of a multiswirl spray combustor. Chapter 10 (Section 1) in this book, pp. 97–110. (*Editor's
remark.*)

Table 11.1 Investigated nozzle conditions.

TARS geometries and outlet conditions	Case I (LDV) open	Case II (LDV) open	Case III (LDV) open
α	45°	45°	c45°
β	45°	c45°	45°
γ	30°	30°	30°
S	0.39	0.25	0.50
R	< 0.04	< 0.04	< 0.04

are distributed between the second and third annular passages. Fuel is injected into the inner and outer annular passages for efficient mixing. A conventional pressure atomized pilot is located in the central passage.

Typical nozzle conditions investigated are indicated in Table 11.1, where angles α, β, and γ — corresponding to the central, intermediate, and external (radial) swirlers, respectively — are used to parameterize the specific triple-swirler geometries. S and R are evaluated based on circumferentially- and time-averaged velocity data at the nozzle outlet, and the "c" in Cases II and III denotes imposed counter-swirl relative to that of the other swirlers. Radial profiles of the axial and tangential velocities for the cases in Table 11.1 are shown in Fig. 11.5, based on UC Laser Doppler Velocimetry (LDV) data (described more in detail in [6]).

The locations of the boundaries between the three annular passages are shown on the abscissa. Compared to the top-hat inlet velocity profiles discussed above, TARS velocity profiles involve a much more complex structure, with the more noticeable aspect being the annular axial TARS velocities with a characteristic well around the axis — as opposed to the simpler top-hat velocities in Fig. 11.2a. All TARS cases reported in this chapter involve unconfined conditions downstream of the fuel-injector nozzle.

Several studies [23–25] have shown the effect of the combustor inlet conditions on the predicted flame structure, liner temperature, and emissions. These boundary conditions include the axial, tangential, and radial velocities; the turbulent kinetic energy; and associated length scale. To characterize the flow field at the exit of the TARS, two approaches were adopted. First, advanced diagnostic, such as LDV (discussed above), was used to measure the flow field distribution at the exit of the swirler. The data collected are used for inlet boundary conditions for the LES and database for numerical model validation. Second, a RANS model was used to study mixing and turbulence parameters in the TARS swirler [6].

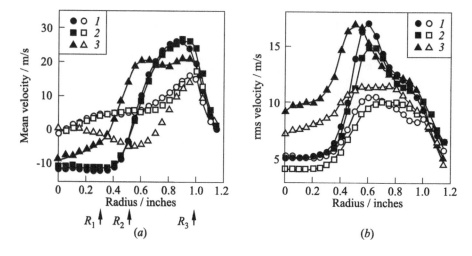

Figure 11.5 Circumferentially- and time-averaged mean (*a*) and rms (*b*) velocity profiles (open symbols — tangential, and filled symbols — axial) at the TARS outlet plane from UC LDV data [6]. Indicated radii ($R_i = D_i/2$) characterize the approximate locations of air-flow passages, cf. Fig. 11.4: *1* — 304545; *2* — 304545c; and *3* — 3045c45.

Although axisymmetric features can typically dominate within the TARS (e.g., Case I, Fig. 11.6), this is not so in general, particularly, when intermediate counter-swirl is present (Case II, Fig. 11.6); in the latter case, the presence of significant azimuthal inhomogeneities at the TARS outlet is apparent, leading to much larger turbulence intensities near the axis (Fig. 11.7). The four peaks in the axial mean and root-mean-squared (rms) velocity are emphasized in Case III. Figure 11.7 compares the main flow features for unconfined Cases I to III in a plane along the axis. The counter-swirl in both Case II and III shortens the recirculation zone and broadens it. Case II has a very short and irregular recirculation zone whose length is about one half of the fuel-injector diameter. The annular jet for this case extends further downstream relative to the other two cases. The annular jet produces two shear layers in the streamwise x–r plane with the internal and external recirculating regions. In addition, due to the counter-rotating swirling flows, two more tangential shear layers develop in the cross-sectional plane. This combination of orthogonal shear layers contributes to the high nonisotropic turbulence level. Because the laboratory velocity data acquisition was carried out at a small distance downstream from the TARS outlet ($x = 0.05R$), the measured velocity (and turbulence intensity) distributions had a chance to be affected by sudden expansion effects.

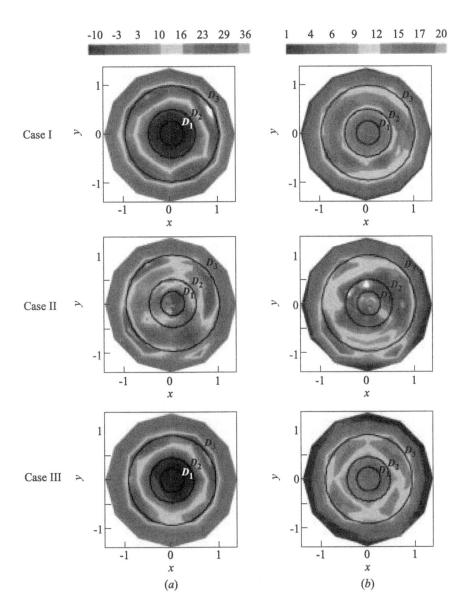

Figure 11.6 Mean (a) and rms (b) axial velocity distributions for three cases in Table 11.1 at a cross-stream plane near the TARS outlet (UC LDV data [6]); levels: mean velocity between −10 m/s (blue) and 36 m/s (white); rms velocity between 1 m/s (blue) and 20 m/s (white); more details are in [6]. (Refer color plate, p. X.)

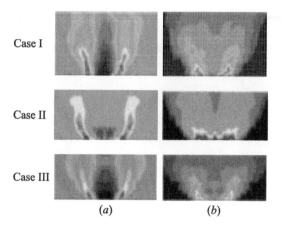

Case I

Case II

Case III

(a)   (b)

Figure 11.7 Caption is the same as in Fig. 11.6, at plane passing through the combustor axis; levels as in Fig. 11.6. (Refer color plate, p. XI.)

The closest distance to the mixer outlet at which the UC LDV measurements could be performed ($x = 0.05R$) was determined by the slanted angle of the intersecting laser beams (practical constraint imposed by the TARS geometry, shown in Fig. 11.4). This is in contrast with the setup in the simulation model, which expects velocity data specified at the inflow boundary of a very short but finite-length inlet ($l < 0.1R$). The *ad hoc* approach used in this early phase of joint hybrid simulations was to linearly scale-back the laboratory velocity information to approximately correct for the expansion effects.

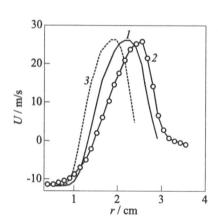

Figure 11.8 Axial velocity profiles near the inlet (nozzle configuration 304545): *1* — LDV, $z = 0.05R$; *2* — LDV, $z = 0.125R$; *3* — LES, and $z = 0$ (inlet boundary conditions).

This approach to defining the inlet velocity profiles does not account for important effects of the sudden expansion and adverse pressure gradients due to swirl, namely, the rapid decay of the axial velocity magnitude (e.g., Fig. 11.2a) and significant reduction of the inner slope of the axial velocity profile (e.g., Fig. 11.8). A challenging additional difficulty relates to emulating (even at axisymmetric profiles) the turbulent intensities in Fig. 11.5b:

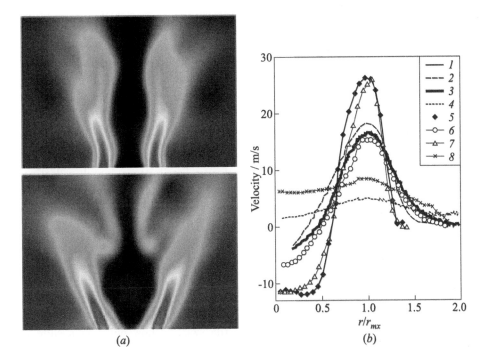

(a)   (b)

Figure 11.9 (*a*) Mean axial velocity distributions in a streamwise plane from LES; levels chosen as in Figs. 11.6 and 11.7 for Case I (top) and Case III (below). (*b*) Mean and rms axial velocity profiles at selected streamwise locations from LES and LDV for Case I; radial coordinate is normalized with local radius, r_{mx}, at which maximum of the corresponding profile is attained. Simulations: *1* — mean velocity at $x = 0$ (inlet BC); *2* — at $x = 0.350R$; *3* — at $x = 0.640R$; and *4* — rms velocity at $x = 0.640R$. Laboratory experiments: *5* — mean velocity at $x = 0.05R$; *6* — at $x = 0.125R$; *7* — at $x = 0.625R$; and *8* — rms velocity at $x = 0.625R$. (Refer color plate, p. XI.)

the potentially significant coherent-structure contribution to the inlet velocity fluctuations suggested by the relatively large peak fluctuation levels (as high as 50% of the peak velocities) could not be assessed at the time of the present joint studies between the NRL and UC. As a preliminary test of the hybrid simulation approach concept, turbulence intensities were neglected, and radial profiles of the axial and tangential velocity used as inlet boundary conditions for the LES were based on the mean (time- and circumferentially-averaged) LDV measurements at the TARS outlet (Fig. 11.5*a*).

Figure 11.9*a* shows mean axial velocity distributions based on LES for Cases I and III, quite similar to the corresponding laboratory visualizations

(Figs. 11.7a and 11.7c). In particular, the comparisons clearly show similar trends in the distributions, such as relatively thinner mixing layers, a shorter recirculation region, and more pronounced expansion effects in Case III.

The latter cases were selected because they exhibited the more approximately axisymmetric TARS outlet features. The simulations are capable of capturing the main features of the flow, including the initial shape of the recirculation zone, its lateral extent, and the development of the annular jet. More detailed comparisons between LES and LDV for Case I are shown in Fig. 11.9b, in terms of radial profiles of the mean value and rms axial velocity at selected cross-stream locations. Disagreements between LDV and LES velocity data are more noticeable in terms of the rms axial velocity distributions, largely reflecting on the neglected inlet turbulent intensities in the simulations, as well as on the needed improved emulation of the laboratory inlet conditions with regards to azimuthal mean velocity variations, turbulence statistics, and spectral content.

11.5 CONCLUDING REMARKS

The aforementioned behavior of the various swirlers emphasizes the importance of several parameters on the behavior of the swirling flow entering a sudden expansion area, namely, swirl and radial numbers; inlet length; and specific velocity profiles of the axial, radial, and tangential components. The flow is driven by the strong interaction between swirling shear-layer instabilities, on the one side, and flow instabilities driven by the sudden expansion and geometry of the combustion chamber, on the other. The coupling of these governing instabilities depends on the swirl magnitudes and the relative length of the inlet; swirl of sufficient strength produces an adverse pressure gradient which can promote flow reversal or vortex breakdown.

Swirl conditions of various degrees of complexity were considered to initialize the simulations at the inlet. This included using top-hat (swirling-pipe and LM-6000) as well as more complex (TARS) velocity profiles. The flow is very sensitive to the detailed nature of the radial profiles of the velocity components. A peak tangential velocity component located farther away from the axis and a more moderate radial gradient of the axial velocity (LM-6000 vs. swirling-pipe conditions) can result in the swirl flow being dominated by the axisymmetric zone at the sudden expansion and in the adverse pressure gradient region necessary for occurrence of vortex breakdown phenomenon not being present. Compared to the top-hat inlet velocity profiles, TARS velocity profiles involve a much more complex structure, with the more noticeable aspect being the annular axial TARS velocities with a characteristic well around the axis.

The ability to control the orientation and magnitude of the individual swirlers allows effective control on the mean velocity field via the vortex breakdown dependency on the overall swirl number and on the turbulence level via the shear layers between the three swirling flows. The annular jet surrounding the internal recirculating zone produces two shear layers in the axial plane with the internal and external recirculating flows, while the counter-rotating swirling flows add two tangential shear layers in the cross-sectional plane. This combination of orthogonal shear layers contributes to the highly nonisotropic turbulent fluctuations.

Ongoing studies address appropriate ways to improve the specification of inlet boundary conditions to more closely emulate those in the UC experiments; important issues to be elucidated relate to:

(1) selecting an appropriate model of the turbulence intensities at the inlet, and

(2) using suitable laboratory data acquisition to provide the necessary turbulence intensity information to close the model.

In the case of the triple-swirler TARS configuration, the circumferential distribution of the mean and turbulent velocity components is generally not axisymmetric, and this feature may have a significant additional effect on the flow evolution. Detailed plane (and possibly large-scale unsteady) velocity and turbulence intensity distributions should likely be used to initialize the simulations at the inlet — an approach which will be systematically pursued in future studies. Previously proposed deterministic (e.g., [26]) and semi-deterministic (e.g., using Proper Orthogonal Decomposition analysis [27]) approaches for formulating the turbulent inlet boundary conditions are currently being investigated in this context.

A systematic study addressing the competing effects of the dominating flow controlling mechanisms and the interaction between combustion and flow dynamics is the subject of the collaborative research project with UC.

ACKNOWLEDGMENTS

Support of this work was provided by NRL and ERDC, and is greatly appreciated.

REFERENCES

1. Bray, K. N. C. 1996. The challenge of turbulent combustion. *26th Symposium (International) on Combustion Proceedings*. Pittsburgh, PA: The Combustion Institute. 1–26.

2. Bray, K. N. C., and P. A. Libby. 1994. Recent developments in the BML model of premixed turbulent combustion. In: *Turbulent reacting flows*. Eds. P. A. Libby and F. A. Williams. London: Academic Press. 115–52.

3. Kim, W.-W., S. Menon, and H. C. Mongia. 1999. Large-eddy simulation of a gas turbine combustor flow. *Combustion Science Technology* 143:25–62.

4. Fureby, C., K. Kailasanath, and F. F. Grinstein. 2000. Large eddy simulation of premixed turbulent flow in a rearward-facing-step combustor. AIAA Paper No. 2000-0863.

5. Spalart, P. R., W. H. Jou, M. Strelets, and S. R. Allmaras. 1997. Comments on the feasibility of LES for wings, and on hybrid RANS/LES approach. In: *Advances in DNS/LES. 1st AFOSR International Conference in DNS/LES Proceedings*. Columbus: Greyden Press.

6. Grinstein, F. F., T. R. Young, E. J. Gutmark, G. Li, G. Hsiao, and H. Mongia. 2002. Flow dynamics in a swirl combustor. *J. Turbulence* 3:030.

7. Grinstein, F. F., and C. Fureby. 2003. LES studies of the flow in a swirl gas combustor. AIAA Paper No. 2003-0484.

8. Galperin, B., and S. A. Orszag. 1993. *Large eddy simulation of complex engineering and geophysical flows*. Cambridge: Cambridge University Press.

9. Boris, J. P., F. F. Grinstein, E. S. Oran, and R. L. Kolbe. 1992. New insights into large eddy simulation. *Fluid Dynamics Research* 10:199–227.

10. Fureby, C., and F. F. Grinstein. 1999. Monotonically integrated large eddy simulation of free shear flows. *AIAA J.* 37:544–56.

11. Fureby, C., and F. F. Grinstein. 2000. Large eddy simulation of high Reynolds number free and wall bounded flows. *J. Computational Physics* 181:68–97.

12. Landsberg, A. M., T. R. Young, and J. P. Boris. 1994. An efficient parallel method for solving flows in complex three-dimensional geometries. AIAA Paper No. 94-0413.

13. Mason, P. J., and N. S. Callen. 1986. On the magnitude of the subgrid-scale eddy coefficient in large-eddy simulations of turbulent channel flow. *J. Fluid Mechanics* 162:439–62.

14. Syred, N., and J. M. Beer. 1974. Combustion in swirling flows. *Combustion Flame* 23:143–201.

15. Paschereit, C., E. Gutmark, and W. Weisenstein. 2000. Excitation of thermoacoustic instabilities by interaction of acoustics and unstable swirling flow. *AIAA J.* 38:1025–34.

16. So, R. M. C., S. A. Ahmed, and H. C. Mongia. 1985. Jet characteristics in confined swirling flow. *Experiments in Fluids* 3:221–30.

17. Gouldin, F. C., J. S. Depsky, and S.-L. Lee. 1985. Velocity field characteristics of a swirling flow combustor. *AIAA J.* 23:95–102.

18. Sarpkaya, T. 1971. Vortex breakdown in swirling conical flows. *AIAA J.* 9:1792–99.

19. Faler, J. H., and S. Leibovich. 1977. Disrupted states of vortex flow and vortex breakdown. *Physics Fluids* 20:1385–400.

20. Pierce, C. D., and P. Moin. 1998. Method for generating equilibrium swirling inflow conditions. *AIAA J.* 36:1325.

21. Grinstein, F. F. 1994. Open boundary conditions in the simulation of subsonic turbulent shear flows. *J. Computational Physics* 115:43–55.

22. Pritchard, B. A., A. M. Danis, M. J. Foust, M. D. Durbin, and H. C. Mongia. 2002. Multiple annular combustion chamber swirler having atomizing pilot. European Patent No. EP1193448A2.

23. Danis, A. M., B. A. Pritchard, and H. C. Mongia. 1996. Empirical and semi-empirical correlation of emissions data from modern turbopropulsion gas turbine engines. ASME Paper No. 96-GT-86.

24. Danis, A. M., D. L. Burrus, and H. C. Mongia. 1997. Anchored CCD for gas turbine combustor design and data correlation. *J. Engineering Gas Turbines Power* 119:535–45.

25. Hura, H. S., N. D. Joshi, H. C. Mongia, and J. Tonouchi. 1998. Dry low emissions premixer CCD modeling and validation. ASME Paper No. 98-GT-444.

26. Lund, T. S., X. Wu, and K. D. Squires. 1998. On the generation of turbulent inflow conditions for boundary-layer simulations. *J. Computational Physics* 140:233–58.

27. Bonnet, J. P., J. Delville, P. Druault, P. Saugat, and R. Grohens. 1997. Linear stochastic estimation of LES inflow conditions. In: *Advances in DNS/LES. 1st AFOSR International Conference in DNS/LES Proceedings.* Columbus: Greyden Press. 341–34.

Comments

Roy: Why did you choose the LM 6000 (GE land-based gas turbine) which has no relevance to the Navy propulsion need?

Grinstein: The LM 6000 case was used for model validation purposes; it was chosen as a convenient reference configuration for which data is available from both laboratory and previous LES studies.

Seiner: In your presentation, only velocity profiles have been discussed. How will your computational approach work with combustion?

Grinstein: We plan to measure and/or model the combustion data (as practical).

Seiner: As you use Favre-averaged variables, how will you evaluate the impact of pressure and mass fluctuation correlations?

Grinstein: Because our (MILES) simulation approach does not require explicit filtering, we do not need to use Favre-averaged variables; the impact of pressure and mass correlations can be thus evaluated directly (at least in principle) based on the database generated in the simulations.

Lindstedt: Do the models work well at predicting the dominant frequencies?

Grinstein: The ability of practical LES models to capture the dominant frequencies has been demonstrated in previous dump combustor studies (including also a few of our own studies at NRL); we expect our simulation approach to be effective in that respect also for the TARS combustor configuration presently under study; suitable data analysis addressing this issue will be included in a forthcoming report.

Lindstedt: How do you implement the two-step chemistry?

Grinstein: In our most recent reacting swirl flow studies (reported elsewhere [7]), we use global (Westbrook–Dryer) chemistry; subgrid fluctuations are neglected; and instantaneous evaluation of quantities such as diffusivities, viscosities, and fuel burning rates are performed in terms of unfiltered variables. Such an approach is effective in the limit of high-Damköhler number flows such as driven by the highly-swirling flow regimes considered. Direct assessments of our MILES reduced-chemistry combustion model for such cases were reported based on direct comparisons of results obtained with other LES models [3, 7]; the comparisons indicate that the two-step MILES model can provide fairly reasonable approximations. Several approaches to improving the combustion modeling are the subject of our ongoing investigations.

Chapter 12

CONTROL OF FLAME STRUCTURE IN SPRAY COMBUSTION

A. K. Gupta, B. Habibzadeh, S. Archer, and M. Linck

In order to develop comprehensive passive combustion control techniques, the effects of radial distribution of swirl and combustion airflow on spray flame characteristics are examined using a double concentric swirl burner. The emphasis of the present work is on the use of shear forces in the swirling air to further reduce the size of the fuel droplets and to transport these droplets to more desirable locations in the spray flame. Flow and droplet characteristics (droplet size, velocity, and number density) and species distribution within the flame have been examined. The results show that the radial distribution of swirl and airflow conditions had a significant effect on droplet size and distribution in the sprays and spray flames. Data shows the direct effect of high-shear in the flow on secondary droplet atomization. The Intensified Charge Coupled Device (ICCD) images of OH and CH species distributions in gas and spray flames are also presented. Particle Image Velocimetry (PIV) images provided airflow characteristics associated with the high-shear regime of the flow. Airflow distribution and shear are shown to have a significant effect on droplet size, flame plume features, and spray flame structure. These features can be used to manipulate passive flame control of swirl-stabilized spray combustion systems.

12.1 INTRODUCTION

In order to control combustion efficiency, intensity, flame signatures, emissions of trace pollutants, and combustion instability, the flame structure must be controlled. Combustion characteristics can be controlled using either active or passive techniques. Even though the active control techniques may be more effective than passive, the active techniques are more expensive, complicated, and less reliable in harsh combustion environments. They also require the use of some external energy to control the instability in practical combustion systems. Passive control techniques can provide a more reliable and economical means to control flame characteristics, since there would be fewer control devices involved, and no external energy supply is necessary.

Smaller fuel droplet size in spray flames is more desirable, since smaller droplets enhance flame stability and combustion intensity and lead to rapid response to input operational conditions [1–9]. The focus of the present work is to reduce the droplet size by creating high-shear regions within the sprays.

Different combustion airflow regimes also have an effect of relocating the droplets within the flame, which, in turn, affects the flame structure. The relocation of droplets to specific regions in the flame can have a significant effect on combustion instability, efficiency, and emissions. If the local heat-release distribution in a flame can be changed, then it may be possible to eliminate the combustion instability at the desired operational conditions without any need to use external energy, instrumentation, or control devices.

A double concentric swirl burner is used with a commercially available air-assist nozzle to atomize the fuel. Nozzle characteristics, atomization airflow rate, and fuel flow rate were maintained constant throughout the experiments.

Information on droplet characteristics in sprays and spray flames is presented. The distribution of OH and CH species in spray and gas-fueled flames is presented in order to decouple the effect of droplet vaporization. As the droplet size in the spray flame becomes smaller, its signature should begin to resemble that of the gas-fueled flame, due to the extremely short evaporation time of the droplets.

12.2   EXPERIMENTAL FACILITY

The double concentric swirl burner used here has a centrally located fuel nozzle surrounded by two annular passages through which combustion air is supplied to the burner. Airflow and swirl vane angle (swirl strength) can be varied independently in each annulus. A swirl blade cascade of any blade angle setting can be placed in each annulus of the burner to provide the desired co- or counter-swirl arrangement. The facility therefore allows the examination of different swirl strengths and directions in the inner and outer annuli of the burner.

A schematic diagram of the experimental burner is shown in Fig. 12.1. It features an air-assist fuel nozzle, nominally rated for 0.5 gallons per hour. In the present study, kerosene has been used as the fuel and air has been used as the atomization gas.

A Phase Doppler Interferometer (PDI) was used to characterize the fuel spray under nonburning and burning conditions. The PDI measures droplet

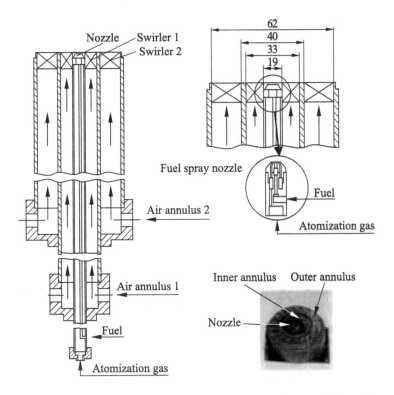

Figure 12.1 The experimental double concentric spray burner. Dimensions are in mm.

characteristics in a given volume in space, so that the burner designed to move in three dimensions allowed data to be taken throughout the flame with fixed optics. Data were taken at various radial positions at known axial positions downstream from the burner exit. Particle image velocimetry was used to measure instantaneous images of the flowfield of the carrier phase and droplets. This provided information on the mean and root-mean-squared (rms) velocity of axial, radial, and tangential components; velocity vectors; vorticity; and strain rate. A three-dimensional (3D) PIV system, utilizing two cameras and mechanical shutters, was used for taking measurements in spray flames. In the present study, the flow velocity data was obtained at the vertical cross-section of the burner by passing a laser beam through the longitudinal axis of the burner. The procedure provided instantaneous velocity and direction of seeded particles in the airflow and/or droplets.

An ICCD camera equipped with 430-, 515-, and 307-nanometer narrow bandpass filters was used to determine the distribution of OH, CH, and C_2 species in the gas and spray flames.

12.3   RESULTS

The effect of swirl distribution in the burner on droplet size and velocity distribution, flow, and flame characteristics is reported. Three examined swirl distributions are $50°/30°$, $50°/-30°$, and $65°/30°$, where the first number indicates the swirl vane angle of the inner annulus and the second number refers to that at the outer annulus. Three airflow distributions of $25\%/75\%$, $50\%/50\%$, and $75\%/25\%$ were examined for each swirl distribution. The first percentage number indicates the fraction of the total combustion airflow through the inner annulus, while the second number refers to that at the outer annulus. The results presented here are for swirl distributions of $65°/30°$ and $50°/30°$ only, where the total airflow was held constant to maintain a fixed equivalence ratio of 0.4.

Global features of the flames were examined using direct image photography. The change in swirl in the outer annulus from co- to counter-swirl resulted in a thinner flame. The co-swirl distribution resulted in a more compact flame as compared to the counter-swirl case. The combustion airflow distribution in the burner had a significant effect on the flame plume configuration.

The droplet size, number density, and volume flux data under nonburning and burning cases, with swirl distributions of $65°/30°$ and $50°/30°$, are presented in Figs. 12.2a and 12.2b. They show the effect of low- and high-shear air distributions on droplet mean size, number density, and volume flux at 5 and 30 mm downstream from the fuel-nozzle exit for the nonburning and burning cases. Data at other locations in the flame were also obtained. Stronger swirl in the inner annulus provides a secondary breakup of the droplets at the region of high-shear under nonburning conditions. One observes smaller-size droplets shown in Fig. 12.2a for the $65°/30°$ case as compared to the $50°/30°$ case at a radial location of about 10 to 15 mm at $X = 5$ mm. In contrast, under burning conditions, the droplet size with the $65°/30°$ case is larger due to rapid depletion of the smaller-size droplets under combustion conditions.

The results show that in the nonburning case high-shear created by the $65°/30°$ swirl combination produces smaller-size droplets as compared to the low-shear case with $50°/30°$ swirl distribution. The increase in droplet number density occurs in the same region where a decrease in droplet size is observed. This then suggests that there must be secondary breakup of the droplets with the high-shear interface between the flows from the inner and outer annular passages of the burner. The increase in number density coincides with the high-shear region at all locations downstream from the fuel-nozzle exit. Smaller droplets evaporate more quickly. The droplet mean size for the high-shear case ($65°/30°$) is smaller than that observed in the low-shear ($50°/30°$) case. Approximate estimates of the Weber number (We) for the high-shear case showed it to be near the critical value. This suggests that the droplets in the high-shear case must have undergone some secondary breakup. A decrease in droplet size by as much as 50% can be seen at some locations in the spray for the high-shear

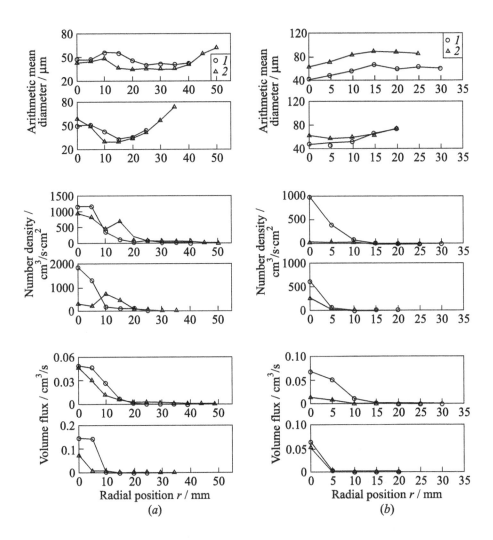

Figure 12.2 Effects of low- and high-shear swirl distribution on arithmetic mean diameter, number density, and volume flux for (a) nonburning and (b) burning cases for equal combustion airflow distribution between inner and outer annulus of the burner: 1 — 50°/30°; and 2 — 65°/30°. The upper panels correspond to $X = 30$ mm, and the lower panels correspond to $X = 5$ mm.

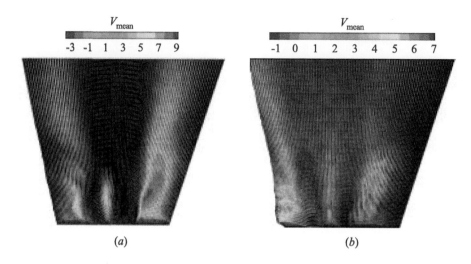

Figure 12.3 Contours of mean axial velocity (m/s) for 65°/30° (*a*) and 50°/30° (*b*) swirl distribution in the burner with equal flow distribution under nonburning conditions. (Refer color plate, p. XII.)

case [8]. This breakup is caused by the velocity difference between the two adjacent air streams. These velocity differences are more pronounced in the high-shear case. An examination of the flowfield was made using PIV. The axial component of droplet velocity in high-shear is smaller than for the low-shear case, see Figs. 12.3*a* and 12.3*b*. Since the total amount of combustion air fed to the burner was held constant during the experiments, the decrease in axial velocity indicates that the tangential component of the velocity is increased. Indeed, an examination of the tangential velocity revealed this to be the case. Thus, a sudden increase in tangential velocity promotes secondary breakup of the droplets.

It can be seen from Fig. 12.2*a* that in the nonburning case, secondary breakup of the droplets occurs at the flow interface region between the inner and outer annulus of the burner. The shear forces in this region are greater for the high-shear case and result in greater secondary breakup of the droplets. The arithmetic mean diameter graph shows smaller droplets, while the number density graph shows an increase in the number of droplets. These results indicate that bigger droplets under high-shear forces disintegrate into a large number of smaller-size droplets. However, in Fig. 12.2*b*, which deals with the burning case, it can be seen that the arithmetic mean diameter of the droplets for the higher-shear case is bigger. This result can be explained by corresponding these droplets with the number-density results at the same location. The number-density graphs show that there is a smaller number of droplets for the

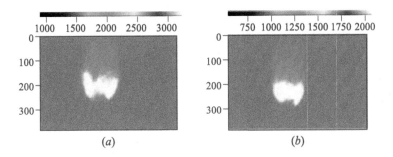

Figure 12.4 Measured OH distribution in the propane flames. Flow distribution: 25% inner and 75% outer annulus. Swirl combinations: 55°/30° (a) and 65°/30° (b). (Refer color plate, p. XIII.)

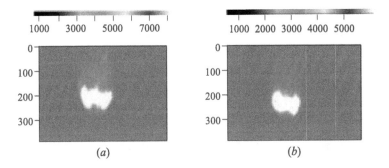

Figure 12.5 Measured CH distribution in the propane flames. Flow distribution: 25% inner and 75% outer annulus. Swirl combinations: 55°/30° (a) and 65°/30° (b). (Refer color plate, p. XIII.)

high-shear case as compared to the low-shear case. From the nonburning case one expects smaller-size droplets occurring in greater number in the high-shear case. During combustion, however, the smaller droplets quickly vaporize and burn, so the remaining droplets are larger in size and smaller in number.

Figures 12.4 and 12.5 show the distribution of OH and CH in swirl-stabilized propane–air flames using 25%/75% combustion air distribution in the inner and outer annulus of the burner, respectively. The results show that the reaction zone moves upstream towards the burner with high-shear, see Fig. 12.4. The distribution of CH shows a measure of the heat-release rate in flames. The heat-release rate is slow with the low-shear case to give a wider and longer flame (see Fig. 12.5). Figure 12.6 shows the distribution of OH in a kerosene flame with 55°/30° swirl distribution and 25%/75% combustion air (a) and 75%/25% combustion air distribution (b) in the two annular passages of the burner. Stronger

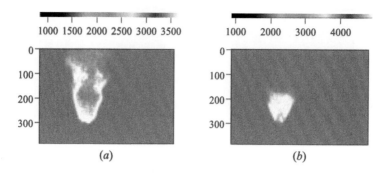

Figure 12.6 Measured OH distribution in the kerosene spray flames. Swirl combination: 55°/30°. Flow distributions: 25% inner and 75% outer annulus (a); and 75% inner and 25% outer annulus (b). (Refer color plate, p. XIV.)

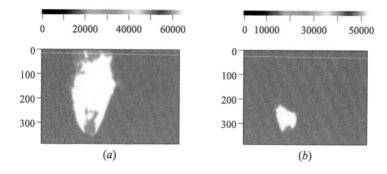

Figure 12.7 Measured CH distribution in the kerosene spray flames. Swirl combination: 55°/30°. Flow distributions: 25% inner and 75% outer annulus (a); and 75% inner and 25% outer annulus (b). (Refer color plate, p. XIV.)

swirl in the inner annulus results in spray reaction zone features that are similar to those observed in gas flames (compare 65°/30° swirl with 25%/75% air case using propane as the fuel case with 55°/30° swirl and 75%/25% air flow spray flame case). Similarly, the features observed from the distribution of CH from spray flames, shown in Fig. 12.7, can be correlated with gas-fueled flames. As the images show, under the right conditions, the reacting region of the kerosene flame can be shaped to resemble that of the propane flames. This indicates that the smaller-size droplets caused by the high-shear forces are small enough for the evaporation step to be negligible and no longer a rate-controlling parameter in the combustion reactions. Under other sets of swirl and airflow conditions, the kerosene flame may not resemble the propane flame.

12.4   CONCLUDING REMARKS

The results show that the radial distribution of swirl in the burner has an important effect on the spatial distribution of droplet size, velocity, and number density. Using swirlers with a sudden large velocity difference in the flow through which the droplets pass can produce smaller droplets. This effect can be explained by the higher-shear forces produced by greater swirl differential in the flow to break up the larger droplets. The tangential velocity component of the droplets is higher in the high-shear case, which helps to break up the large-size droplets.

A co-swirl distribution in the burner increases both the droplet size and the axial mean velocity at all positions in the spray, including the longitudinal axis through the spray centerline and all radial positions downstream of the burner exit. This effect becomes more pronounced as the axial distance from the burner exit increases. The combustion airflow distribution in the burner has also been shown to have a significant effect on the size and velocity distribution of droplets in the spray flames. Since it is possible to change the droplet size, velocity, and number-density distribution in the spray flame using these techniques, it becomes possible to control the flame characteristics for any set of operating conditions of the combustor.

ACKNOWLEDGMENTS

This work was performed under ONR contract N00014-99-1-0491.

REFERENCES

1. Presser, C., A. K. Gupta, and H. G. Semerjian. 1993. Aerodynamic characteristics of swirling spray flames. *Combustion Flame* 92:25–44.

2. Presser, C., A. K. Gupta, C. T. Avedisian, and H. G. Semerjian. 1994. *Atomization Sprays* 4:207–22.

3. Presser, C., A. K. Gupta, H. G. Semerjian, and C. T. Avedisian. 1994. *J. Propulsion Power* 10(5):631–38.

4. Aftel, R., A. K. Gupta, C. Presser, and C. Cook. 1996. Gas property effects on droplet atomization and combustion in an air-assist atomizer. *26th Symposium (International) on Combustion Proceedings*. 1645–51.

5. Gupta, A. K., T. Damm, C. Cook, S. R. Charagundla, and C. Presser. 1997. Effect of oxygen-enriched atomization air on the characteristics of spray flames. *35th Aerospace Sciences Meeting Proceedings*. Reno, NV. Paper No. 97-0268.

6. Gupta, A. K., and M. Megerle. 1997. Effect of steam assisted atomization on spray flame characteristics. *33rd AIAA/ASME/SAE/ASEE Joint Propulsion Conference Proceedings.* Seattle, WA. Paper No. 97-2839.

7. Gupta, A. K., M. Megerle, S. R. Charagundla, and C. Presser. 1998. Spray flame characteristics for high temperature gas-assisted atomization. *ILASS-98 Proceedings.* Sacramento, CA. 354–58.

8. Habibzadeh, B., and A. K. Gupta. 2001. Passive control of kerosene spray flame structure in a swirl burner. *Conference (International) on Advanced Energy and Related Technologies, RAN 2001 Proceedings.* Nagoya, Japan.

9. Mehresh, P., B. Habibzadeh, and A. K. Gupta. 2002. Control of spray flame characteristics using high-shear in a double concentric swirl burner. *ASME International Joint Power Generation Conference (IJPGC) Proceedings.* Scottsdale, AZ.

Comments

Mashayek: How many data sets were averaged in your PIV results and is there any criterion?

Gupta: It depends on the confidence limit you want on your data. We take the ensemble average of many instantaneous PIV velocity fields. Typically, 500 images are required for statistical convergence.

Santoro: How are you able to separate the flow seeding from the fuel droplets?

Gupta: At the present time, the PIV simply captures the velocity dynamics of the droplets. For nonreacting flows, we seed the two systems separately.

Strykowski: You indicate the presence of strong recirculation. Have you evaluated the impact of global instability concepts to the spray dynamics?

Gupta: There is circumstantial evidence for global instability, but no hard evidence that we have identified.

Chapter 13

POROUS MEDIA BURNERS FOR CLEAN ENGINES

J. J. Witton and E. Noordally

Combustion in porous media offers some advantages for propulsion devices, notably, wide range, low emissions, and very good spatial temperature control. The study will examine these characteristics at operating conditions of pressure and air-inlet temperature representing typical high-performance propulsion devices. Initial studies will be run on gaseous fuels, with extension to liquid fuels if results are promising. This chapter describes the experimental setup for this newly started work.

13.1 INTRODUCTION

The objective of this study is to examine the porous media combustion concept at high pressure and thus to indicate its potential in propulsion devices. The technique of establishing combustion in and over a porous matrix has certain well-documented advantages relating to the flammability range over which combustion can be maintained and the quality of gaseous emissions from the combustor [1–3]. Hitherto, these characteristics have been demonstrated widely at atmospheric pressure, see, for example, [4], but only in general terms at high pressures [5]. In addition, very good spatial control of the gas temperature might be expected, coupled with low-acoustic emission from the combustion process. This study will concentrate on performance and characterization of such a combustor over a range of pressures, typically ambient to approximately 2500 kPa, with reactant preheat to about 750 K, and will use a variety of fuels to examine the effects of fundamental combustion properties such as flame speed.

13.2   EXPERIMENTAL SETUP

The study is based on a small combustor, approximately 50 mm in diameter, using several different types of porous media including ceramics and metals of different porosities. Services such as fuel and instrumentation will be common throughout the study wherever possible. Parameters and measurements made will include at least the following:

- Fuel and air flow rates;

- Rig operating pressure and temperature;

- Pressure loss across the porous matrix elements;

- Fuel composition;

- Temperature of the porous media and its evolution through the combustor;

- Burnt gas composition using either dedicated single-species analyzers or multicomponent methods such as Fourier transform infrared analyzer (FTIR) and mass spectrometer (MS); and

- Radiant emissions from the matrix in certain cases.

The basis of the design is a simple laboratory-scale unit [4]. The high-pressure variant scheme is illustrated in Fig. 13.1. The enclosure for the media is a refractory tube, providing a quasi-adiabatic environment and enabling small-sized experiments to be used while limiting radial heat losses. Figure 13.2 shows the tube and two varieties of media. In these combustors, as in catalytic combustors, radiant exchange from the exposed ends of the media is an important energy transfer mechanism. An analysis based on view factor exchange has been used to assess this and to guide the initial design. Detailed numerical modeling has shown this effect — see companion chapter from Ellzey, Barra, and Diepvens*.

Two combustors have been prepared, one for predominantly high-pressure operation and the other for low-pressure work, especially ignition and fuel placement studies. The first set of experiments, both at elevated and normal pressure, uses a small (approximately 0.5 mm in diameter) electrically heated igniter located in the porous medium. The propulsion application puts a premium on reliable, simple ignition systems.

The atmospheric rig is also used for fuel placement studies. Initially, these will be conducted using an inert tracer — Helium — to label a surrogate fuel flow, with mass spectrometry as the measurement method. The local fuel distribution can be applied to the numerical model to assess the effects of distortion in fuel–air

---

*See Ch. 14 (Section 1) in this book, pp. 145–156. (*Editor's remark.*)

mixedness on the reaction-zone performance and temperature evolution
within the porous media.

In both experiments, the fuel–air
mixture is prepared by separate injection of the reactants and in-line mixing
using a proprietary static mixer (type
SMV, Sulzer Co.). Other experiments
have used this method with good results, but it is not seen as other than
a convenience for the laboratory work.
Other techniques will apply, for example, to liquid fuels, which will follow.

Both rigs use existing Cranfield facilities for reactant supply and metering, rig pressure control, and gas analysis. The gas fuel and combustion air
are fed from bottled storage through
PC-controlled mass-flow controllers.
For the high-pressure rig, the combustion air passes through a high-pressure
electric heater capable of 800 K and
is blended with the fuel immediately

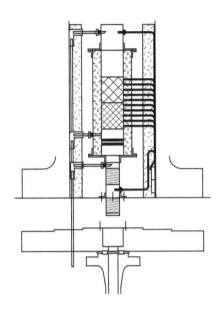

Figure 13.1 Assembly of high-pressure combustor.

prior to injection. For some of the fuels to be used in the study, the higher
feed temperatures present a potential spontaneous ignition hazard. The average
mixing residence time, local temperatures, and methods will be changed during
parts of the work to control this hazard. Within the combustor, a long length
(50 mm) of fine-pore material has been incorporated to act as a suppressor.
This draws upon other work on catalytic combustion that has a similar constraint and has successfully used a comparable arrangement. Exhaust products
from the media stack pass into a film-cooled tube representing the dilution zone
of a combustor. This is a metal tube with film cooling at the inlet end and
dilution air jets fed through ports in the wall. In addition to the film cooling,
the dilution air will be fed cold initially, and an additional measure of convective
cooling is obtained on the backside tube face. At this stage, there is no plan to
configure the preliminary experiments to fully reproduce the major features of a
turbine combustor or temperature profile control that would normally be found
in a practical geometry: if the work is promising, such additions will be made
later, incorporating representative feed temperatures, pressure losses, and flow
splits.

Rig pressure control occurs via a simple exhaust throttle valve, with water
spray to reduce the bulk gas temperature. The rig exhaust passes through a
silencer. Gas for exhaust analysis is drawn by a probe that can axially traverse,

Figure 13.2 Porous media and refractory liner before build.

mounted on the duct centerline, and by fixed probes at 1/3 insertion radial positions at various locations along the flow path. Gas analysis is obtained from PC-controlled analyzers, both single-species and multispecies. Single components for CO, CO_2, unreacted hydrocarbons (as methane or propane, depending on fuel fed), NOx, and O_2 are available. An FTIR spectrometer can also be applied. It is possible to add a window to the end-plate of the high-pressure (HP) casing to give optical access later.

Measurements of the interface temperature between the media and refractory liner are made using a combination of B- and N-type sheathed thermocouples. Thermocouples are positioned through radial drillings and sealed into the refractory with high-temperature cement. Lead-throughs pass out to the logging computer via the mounting plate of the combustor. Pressure measurements across the entire media stack, both coarse and fine, will be used to indicate combustor pressure loss. It is likely that the losses due to the individual media will be obtained using the atmospheric rig, which is easier to rebuild and instrument to the necessary accuracy.

One of the key issues in the porous media burner is where flame stabilization occurs. Numerical modeling, backed up by atmospheric pressure laboratory experiments, suggests that when stabilization is obtained at the interface between the media, operating flexibility and heat release is optimal. The numerical model results from the University of Texas (Prof. J. L. Ellzey) have been used to guide the positioning of thermocouples in the HP rig wall that will measure the rapid temperature rise adjacent to the interface, both axially and circumferentially. This is an important issue in the "excess enthalpy" concept on which this type of combustor is based [5, 6] and which may be expected to be important at high operating pressures. A separate concern that has been expressed is that the flame front may not be perpendicular to the bulk flow direction at the relatively low speeds through the combustor that will be present in early work. The HP experiments will be conducted with the burner axis horizontal, and additional instrumentation will measure any gross reaction-zone distortion. Later in the program, methods other than thermocouples may be employed to examine solid and possibly gas temperatures.

The design and build phase is completed for the initial experiments, which are now establishing a reliable ignition envelope for the combustor and obtaining preliminary information at pressure. The experiments are proceeding, via a coarse and fine sift, to define operational procedures and regions. Initially, the work is being done on methane as much of the open-domain literature has employed this fuel, and the modeling support is available from the companion program at the University of Texas with a full reaction scheme. Subsequently, work will embrace propane and other fuels having different fundamental properties such as flame speed; this is expected to overlap from the second to the third year of the project, when the detailed performance mapping will continue.

13.3   CONCLUDING REMARKS

The potential attraction of the method in controlling combustor exit gas temperature profile and emissions, coupled with the ability to burn at lean equivalence ratios in simple and inexpensive geometries, offers simplicity of design. At this stage, it is not possible to predict the benefits and limitations of the technique. Two clear requirements are to increase the combustion intensity and, with practical devices in mind, to establish simple ignition methods. The type and range of fuels that can be accommodated by the technique will be explored during later parts of the work as part of the overall characterization; this will include temperature profile, gas composition, and operable equivalence ratio range, all obtained at representative propulsion cycle gas turbine conditions. It is expected that some useful information on matrix materials will be obtained also, although a systematic study of the materials issue is beyond the scope of the work.

ACKNOWLEDGMENTS

The work was performed under the U.S. Navy's Office of Naval Research Combustion Control Research Program; award number N00014-01-0393. Funds found from the Naval International Cooperative Opportunities Science and Technology Program for part of the Cranfield work are gratefully acknowledged.

REFERENCES

1. Sathe, S. B., M. R. Kulkarni, R. E. Peck, and T. W. Tong. 1990. An experimental and theoretical study of porous radiant burner performance. *23rd Symposium (International) on Combustion Proceedings*. Pittsburgh, PA: The Combustion Institute. 1011–18.

2. Evans, W. D., J. R. Howell, and P. L. Varghese. 1991. The stability limits of methane combustion inside a porous ceramic matrix. AIAA Paper No. 91-1966.

3. Howell, J. R., M. J. Hall, and J. L. Ellzey. 2001. Combustion of hydrocarbon fuels within porous inert media. *Progress Energy Combustion Science* 22:121–45.

4. Khanna, V., R. Goel, and J. L. Ellzey. 1994. Measurements of emissions and radiation for methane combustion within a porous burner. *Combustion Science Technology* 99:133–42.

5. Babkin, V. S., A. A. Korzhavin, and V. A. Bunev. 1991. Propagation of premixed gaseous explosion flames in porous media. *Combustion Flame* 87:182–90.

6. Hardesty, D. R., and F. J. Weinberg. 1974. Burners producing large excess enthalpies. *Combustion Science Technology* 8:201–14.

Comments

Gupta: Why are you using methane?

Witton: Methane was chosen because a database is available at the University of Texas and to coordinate with Prof. Ellzey's computational results.

Soto: Are you planning to look at exhaust for emission and acoustic characteristics?

Witton: Yes, we are. Standard analysis of pollutants' emission will be made (CO_2, CO, NO_x). We also have the capability to measure noise, but we expect the burner will be relatively quiet.

Smirnov: How durable are the porous materials?

Witton: Zirconium-containing material would seem to be the best. Aluminum oxides are also promising. In addition, manufacturers have indicated their flexibility in producing new materials.

Soto: Are you planning to vary the pore size?

Witton: It will depend on what is available from the manufacturer.

Chapter 14

SIMULATIONS OF A POROUS BURNER FOR A GAS TURBINE

J. L. Ellzey, A. J. Barra, and G. Diepvens

Porous burners in which the flame is stabilized within the matrix of a porous solid offer various advantages such as low emissions and fuel flexibility. A major issue in these burners, however, is the stabilization of the flame within the matrix. A promising design that has emerged from recent research consists of two sections of porous medium with different characteristics. Since the effective flame speed within the matrix is determined by the porous medium properties, the interface between the two sections acts as a flame holder preventing flashback for a range of conditions. In this chapter, a computational study of a two-section porous burner is presented. The values of the matrix conductivity, extinction coefficient, and volumetric heat-transfer coefficient were varied, and the range of stable flow rates for which a flame stabilized at the interface was determined. For each property, the optimum values are discussed with respect to maximum velocity and maximum stable range.

14.1 INTRODUCTION

Porous burners in which the flame is stabilized within the matrix rather than at the surface have received considerable attention in the last two decades due to their low emissions, fuel flexibility, and wide operating range [1, 2]. Interest in combustion in porous media grew out of early work on excess enthalpy or superadiabatic flames [3] in which peak temperatures were shown to be higher than the adiabatic flame temperature if heat was recirculated from the hot products to the incoming reactants*. Subsequent work [4] showed that a porous solid inserted into the flow provided the necessary heat recirculation. Various researchers showed that flames in porous media exhibit higher effective flame speeds than free flames [5, 6].

Much of the recent research in combustion in porous media has focused on developing a porous burner as a radiant heater [7–13]. This is an attractive application because the porous solid is an efficient radiator while still permitting the use of a clean fuel such as methane. One design that has shown promise is a burner consisting of two sections of porous medium with different characteristics [2, 9, 10]. This design is based on the idea that the effective flame speed within the matrix is determined by the porous medium properties such as solid conductivity, porosity, and pore diameter. The interface between the two sections of porous media acts as a flame holder preventing flashback.

Hsu et al. [9] studied the two-section burner through experiments and computations. As the upstream extinction coefficient increased, the amount of preheating was reduced, resulting in lower peak flame temperatures. Their computational results predicted a smaller range of stable burning velocities than observed in the laboratory, and the sensitivity to various other porous medium properties was not investigated.

In this chapter, computational results of a flame stabilized within a porous burner consisting of two sections of porous media are presented. The range of velocities for which a flame will stabilize at the interface is determined for a burner similar to that described in Khanna et al. [10]. The effects of varying the conductivity, the extinction coefficient, and the volumetric heat-transfer coefficient are examined through a parametric study. An optimum design is discussed with respect to these parameters.

14.2   NUMERICAL METHOD

The computational model follows that reported in Henneke and Ellzey [13]. The following conservation equations for mass, gas energy, solid energy, and gas species are solved in the computational model:

$$\frac{\partial(\rho_g \varepsilon)}{\partial t} + \frac{\partial(\rho_g \varepsilon u)}{\partial x} = 0 \tag{14.1}$$

---

*The ideas of fuel burning with "excess enthalpy" and some schemes of practical realization with external heat exchangers were first published in Weinberg, F. J. 1971. *Nature*. 233; Weinberg, F. J. 1975. *15th Symposium (International) on Combustion Proceedings*. Pittsburg, PA: The Combustion Institute; Jones, A. R., *et al.* 1987. *Proc. Royal Society London A* 360. The use of "internal" heat exchangers in the form of porous filling was proposed in [5]. (*Editor's remark.*)

$$\rho_g C_g \varepsilon \frac{\partial T_g}{\partial t} + \rho_g C_g \varepsilon u \frac{\partial T_g}{\partial x} + \sum \rho \varepsilon Y_i V_i C_{gi} \frac{\partial T_g}{\partial x} + \sum \dot{\omega}_i h_i W_i$$
$$= \varepsilon \frac{\partial}{\partial x} \left(\left(k_g + \rho_g C_g D^d \right) \frac{\partial T_g}{\partial x} \right) - h_v (T_g - T_s) \qquad (14.2)$$

$$\rho_g C_g (1 - \varepsilon) \frac{\partial T_s}{\partial t} = k_s (1 - \varepsilon) \frac{\partial^2 T_s}{\partial x^2} + h_v (T_g - T_s) - \frac{dq^r}{dx} \qquad (14.3)$$

$$\rho_g \varepsilon \frac{\partial Y_i}{\partial t} + \rho_g \varepsilon u \frac{\partial Y_i}{\partial x} + \frac{\partial}{\partial x} (\rho \varepsilon Y_i V_i) - \varepsilon \dot{\omega} W_i = 0 \qquad (14.4)$$

The symbols are defined as follows: ε is porosity; ρ is gas density; u is gas velocity; t is time; C_g is specific heat of the gas; T is temperature of the gas; x is distance; Y_i, V_i, C_{gi}, w_i, h_i, and W_i are the mass fraction, diffusion velocity, specific heat, molar rate of production, molar enthalpy, and molecular mass of species i, respectively; k_g is gas thermal conductivity; D^d is the thermal disperse coefficient; h_v is the volumetric heat-transfer coefficient between the porous medium and the gas; T_s, C_s, and k_s are the temperature, specific heat, and thermal conductivity of the porous medium, respectively; and q^r is the radiative heat flux in the x-direction.

Gas densities are computed from the ideal gas equation of state for a multi-component mixture,

$$P = \frac{\rho_g R T_g}{\overline{W}} \qquad (14.5)$$

where $\overline{W}$ is the mean molecular mass, and R is the gas constant. The diffusion velocities are given by the mixture-averaged formulation

$$V_i = - \left(D_{im} + D_m^d \right) \frac{1}{X_i} \frac{\partial X_i}{\partial x} \qquad (14.6)$$

with

$$D_{im} = \frac{1 - Y_i}{\sum (X_j / D_{ij})} \qquad (14.7)$$

where X_i is the mole fraction, and D_{ij} is the binary diffusion coefficient. The radiation field q^r is represented with the P_3 approximation with scattering [14]. Gas-phase thermochemical and transport properties are obtained from the Chemkin [15] and Tranfit [16] packages. The GRI 1.2 mechanism for methane combustion [17] was used to calculate the production rates $\dot{\omega}$.

Values for conductivity, extinction coefficient, and albedo were obtained from [18] for porous partially stabilized zirconia (PSZ). The correlation for Nusselt number [19] is given by

$$\mathrm{Nu}_v = C\mathrm{Re}^n \tag{14.8}$$

where C and n take on various values depending on the number of pores per centimeter (ppc). An inverse relationship between pore diameter and extinction coefficient [18, 20] is valid for pore sizes larger than 0.6 mm:

$$\kappa = \frac{3}{d_p}\left(1 - \varepsilon\right) \tag{14.9}$$

In addition, the properties for the two sections of the burner are smoothed across the interface using an error function in order to avoid a discontinuity.

The computational domain is shown in Fig. 14.1. The domain was discretized into 300 grid points with a cluster located in regions of high-temperature gradient. The burner consists of two sections of porous media. For the reference case, the burner consisted of an upstream section of porous medium with 25.6 ppc and a downstream section with 3.9 ppc. The upstream and downstream porosities were 0.84 and 0.87, respectively. The upstream and downstream pore diameters were 0.029 and 0.152 cm, respectively. Methane and air, with an equivalence ratio of 0.65, a temperature of 298 K, and a specified velocity, flow into the domain. Hot products exit at the downstream end of the domain. Both downstream and upstream boundaries radiate to a black body at 298 K.

14.3   RESULTS

General Behavior

The computed gas temperatures for various inlet velocities at an equivalence ratio of 0.65 are shown in Fig. 14.2. For velocities from 20 to 45 cm/s, the flame stabilized at $x \approx 0.2$ cm, near the upstream face of the burner. From 48 to 74 cm/s, the flame stabilized near the interface ($x = 3.5$ cm). From 75 to 80 cm/s (not shown in Fig. 14.2), the flame stabilized near the exit of the burner. At higher velocities, a stable solution was not obtained within the computational domain. The laminar flame speed for a methane/air mixture with an equivalence ratio of 0.65 is 15.5 cm/s. All of the computed results for stable flames show effective flame speeds greater than 15.5 cm/s.

Experimental results from Khanna et al. [10] show that flames stabilized between approximately 15 and 50 cm/s for an equivalence ratio of 0.65. These velocities are lower than predicted by the authors' computations. Heat losses tend to reduce the operating range by decreasing the peak temperature and, hence, the effective flame speed. In the experimental burner, there were radial heat losses, whereas in the authors' computations, these losses were not modeled.

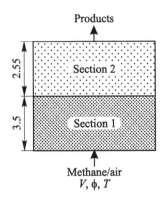

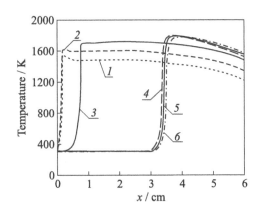

Figure 14.1 Computational domain (dimensions are in cm).

Figure 14.2 Gas temperature profiles [21]: *1* — 20; *2* — 30; *3* — 45; *4* — 50; *5* — 60; and *6* — 71 (dimensions are in cm/s).

As shown in Fig. 14.2, the peak temperature is lower for the flames that stabilize near the upstream face due to the increased heat losses. The peak temperature for the flames stabilized just downstream of the interface are slightly higher (by ~ 50 K) than the adiabatic flame temperature (1753 K) for a free flame at these conditions, thus demonstrating superadiabatic combustion.

Effect of Solid Conductivity

Heat recirculation from the postflame zone to the preflame zone occurs due to radiation and conduction. If either process is enhanced, then the effective flame speed in the matrix is increased. In order to test the sensitivity of flame stabilization to conductivity, the conductivities in the upstream and downstream porous matrices were changed and the minimum and maximum velocities for which a flame would stabilize at the interface were determined. These results are presented in Table 14.1. For the base case, the conductivities are those for a Khanna burner [10].

In Case A, the conductivities of both the downstream and upstream sections were divided by a factor of 10. Decreasing the conductivity reduces the heat recirculation from the postflame to the preflame zone. This, in turn, reduces the effective flame speed. In Case A, the maximum velocity is decreased slightly and the minimum velocity is decreased more substantially. The reverse effect is observed when both conductivities are increased (Case B). Heat recirculation due to conduction is enhanced, and both the minimum and maximum velocities are increased.

Table 14.1 Effect of solid conductivity k (W/(m·K)) on the flame stability range.

Cases	Base	A	B	C	D
Upstream	$k_1 = 0.2$	$k_1/10$	$10k_1$	$10k_1$	k_1
Downstream	$k_2 = 0.1$	$k_2/10$	$10k_2$	k_2	$10k_2$
V_{max}, cm/s	74	72	100	74	92
V_{min}, cm/s	48	40	79	65	48
Stable range: $V_{max} - V_{min}$, cm/s	26	32	21	9	44

In Case C, the upstream conductivity is multiplied by a factor of 10, while the downstream conductivity is the same as in the reference case. Lower velocity flames stabilize close to the interface, and the minimum stable velocity is primarily determined by the heat-transfer characteristics of the upstream section. Increasing the upstream conductivity increases preheating in this section. As a result, the minimum velocity for which a flame can stabilize at the interface is higher for Case C than the base case. The maximum velocity is primarily determined by the conductivity in the downstream section because the higher velocity flames stabilize further away from the interface and preheating takes place entirely within the downstream section. Increasing the conductivity in this section while leaving the upstream conductivity unchanged (Case D) increases the maximum stable velocity. The minimum velocity is similar to that for the base case. Case D represents the optimum combination: a high downstream conductivity and low upstream conductivity. This has the highest maximum velocity and the widest stable range of all cases studied.

Effect of Extinction Coefficient

The radiation parameters included in the present model are the extinction coefficient κ (the inverse of the mean radiative path length) and the scattering albedo which is defined as σ_s/κ, where σ_s is the scattering coefficient. In general, a porous medium with a large pore diameter has a longer radiative path length and a smaller extinction coefficient. A small extinction coefficient allows more radiation from the flame zone to escape to the surroundings. For this study, the albedo is kept constant and the extinction coefficient is varied, as indicated in Table 14.2. If the extinction coefficient is large (small pore diameter), then radiative transfer takes place over a distance that is insufficient to preheat the incoming mixture. If the extinction coefficient is small (large pore diameter), then the radiation is spread over a large distance and the incoming mixture is

Table 14.2 Effect of extinction coefficient κ on flame stability range.

Cases	Base	E	F	G	H	I
Upstream	κ_1	$6\kappa_1$	$\kappa_1/6$	$\kappa_1/6$	κ_1	$6\kappa_1$
Downstream	κ_2	$6\kappa_2$	$\kappa_2/6$	κ_2	$\kappa_2/6$	κ_2
V_{max}, cm/s	74	55	58	75	57	72
V_{min}, cm/s	48	36	43	57	34	40
Stable range: $V_{max} - V_{min}$, cm/s	26	19	15	18	23	32

again not preheated effectively. The results for varying the extinction coefficient are shown in Table 14.2.

The correlation from Eq. (14.9) is used to determine the extinction coefficient, and although extinction coefficients are considered out of the range of validity of the correlation, it is still possible to interpret the results in terms of the pore diameter. For the base case, the upstream and downstream sections have pore diameters of 0.029 and 0.152 cm, respectively. The extinction coefficients for the upstream and downstream sections in the base case are 17.07 and 2.57 cm^{-1}, respectively.

For Case E, the extinction coefficient in both sections is increased by a factor of 6. According to Eq. (14.9), this is equivalent to decreasing the pore diameter by the same factor. Case E has a lower maximum velocity and a smaller stable range. Hsu *et al.* [9] modeled a burner similar to that in this study and consistent with these results. They showed that increasing the upstream extinction coefficient decreased the amount of preheating and reduced the peak temperature. For Case F, the extinction coefficient in both sections is decreased by a factor of 6, which is equivalent to increasing the pore diameter. Similar to Case E, the maximum stable velocity is lower and the stable range is narrower than for the base case. In Cases G and H, the extinction coefficient is decreased in either the upstream or the downstream section, while the other section remains at the reference value. Changing only the upstream value has little effect on the maximum velocity. Higher velocity flames stabilize further downstream of the interface and are less affected by the upstream properties. Decreasing the downstream extinction coefficient, however, significantly decreases the maximum and minimum stable velocities. An optimum case is Case I in which the upstream extinction coefficient is increased and the downstream coefficient remains at the base case value. Case I has a maximum velocity close to that of the base case, but a wider stable range. This extinction coefficient corresponds to a decrease in the upstream pore diameter. The upstream section in the base case has a pore diameter of 0.029 cm. Although smaller pore diameters are optimal according to the present study, reticulated ceramics with smaller pore diameters are not currently available.

The results obtained by varying the extinction coefficients by factors of 2 were also compared. These show all of the same trends as those reported for changing the diameters by factors of 6, but the changes were modest.

Effect of Volumetric Heat-Transfer Coefficient

Unlike conductivity that can be changed through the material properties of the solid, the heat-transfer coefficient is primarily a function of the pore diameter and geometry. Pore diameter affects the volumetric heat-transfer coefficient through the surface area-to-volume ratio. A smaller pore diameter means greater surface area per volume and higher heat transfer. The effect of the heat-transfer coefficient is shown in Table 14.3. The volumetric heat-transfer coefficient was calculated from Eq. (14.8) for different pore diameters.

Table 14.3 Effect of volumetric heat-transfer coefficient h_v ($W/(K \cdot m^3)$) on flame stability range.

Cases	Base	J	K	L
Upstream	$h_{v1}(d_1)$	$h_{v1}(2d_1)$	$h_{v1}(d_1/2)$	$h_{v2}(d_2)$
Downstream	$h_{v2}(d_2)$	$h_{v2}(2d_2)$	$h_{v2}(d_2/2)$	$h_{v2}(d_2/2)$
V_{max}, cm/s	74	64	82	82
V_{min}, cm/s	48	40	55	52
Stable range: $V_{max} - V_{min}$, cm/s	26	24	27	30

Case J represents a burner in which the pore diameter in both sections was doubled, thus decreasing the heat-transfer coefficient throughout the burner. Consistent with the results on conductivity, decreasing the heat-transfer coefficient in both sections decreases both the minimum and the maximum stable velocities. In Case K, the pore diameter was halved in both sections, thus increasing the heat-transfer coefficient. In this case, both the minimum and maximum stable velocities are increased. Similar to the results on conductivity, the optimum case is for a high heat-transfer coefficient in the downstream section and lower one in the upstream section (Case L).

Viskanta and Gore [2] studied a two-section burner which was substantially thinner than the current study and was constructed with different materials. In agreement with their results, the temperature profiles (not shown) for different heat-transfer coefficients indicated that for higher values of h_v the peak solid temperature increased due to more effective gas-to-solid heat transfer. This then promoted higher radiative flux from the high-temperature zone. The maximum gas temperature was not, however, significantly affected.

14.4 CONCLUDING REMARKS

In this study, computations of a flame stabilized at the interface of two sections of porous media are presented. The results indicate that the range of stable velocities is significantly affected by the conductivity, the extinction coefficient, and the volumetric heat-transfer coefficient. By proper selection of the porous media for the two sections, an optimum design may be achieved. For the broadest operating range and highest maximum velocity, the upstream section should have a small conductivity, while the downstream section should have a large conductivity. A large extinction coefficient is desired in the upstream section in order to minimize the radiative heat transfer. This corresponds to a material with a small pore diameter. The selection of the pore diameter, however, needs to be balanced with the effect on the heat-transfer coefficient.

ACKNOWLEDGMENTS

This work was performed under ONR contract N00014-01-1-0207 and was also supported by the Independent Research and Development Program at the Applied Research Laboratories, University of Texas at Austin. The authors acknowledge support through the National Science Foundation Fellowship Program.

REFERENCES

1. Howell, J. R., M. J. Hall, and J. L. Ellzey. 1996. Combustion of hydrocarbon fuels within porous inert media. *Progress Energy Combustion Science* 22:121.

2. Viskanta, R., and J. P. Gore. 2000. Overview of cellular ceramics based porous radiant burners for supporting combustion. *Environ. Combustion Technology* 1:167.

3. Weinberg, F. J. 1971. Combustion temperatures: The future. *Nature* 233:239.

4. Echigo, R., Y. Yoshizawa, K. Hanamura, and T. Tomimura. 1986. Analytical and experimental studies on radiative propagation in porous media with internal heat generation. *International Heat Transfer Conference* 8:827.

5. Takeno, T., and K. Sato. 1979. An excess enthalpy flame theory. *Combustion Science Technology* 20:73.

6. Babkin, V. S., A. A. Korzhavin, and V. A. Bunev. 1991. Propagation of premixed gaseous explosion flames in porous media. *Combustion Flame* 87:182.

7. Sathe, S. B., M. R. Kulkarni, R. E. Peck, and T. W. Tong. 1990. An experimental study of porous radiant burner performance. *23rd Symposium (International) on Combustion Proceedings*. Pittsburg, PA: The Combustion Institute. 1011.

8. Sathe, S. B., R. E. Peck, and T. W. Tong. 1990. Flame stabilization and multi-mode heat transfer in inert porous media. *Int. J. Heat Mass Transfer* 33:1331.

9. Hsu, P.-F., W. D. Evans, and J. R. Howell. 1993. Experimental and numerical study of premixed combustion within nonhomogeneous porous ceramics. *Combustion Science Technology* 90:149.

10. Khanna, V., R. Goel, and J. L. Ellzey. 1994. Measurements of emissions and radiation for methane combustion within a porous medium burner. *Combustion Science Technology* 99:133.

11. Rumminger, M. D., R. W. Dibble, N. H. Heberle, and D. R. Crosley. 1996. Gas temperature above a porous radiant burner: Comparison of measurements and model predictions. *26th Symposium (International) on Combustion Proceedings*. Pittsburg, PA: The Combustion Institute. 1755.

12. Mital, R., J. P. Gore, and R. Viskanta. 1997. A study of the structure of submerged reaction zone in porous ceramic radiant burners. *Combustion Flame* 111:175.

13. Henneke, M. R., and J. L. Ellzey. 1999. Modeling of filtration combustion in a packed bed. *Combustion Flame* 117:832.

14. Siegel, R., and J. R. Howell. 1992. In: *Thermal radiation heat transfer*. 3rd ed. Washington, D.C.: Hemisphere Publishing. 771.

15. Kee, R. J., F. M. Rupley, and J. A. Miller. 1989. CHEMKIN-II: A FORTRAN chemical kinetics package for the analysis of gas phase chemical kinetics. Livermore, CA: Sandia National Laboratory. SAND89-8009B.

16. Kee, R. J., G. Dixon-Lewis, J. Warnatz, M. E. Coltrin, and J. A. Miller. 1986. A FORTRAN computer package for the evaluation of gas-phase, multicomponent transport properties. Livermore, CA: Sandia National Laboratory. SAND86-8246.

17. Frenklach, M., H. Wang, C.-L. Yu, M. Goldenberg, C. T. Bowman, R. K. Hanson, D. F. Davidson Chang, E. J. Smith, G. P. Golden, W. C. Gardiner, and V. Lissianski. http://www.me.berkeley.edu/gri-mech/version12/test12.html.

18. Hsu, P.-F., and J. R. Howell. 1992. Measurements of thermal conductivity and optical properties of porous partially stabilized zirconia. *Experimental Heat Transfer* 5:219.

19. Younis, L. B., and R. Viskanta. 1993. Experimental determination of the volumetric heat transfer coefficient between stream of air and ceramic foam. *Int. J. Heat Mass Transfer* 36:1425.

20. Mital, R., J. P. Gore, and R. J. Viskanta. 1996. Measurements of radiative properties of cellular ceramics at high temperatures. *J. Thermophysics Heat Transfer* 10(1):33–38.

21. Ellzey, J. L., G. Diepvens, W. Mathis Jr., and P. Elverum. 2001. Porous burners for clean gas turbine engines. *14th ONR Contractors Meeting Proceedings*. Chicago, IL. 93–98.

Comments

Singh: The flame structure is quite different for different materials. It will be very difficult to stabilize the flame.

Ellzey: Our results show that flame stabilization is affected by material properties. Proper selection of the porous materials is an important consideration.

Edwards: What is the radiation out compared to the conduction heat transfer?

Ellzey: High-end of the radiation out is about 20%, and we don't have the conduction measurement yet. It is a radiation-dominated problem, but conduction affects the process.

Edwards: What is the pressure drop associated with a porous matrix inserted in the realistic combustor?

Ellzey: I don't know at present.

Singh: How do you avoid liquid fuel from carbonizing?

Ellzey: For liquid fuels, the thickness of the upstream section will be critical for prevaporization.

Smirnov: There may be unsteady propagation when the flame moves to a new location. Have you observed such effects?

Ellzey: We have not observed that in computation[*].

---

[*]Regarding this issue, there are several points to be mentioned. First, in real porous matrices, there is always a pressure gradient as mentioned by Edwards. Second, combustion in the matrix generates radiation heat fluxes and pressure waves that make the conditions ahead of the flame different from undisturbed conditions at infinity (usually assumed in calculations). Third, the matrix considered here is of finite length that is somewhat comparable in size with the flame zone. Fourth, if turbulence is taken into account, a whole spectrum of combustion regimes can be realized as reported by Korzhavin A. A. *et al.* (1999. In: *Gaseous and heterogeneous detonations: Science to applications*. Eds. G. D. Roy, S. M. Frolov, K. Kailasanath, and N. N. Smirnov. Moscow, Russia: ENAS Publ. 255–68). In view of these, the validity of calculations may become questionable, in particular, when flame stabilizes near the edge of the matrix (as in Cases 1 to 3 in Fig. 14.2).

Further elaborating the problem, it can happen that the range of anchored flames in porous matrices will be so different that nonstationary propagating flames characteristics may be predicted (question of Smirnov). (*Editor's remark.*)

O'Connell: With respect to liquid fuel, carbon is formed at a certain temperature, so it needs to be above 500 °C. Then, you cannot have the upstream section very thick because you need the heat radiated.

Ellzey: Yes, you need to select the proper upstream length for liquid fuel. It serves as a stabilizer and fuel vaporizer.

Chapter 15

CHARACTERISTICS AND CONTROL OF COMBUSTION INSTABILITIES IN A SWIRL-STABILIZED SPRAY COMBUSTOR

S. Acharya and J. H. Uhm

Active combustion control in a swirl-stabilized combustor is being investigated to reduce combustion instabilities and to control mixing in order to enhance certain key performance metrics (pattern factor, emissions, and volumetric release). For instability control, it has been demonstrated that experimental model-based controllers can provide substantially greater reductions in pressure oscillations relative to time-delay controllers. Phase-locked CH measurements are presented to provide improved understanding of the heat-release dynamics. It has also been shown that active modulation of dilution air jets can be utilized to control the heat-release and temperature distributions.

15.1 INTRODUCTION

The focus of the ongoing research is on active control of the combustion processes in a swirl-stabilized spray combustor [1–7]. Swirl can significantly alter the flow and combustion dynamics, and it is therefore important to develop control strategies that are optimal for swirling flows. Active control has the potential of providing high performance levels over a wide range of operating conditions and being adaptable to a wide variety of fuel-powered systems with minimal changes in design.

Primary emphasis of the present work is on developing more effective feedback control strategies (using model-based controllers [8]) for reducing pressure oscillations and on using active control through dilution-jet modulation to control temperature distributions or pattern factor. Efforts have been also directed

157

at improved understanding of the heat-release mechanisms using phase-locked CH imaging.

This chapter outlines key results and efforts. Section 15.2 deals with the work on instability control and heat-release mechanisms. Section 15.3 deals with dilution-jet modulation for pattern factor control.

15.2   EXPERIMENTAL SETUP

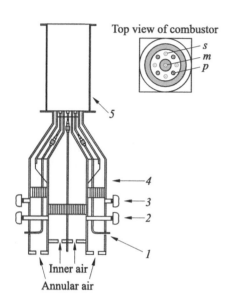

Figure 15.1 Coaxial jet spray combustor: *1* — primary and secondary fuel lines; *2* — primary speaker; *3* — secondary speaker; *4* — inlet section; *5* — combustion section; *p* — primary fuel nozzle; *s* — secondary fuel nozzle; and *m* — fuel-modulation nozzle.

The experiments were carried out in a multifuel-feed combustor operating at nearly 150-kilowatt heat release. The combustor configuration, shown in Fig. 15.1, consists of two concentric pipes that serve as settling chambers for the coaxial jets. Swirl vanes with a 45-degree angle are placed in the inner and outer air streams at the exit of the settling chambers that are accelerated through concentric large area-ratio nozzles. Each settling chamber can be acoustically forced by an array of eight 75-watt loudspeakers (Sanming 75-A) driven by four audio amplifiers (Radio Shack MPA-250) and mounted at equal polar angles about the circumference of the chambers. The air supply lines piped from a reservoir pressurized to 150 psi are used to deliver the primary and secondary air. The airflow rates are measured using Dwyer VFC series rotameters. Provision for the placement of eight liquid fuel atomizers (Arizona Mist with 0.012″ orifices) equally spaced between the coaxial jets is provided. Pressure-fed nozzles provide the atomization of the liquid spray. A fuel nozzle coupled to an automotive fuel injector is located at the geometric center of the inner air stream. Independent fuel supply lines are provided to the central injector and eight unmodulated circumferential fuel nozzles. Ethanol is used as the liquid fuel and is pressurized to the maximum of 250 psi in three fuel tanks by high-pressure nitrogen.

The combustor is equipped with a square shell with $7^5/_8$-inch sides and windows that are either stainless-steel or quartz for optical access. A high-sensitivity, water-cooled pressure transducer (Kistler 6061B) is mounted at a normalized axial distance of 6 cm from the expansion plane to measure pressure oscillation and to provide the feedback signal to the control actuators. Light emission recorded at a CH radical wavelength using a photodiode (Melles Griot) with a 430-nanometer bandpass optical filter was taken as a measure of the heat-release fluctuations from the flame. These signals were then processed in real time using a digital signal processor (DS1103, DSPACE, 333 MHz Motorola Power PC) to be used in active control. The spatial and temporal variations in the heat release were visualized using a Princeton Instruments PI-MAX 512×512 Intensified Charge Coupled Device camera with a UV lens (Electrophysics) and a bandpass filter (DIOP) at 430 nm for CH-chemiluminescence imaging. The images could be triggered with respect to the pressure oscillations in the combustor to observe the fluctuations in heat release at different instances of the oscillation cycle.

15.3   RESULTS AND DISCUSSIONS

Characteristics of Combustion Oscillations

The trace of the pressure signal at two locations (upstream of the dump plane and 6 cm downstream of the dump plane) is shown in Fig. 15.2 for flow conditions corresponding to 60 scfm of air and 4 gph of fuel. At both locations, a dominant frequency of 213 Hz (see pressure spectra in Fig. 15.3) modulated by a low frequency of 12 Hz is observed. The low-frequency modulation is attributed to the acoustics of the settling chamber and, as will be discussed later, plays a role in the effectiveness of the control strategy.

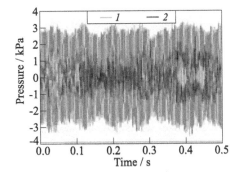

Figure 15.2 Time trace of the pressure signals (*1* — combustor pressure and *2* — inlet pressure). Inner and annular airflow rates: 16 and 44 cfm, respectively. Primary and secondary fuel flow rates: 2.0 gph each. (Refer color plate, p. XV.)

Figure 15.3 shows the pressure signal for one cycle of the oscillations and its spectrum and the CH images phase-locked with respect to the pressure signal. The flow conditions represented by Fig. 15.3 correspond to fuel injection from only the four fuel injectors (secondary fuel injectors in Fig. 15.1) that are facing the flat sides of the square

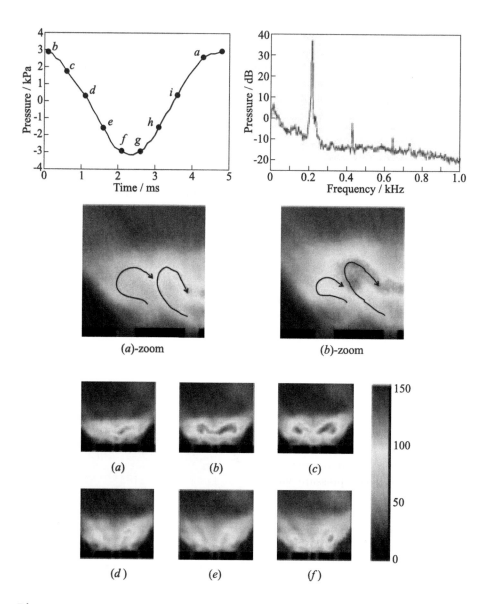

Figure 15.3 Pressure signal and spectrum (top row), and heat-release distribution of CH radical intensity synchronized with the pressure signal of the instability cycle (bottom row *a* to *f* corresponding to the points at the top left plot) (213 Hz). Secondary fuel supply of 2.5 gph. Primary and secondary air: 11 and 55 cfm. (Refer color plate, p. XVI.)

combustion chamber. The CH images are shown at several instances during the pressure cycle (denoted by $a-i$). The heat release may be represented by the measured gray level of CH-chemiluminescence intensity. The region of maximum heat release is seen to be located downstream of the expansion plane and in the annular mixing-layer region between the primary and secondary air streams. The dynamics of this high heat-release region appears to be in phase with the pressure oscillations, with high CH levels and a compact lifted flame at peak pressures (Fig. 15.3a) and low CH levels with a broader attached flame at minimum pressures (Fig. 15.3e). Thus, with the changes in the pressure levels, the flame attaches or lifts-off at the dump plane, and this flame movement introduces dynamics (heat-release variation) close to the dump plane. However, these dynamics associated with the flame movement are out of phase with the pressure (see region near the dump plane, where CH levels are low at high pressures and high at low pressures) and therefore do not contribute to the instability. Thus, the primary heat-release dynamics of interest which contribute to the instability are associated with the high heat-release regions.

In order to explain the heat-release dynamics shown in Figs. 15.3a–15.3f, an enlarged view of Figs. 15.3a and 15.3b is also shown, with postulated flow structures indicated. It is postulated that with the high degree of swirl present, both the primary and secondary air streams lead to a vortex-core structure with a strong central recirculation and a weaker outer recirculation. These vortical flow structures have inherent dynamics or transverse movements that lead to the heat-release variations. These transverse movements can be seen by comparing the locations of high CH values in Figs. 15.3a–15.3f. In Figs. 15.3a–15.3c (high pressures), the high intensity regions are well defined and appear to correlate with vortical structures. In Figs. 15.3e–15.3f, the pressure and CH levels are low, and the flow structures are not as evident. Flow measurements are being planned to better understand these dynamics.

Figure 15.4 shows the plot of the Rayleigh index (integral of the product of the pressure and heat-release fluctuations). These plots indicate that the pressure and heat release are out of phase with the pressure close to the dump plane (at location marked "b"), where a negative Rayleigh index is obtained, and in phase at location marked "a," where a positive Rayleigh index is obtained. Instability arises from the regions of positive Rayleigh index, and for control, heat-release variations in these regions must be altered.

The existence of low-frequency combustion oscillations superimposed on the primary instability has been reported by De Zilwa et al. [9] for premixed gaseous systems. As noted earlier, the present pressure and CH-photo-diode measurements also reveal low-frequency oscillations at around 12 Hz. This low-frequency mode is due to the cavity between the acoustically-closed upstream inlet end and the constricted exit nozzle of the air-delivery inlet that leads to a bulk mode oscillation in the flame. The existence of these oscillations can be clearly seen in Figs. 15.2, 15.5, and 15.6, where the time-variations in the peak pressure and

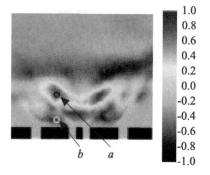

Figure 15.4 Rayleigh index at the same flow condition as that in Fig. 15.3. (Refer color plate, p. XV.)

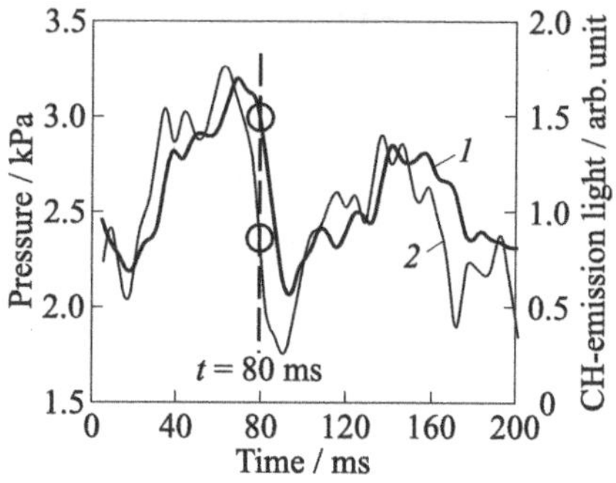

Figure 15.5 Variation of the peak in the pressure (*1*) and CH-emission (*2*).

CH levels are plotted. The low-frequency (12 Hz) flame movement makes the control of the combustion oscillations difficult due to changes in the heat-release distribution from one cycle of thermoacoustic instability (at 213 Hz) to another. It is evident in Fig. 15.5 that there is a time lag (of the order of 2–5 ms) between the peak in the pressure oscillation and the peak in the CH level. The variability in the time lag is presumably associated with the cycle-to-cycle variations in the flame movement.

Figure 15.6 shows the normalized peak pressures, CH-chemiluminescence, and pixel-averaged values from the CH images (top figure). Note that the peak pressures are modulated by the low-frequency 12-hertz mode, and therefore, the time traces show variations in the peak values of both pressure and CH. As in Fig. 15.5, it is of specific interest to correlate CH values with the locations of maximum pressure. In view of this, the CH imaging was triggered at locations indicated by (*a*) through (*h*) in Fig. 15.6. These selected locations represent local maximum and minima in the pressure peak. The CH images at two of the triggered points (*c* and *h*) are shown in Fig. 15.6 (bottom), while Fig. 15.6 (middle) shows the time traces around the trigger points (indicated by symbols in the time traces). It is evident from Fig. 15.6 that the maximum in the peak pressure does not always correspond to the maximum in the CH value. At location (*c*) there is a close correspondence between the maximum pressure and the maximum heat release (CH image shows large pixel intensity values), while at location (*h*) where the pressure peak has a local maximum, CH intensity values are relatively lower. Thus, for control purposes, in order to introduce secondary fuel that will produce heat-release oscillations out of phase with the baseline or driving heat-release dynamics, the pressure signal should not be used as the control input. A better measure of the control input is the heat-release signal itself.

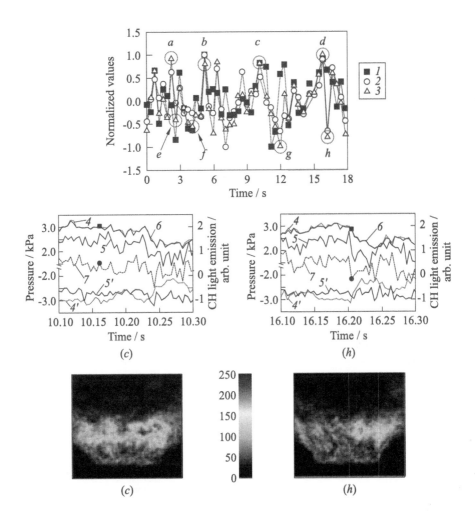

Figure 15.6 Top: normalized values of the peak pressures and corresponding CH signals and pixel-averaged CH image. Typical trigger location for CH imaging is denoted by (a) through (h). Inner and annular airflow rates: 16 and 44 cfm. Primary and secondary fuel flow rate: 2.0 gph. Normalized values by maximum and minimum peaks of 1 — pressure (min: 1.8, max: 3.1 kPa); 2 — CH light emission (arb. unit); and 3 — pixel-averaged CH image (arb. unit). Middle and bottom: expanded time trace and CH images for (c) and (h) instances: 4 and 4′ — maximum and minimum peaks of pressure; 5 and 5′ — maximum and minimum peaks of CH signals; 6 and 7 — pressure and CH values around trigger point. Square and cycle solid symbols correspond to trigger points. (Refer color plate, p. XVII.)

Acoustic- and Fuel-Modulation Control

In this section, the results of active control using both the acoustic-modulation and fuel-modulation strategies are reported. Acoustic modulation alters the flow dynamics and mixing, while fuel modulation directly controls the timing of the secondary fuel injection. With both strategies, the heat-release dynamics is affected and therefore has the potential of controlling instabilities. The unique aspect of the work is in the implementation of model-based control strategies and, in particular, Linear Quadratic Gaussian Control with Loop Transfer Recovery (LQG-LTR-based control). Since the model-based control strategies have been described in earlier publications, additional discussion on the controller methodology is not presented here, and emphasis is placed primarily on the results.

Figures 15.7a and 15.7b show the results of acoustic-modulation and fuel-modulation control, respectively. Controllers used include phase-delay, LQG-LTR, and adaptive posi-cast controllers. With acoustic modulation, phase-delay and LQG-LTR control led to a reduction of about 4 and 26 dB in the amplitude of rms pressure fluctuations. Thus, a significant improvement (22 dB) was achievable with LQG-LTR control. Note that the advantage of acoustic modulation is that a variable gain/amplitude can be dialed into the control signal, and this capability is exploited by the model-based control. The input energy to the loudspeakers with an LTR controller is varied in proportion to the feed-

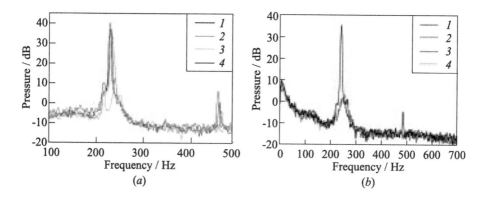

Figure 15.7 Pressure spectra with and without control: (a) acoustic modulation (1 — without control; 2 — with LQG-LTR control; 3 — with phase-delay control at 0 ms; and 4 — with phase-delay control at 3 ms) and (b) fuel modulation (1 — base line; 2 — phase delay; 3 — LQG-LTR; and 4 — adaptive posi-cast). Flow conditions in (a) and (b) correspond to inner and annular airflow rates: 16 and 44 cfm. Primary and secondary fuel flow rates: 2.0 gph.

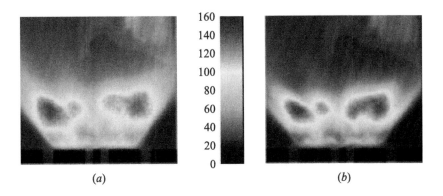

Figure 15.8 Time-averaged heat-release distribution of CH: (*a*) without control and (*b*) with LQG-LTR control. (Refer color plate, p. XV.)

back pressure signal, in contrast to that of the phase-delay controller where the input is a constant. With fuel-modulation control, all controllers appear to be quite effective in reducing the instability level and appear to suppress the pressure oscillations by nearly 30 dB. No significant improvement over the phase-delay control was achieved with either the LQG-LTR control or the adaptive posi-cast control. One key limitation in the current fuel-modulation strategy is the use of an on–off solenoid valve, where the amplitude/gain information from the controller cannot be utilized. Thus, this advantage of a model-based controller over phase-delay controllers cannot be effectively utilized and explains, in part, the lack of improvement with model-based controllers when fuel modulation is utilized. To remedy this drawback, proportional drive valves are needed, and the authors retrofitted their combustor with a proportional drive Moog Valve.

Figure 15.8 shows the time-averaged distributions of heat release with and without control. With strong oscillations and no control, the flame is anchored to the flame holder. Thus, the pressure anti-node is associated with high heat release, a condition that will promote thermoacoustic coupling. With control, the flame is slightly lifted, and heat release at the flame holder or pressure anti-node is considerably lower.

Control of Mixing, Pattern Factor, and Emissions

Active control of mixing and pattern factor is being explored in the present study through modulation of dilution air jets introduced through the combustor shell. The combustor has been retrofitted with four dilution air jets located at four

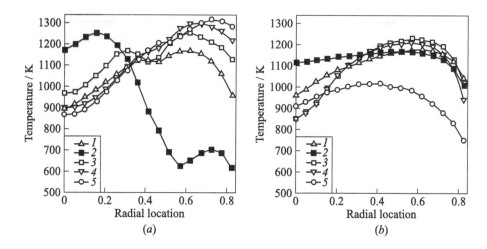

Figure 15.9 Temperature distributions at two axial locations, $z/H = 0.05$ (a) and $z/H = 0.22$ (b). Baseline *1* is with dilution air jet (no modulation), while baseline *2* represents no dilution air jet. The equivalence ratio is the same for all cases. *3* — 185 Hz; *4* — 300 Hz; and *5* — 700 Hz.

circumferential locations. The dilution air jets are modulated by a spinning valve capable of operation up to 1.5 kHz. Studies completed to date have included open-loop forcing and detailed temperature maps at various forcing frequencies. In addition, limited unsteady numerical simulations have been performed in order to get a better understanding of the physical effects of forcing the dilution air jets. Typical jet blowing ratios (jet velocity to crossflow velocity) of 6 have been used in both the experiments and the simulations.

As a first step, the characteristics of the spinning valve have been investigated. The output of a pressure transducer and hot wire located downstream of the modulated jet was recorded for different forcing frequencies, and frequencies were identified where the amplitude of forcing would be nearly the same. Detailed temperature maps were then recorded at these selected frequencies.

Figure 15.9 shows the radial temperature distributions plotted at two axial locations for forcing frequencies of 185, 300, and 700 Hz. These results are compared with two baseline cases where no modulation is applied, but the equivalence ratio is maintained the same. Forcing is clearly seen to have an influence on the temperature distribution. In the near field ($z/H = 0.05$), temperature levels are enhanced with forcing, indicating improved mixing and higher heat-release levels. Further downstream, at 700 Hz forcing, temperature levels are lower, implying a greater degree of completion of the combustion process and indicating the potential for more compact combustors and lower emissions. The higher forcing frequency studied (700 Hz) appears to result in the best performance. Current research explores forcing effects in greater detail and over a broader parametric range.

15.4 CONCLUDING REMARKS

Active control studies on a swirl-stabilized spray combustor are presented. Significant improvements with model-based control over traditional time-delay control is demonstrated in the present work. These improvements are particularly noted with acoustic modulation. Future work in this area is directed toward using a proportional drive spray injector where the full amplitude/phase information from the model-based controller can be exploited.

Modulation of the dilution air jet is shown to control mixing and temperature distributions. Ongoing studies are directed towards more detailed measurements of temperature and emissions over a broader range of forcing frequencies to fully exploit the potential of dilution air-jet modulation.

ACKNOWLEDGMENTS

The work was performed under ONR contract N00014-97-1-0957. The work on model-based control was done in collaboration with Profs. Annaswamy and Ghoneim of MIT. Their help is sincerely appreciated.

REFERENCES

1. Acharya, S., S. Murugappan, E. J. Gutmark, and T. Messina. 1999. Characteristics and control of combustion instabilities in a swirl-stabilized spray combustor. AIAA Paper No. 99-2487.

2. Allgood, D., S. Murugappan, S. Acharya, and E. Gutmark. 2000. Infra-red temperature measurements in an actively controlled spray combustor. *ASME Turbo Expo Proceedings*. Munich, Germany.

3. Murugappan, S., C. Park, S. Acharya, A. Annaswamy, and A. Ghoneim. 2000. Optimal control of combustion instability using system identification approach. *Turbine-2000. Symposium (International) on Gas Turbine Heat Transfer Proceedings*. Cesme, Turkey.

4. Murugappan, S., S. Park, S. Acharya, A. Annaswamy, and A. Ghoneim. 2001. Optimal control of swirl-stabilized spray combustion using system-identification approach. AIAA Paper No. 2001-0779.

5. Campos-Delgado, D., D. Allgood, K. Zhou, and S. Acharya. 2001. Identification and active control of thermoacoustic instabilities. *IEEE Conference on Control Applications Proceedings*. Mexico City, Mexico.

6. Allgood, D., D. Campos-Delgado, S. Acharya, and K. Zhou. 2001. Active control of combustion instabilities using experimental model-based controllers. *26th Annual Dayton-Cincinnati Aerospace Sciences Symposium Proceedings.* Dayton, OH.

7. Campos-Delgado, D. U., K. Zhou, D. Allgood, and S. Acharya. 2003. Active control of a swirl-stabilized spray combustor using model-based controllers. *Combustion Science Technology* 175(1):27–53.

8. Ljung, L. 1999. *System identification: Theory for the user.* 2nd ed. Upper Saddle River, NJ: Prentice-Hall Inc.

9. De Zilwa, S. R. N., I. Emiris, J. H. Uhm, and J. Whitelaw. 2000. Oscillations, extinction and combustor performance. *13th ONR Propulsion Meeting Proceedings.* Minneapolis, MN.

Comments

Smirnov: What is the physical mechanism responsible for control?

Acharya: There are two basic mechanisms at work: the first involves the introduction of heat release out of phase with respect to pressure; the second is the use of high momentum jets to disrupt the vortex dynamics.

Seiner: Did you characterize the acoustics to know the source of instability?

Acharya: Yes, we see a quarter-wavelength mode and the Helmholtz mode. Of course, the dynamics will depend on the facility resonance.

Whitelaw: All the cases shown in your presentation have the primary jets moving downstream. I am not aware of any combustor that works this way. Your configuration does not have enough space for putting in dilution jets and none go into the primary zone. Have you considered the compactness implications of this strategy?

Acharya: We introduce the jets for control purposes and investigate the ways of injecting secondary jets.

Carlin: How relevant is the use of ethanol as a fuel?

Acharya: Just simulate a liquid fuel by using ethanol.

Strykowski: How well do other fuels respond to the control?

Acharya: Jet-A works similar to ethanol. We use ethanol because it is well-characterized and clean.

Frolov: What was the reason to choose ethanol rather than other primary hydrocarbon?

Acharya: With ethanol, optical measurements are easier.

Chapter 16

COMBUSTION AND MIXING CONTROL STUDIES FOR ADVANCED PROPULSION

B. Pang, S. Cipolla, O. Hsu, V. Nenmeni, and K. Yu

Four sets of experimental and analytical studies have been conducted to identify key physical mechanisms for controlling mixing and combustion processes in ramjet and scramjet combustors. The first set of experiments was used to study the interaction between periodic vortex structures and unsteady heat release processes, which may occur under certain operating conditions. The results from these controlled experiments were used to identify the spatial distribution of controller fuel in vortex, necessary to obtain active combustion control. The results are also being tested in the second set of experiments, involving liquid-fueled active combustion control in a larger-scale dump combustor. The third study utilized the existing data on heated inlet experiments to model the liquid-fueled active combustion control process and to identify the amount of controller fuel flux required for control. Lastly, a supersonic mixing-enhancement experiment was conducted to assess the practicality of utilizing passive acoustic excitation as a means to enhance mixing in a scramjet.

16.1 INTRODUCTION

Under the U.S. Office of Naval Research (ONR) combustion control program, several experimental rigs have been constructed and a number of experiments have been conducted to demonstrate the potential of applying active combustion control to advanced propulsion systems. Figure 16.1 shows various rigs that have been used to conduct the ONR studies. The aim was to obtain data under more relevant operating conditions while targeting the experimental effort to a specific goal of providing a detailed database for model validation. Thus, the emphasis was placed on liquid-fuel issues and high-speed combustors.

Figure 16.1 Existing experimental facilities: (*a*) University of Maryland (UMD) liquid-fueled dump combustor; (*b*) UMD supersonic mixing rig; (*c*) Naval Air Warfare Center (NAWC) heated inlet rig; and (*d*) NAWC dual-mode scramjet rig.

Finally, four sets of related studies were conducted to analyze various physical mechanisms concerning the problems in high-speed, air-breathing propulsion systems. The details of the study, as well as the experimental and analytical results, are described in the following sections.

16.2   VORTEX-STABILIZED FLAMES AND HEAT RELEASE

Vortex shedding plays a significant role in maintaining self-sustained combustion instabilities in dump combustors. Vortices may entrain a large amount of hot recirculating products into fresh reactants, causing a substantial modification in

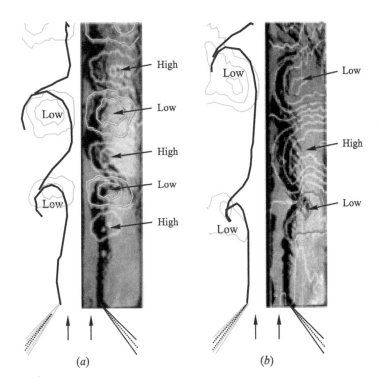

Figure 16.2 Heat release fluctuation with respect to acoustically-stabilized vortex propagation in premixed flames: Re = 60,000; $u'/U \sim 0.1$; $\phi = 0.75$; St = 0.8 (a); and St = 0.4 (b).

the temporally-evolving heat release pattern. Depending on the location and timing of the heat release, it may either amplify or suppress the instabilities by coupling with the acoustic field and becoming a source or sink of acoustic energy. In order to actively control the instabilities, it is desirable to understand the precise timing and location of the heat release cycle associated with the vortex shedding cycle.

Previous studies have shed some light on this issue. Figure 16.2 shows a comparison of local heat release fluctuation with respect to vortex flame structure [1], obtained in the wake of a V-gutter flameholder. Each image was obtained by combining spark shadowgraph with local C_2-chemiluminescence corresponding to the same phase. The illustration on the left half of each image shows where the heat release oscillation goes through a low cycle with respect to vortex. The inlet velocity in this case was 40 m/s and the equivalence ratio was 0.75. The local heat release oscillation in a premixed combustor goes through the low cycle in

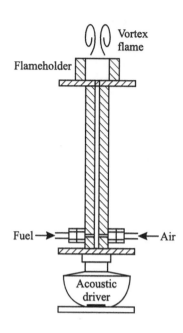

the leading part of the vortex and the high cycle in the trailing part of the vortex. This is caused by the entrainment of the fresh reactants by vortex action, which initially lowers the temperature in the leading part of the vortex. As the entrained reactants mix with the high-temperature products in the trailing part of the vortex, the local heat release increases, reaching a high cycle. While such diagnostics captured the dynamics of oscillatory heat release in burning vortices during a particular instability cycle, these results may not be general for other vortex flames.

In this study, the relationship between oscillatory heat release and large vortex structure was systematically examined as a function of flow and chemistry scaling. Figure 16.3 shows the experimental setup used for producing vortex flames. A premixed propane–air jet was ignited, and the flame was stabilized at the exit downstream of a sudden-expansion flameholder. A 75-watt compression driver was used to apply controlled disturbance and to produce periodic vortices into a jet flame. The frequency response of actuation was evaluated separately

Figure 16.3 Experimental setup for characterizing proper fuel injection location into vortices: $D_{exit} = 1.375$ in. and $L_{inlet} = 10$ in.

to maintain a similar amplitude of acoustic disturbance at the exit plane. The objectives were to identify the location for active fuel injection in the general case and to establish a scaling criterion.

A vortex-stabilized jet flame has been established at several different scales, and temporally-evolving heat release associated with periodic vortices was characterized using a combination of phase-resolved Schlieren technique and CH-chemiluminescence. Figure 16.4 shows a set of Schlieren images showing propagation of periodic vortices in the flame for a typical case. Figure 16.5 shows the fluctuating component of CH-radical chemiluminescence at 431 ± 2 nm corresponding to the respective phases in Fig. 16.4. Since the CH-chemiluminescence is related to local heat release, the relative maps of heat release surplus and deficit at each phase in the cycle can be obtained with respect to the vortex structure. Figure 16.6 shows a composite picture of heat release fluctuation with the corresponding Schlieren image. In the gray region, there is a deficit in local heat release, while a surplus in heat release is found in the white region. The dark-color structure denotes the Schlieren images showing the density gradient.

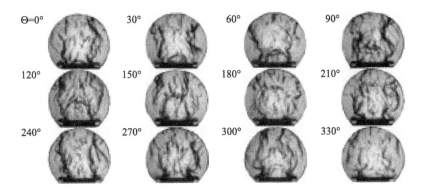

Figure 16.4 Phase-lock Schlieren images of periodic vortices convecting inside a turbulent jet-flame structure during a cycle. The spacing between images is 30°: Re = 7200 and $\phi = 1$.

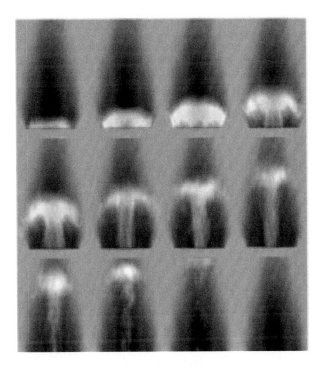

Figure 16.5 CH-radical chemiluminescence measurements corresponding to various phases shown in Fig. 16.4.

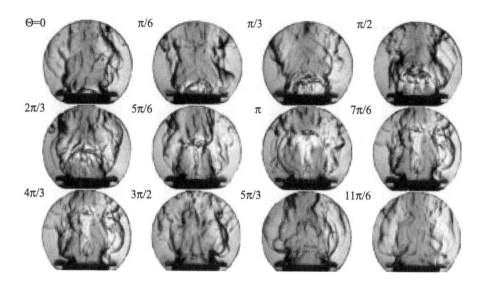

Figure 16.6 Phase-resolved visualization of vortex-driven flame structure and the corresponding heat release fluctuation displayed together. Phase-lock averaged C_2-chemiluminescence fluctuations were overlaid on top of the Schlieren images, with the darker color denoting the region of heat release deficit and the lighter color denoting the surplus region.

Interestingly, the high cycle of heat release or simply "the hot spot" in this case closely follows the vortex development. The main difference from the previous case was that the inlet velocity was much lower, while the equivalence ratio was higher. In other words, the convective time scale was less and the chemical time scale was higher. This suggests that one of the important parameters to consider in identifying the proper location of controlled fuel injection is the Damköhler number. Near the flame blow-off limit, however, one would expect the hot spot to lag the vortices slightly.

16.3   DUMP COMBUSTOR CHARACTERIZATION AND LIQUID-FUELED ACTIVE CONTROL

The experiments were performed in a 102-millimeter diameter axisymmetric dump combustor shown in Fig. 16.1a. The inlet diameter is 41 mm. Ethylene was injected at 24.5 inlet diameters upstream of the dump plane to simulate the prevaporized, premixed reactants entering a ramjet combustor. The cylindrical combustor is 610 mm long and is capped by a 38-millimeter diameter exhaust

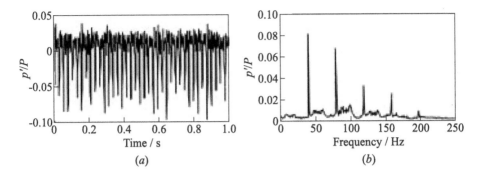

Figure 16.7 Unstable operation at $U = 15$ m/s and $\phi = 0.58$: (a) pressure–time trace, and (b) pressure spectrum.

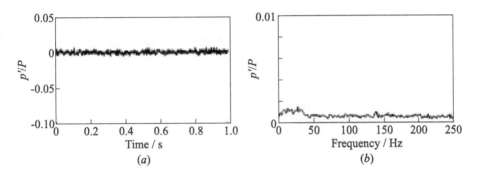

Figure 16.8 Stable operation at $U = 10$ m/s and $\phi = 0.53$: (a) pressure–time trace, and (b) pressure spectrum.

nozzle. Active control is attained by injecting a small amount of ethanol directly into the combustor at controlled frequencies and duration. Previously, the actuator system was characterized as providing pulsed sprays with 40-micrometer Sauter-mean-diameter droplets at these conditions.

The combustor is naturally unstable under certain operating conditions. Figure 16.7 shows combustor pressure oscillations and the Fast Fourier Transform (FFT) spectrum under a typical, unstable operating condition. The fundamental mode at 39 Hz and its higher harmonics were observed. The fundamental-mode frequency corresponds to the inlet quarter-wave mode of acoustic oscillations. During stable operation as shown in Fig. 16.8, the amplitude of pressure oscillations is much less. Also, no significant peak was observed in the pressure spectrum.

Natural operation characteristics of the uncontrolled dump combustor are shown in Fig. 16.9.

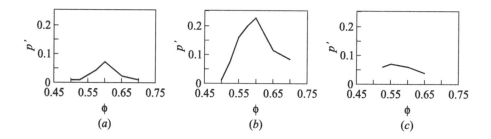

Figure 16.9 Characterization of combustor pressure oscillation near the lean blow-off limit: (a) $U = 10$ m/s; (b) $U = 15$ m/s; and (c) $U = 20$ m/s.

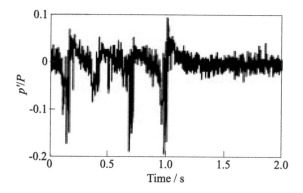

Figure 16.10 Large-amplitude pressure fluctuation leading to flame extinction at $t \approx 1.1$ s, $U = 20$ m/s, and $\phi = 0.53$.

Particularly intense pressure fluctuation can also lead to premature flame extinction as shown in Fig. 16.10. Previously, active combustion control was demonstrated with a phase-lock and delay circuit using a Wavetek Phase Synthesizer, which controlled the timing of fuel injection into the vortices. The new technique, presently being developed, will use PC-104 technology to make the controller more practical while achieving the same physical function.

16.4 HIGH-ENTHALPY INLET EXPERIMENT AND CRITICAL FUEL-FLUX MODEL

The high-enthalpy inlet experiments were conducted using the NAWC combustor shown in Fig. 16.1c. The characteristic features are similar to the UMD combustor except for a vitiated air heater, which allows heating of incoming air upstream. Details of the vitiated heater and its stability map were shown in previous reports [2]. The experiments were performed at three different elevated inlet temperatures, and the results from these experiments were used to calibrate the critical fuel-flux model.

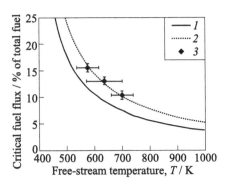

Figure 16.11 Comparison of predicted ($1 - D_{32} = 40$ μm, and $2 - 48$ μm) and measured (3) critical fuel flux.

The essence of the critical fuel-flux model is based on the balance between acoustic energy contained in the oscillation cycle and the attenuation of acoustic energy using properly timed heat release from active control. Using this model, the change in critical fuel flux for sustaining control authority was derived as

$$\frac{\Delta \dot{m}_{f,C}}{\dot{m}_{f,C}} = \frac{\Delta q'}{q'} = 1 - \frac{q'|_{T+\Delta T}}{q'|_{T}}$$

The results from heated inlet experiments were in good agreement with the model prediction (Fig. 16.11).

16.5   PASSIVE CONTROL OF SUPERSONIC MIXING

It has been shown that a wall cavity, strategically placed inside a scramjet combustor, could provide many potential benefits by significantly improving supersonic mixing (Fig. 16.12) or enhancing flameholding characteristics [3, 4]. Measurements in a direct-connect experiment have suggested that a higher volumetric heat release could be obtained when a cavity shape is optimized. Because such a configuration is expected to operate over a wide range of flight speed, it is important to establish the robustness of such a mechanism. Thus, a parametric investigation was performed in which the stagnation pressure was varied from 35 to 120 psi and the acoustic characteristics of various cavities were examined. Spark Schlieren images (Fig. 16.13) of the flow field over cavities revealed a strong interaction between the leading edge expansion wave and boundary layers. The wave–boundary layer interaction was also affected by downstream conditions which changed the Mach number of the flow entering the leading edge of the cavity.

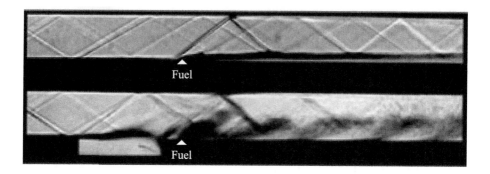

Figure 16.12 Schlieren images of mixing between Mach 2 air stream and transversely injected fuel with and without mixing enhancement cavity.

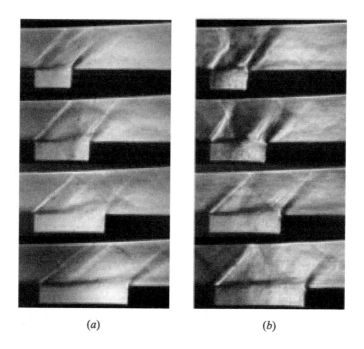

Figure 16.13 Schlieren images of supersonic flow over various aspect-ratio cavities under off-design inlet conditions. Inlet flow stagnation pressure is 35 psi (a) and 120 psi (b).

Table 16.1 Measured oscillation frequencies in kHz.

P_0, psi	L_2, D_1	L_3, D_1	L_4, D_1	L_5, D_1	L_4, D_2	L_6, D_2	L_8, D_2	L_{10}, D_2
30	38.4	15.2	15.2	15.2	21.0	23.6	17.6	8.6
35	37.6	21.4	15.2	15.2	20.4	22.7	17.5	9.1
55	37.1	25.9	15.2	15.2	19.8	21.5	10.1	12.0
75	37.5	25.3	15.2	28.0	19.5	21.3	10.1	12.0
95	36.8	25.5	20.4	27.6	30.4	21.2	9.9	11.8
120	36.8	25.0	20.1	26.7	30.3	21.2	9.9	12.0

Despite such complexity in the upstream flow field, the flow-induced cavity resonance mechanism was fairly robust, producing large-amplitude coherent fluctuations in all cases of selected cavity configurations. For certain cases, however, the dominant frequency of oscillations was shown to change as a function of stagnation pressure (Table 16.1). The effect of such mode hopping behavior on quantitative results of mixing enhancement warrants further investigation.

ACKNOWLEDGMENTS

This work was performed under ONR contract N00014-00-1-0704.

REFERENCES

1. Yu, K., A. Trouve, and S. Candel. 1991. Combustion enhancement of a premixed flame by acoustic forcing with emphasis on role of large-scale structures. AIAA Paper No. 91-0367.

2. Yu, K. H., and K. J. Wilson. 2002. Scale-up experiments on liquid-fueled active combustion control. *J. Propulsion Power* 18(1):53–60.

3. Burnes, R., K. J. Wilson, T. P. Parr, and K. Yu. 2000. Investigation of supersonic mixing control using cavities: Effect of fuel injection location. AIAA Paper No. 2000-3618.

4. Yu, K. H., K. J. Wilson, and K. C. Schadow. 2001. Effect of flame-holding cavities on supersonic combustion performance. *J. Propulsion Power* 17(6):1287–95.

Comments

Roy: I really commend you on following the decisions taken at the mid-term meeting and keeping up the pace. For high-speed mixing enhancement part using flow-induced cavity resonance, it would be interesting to try trapezoidal-shaped cavities.

Yu: We have not done this, but we can try.

Annaswamy: You said the droplet size affects the control. How does it do that?

Yu: The droplet size affects the combustion response as well as the spatial dispersion. The temporal response is affected through the rate of vaporization and the spatial response by the changes in dispersion characteristics.

Mashayek: Have you characterized the Damköhler number and the related time-scales of the problem?

Yu: The Damköhler number ranged between 10 and 20 for the cases presented here. We are also working on ways to conduct experiments at lower Damköhler numbers, and the results will be presented at the upcoming Aerospace Sciences Meeting.

Mashayek: Can you comment on the physical mechanism controlling the droplet dispersion?

Yu: Vortex–droplet interaction is affected not only by the droplet size, but also by the magnitude of slip velocity, which depends on the injection timing with respect to the vortex development phase.

Frolov: Are your results sensitive to the type of fuel injector used? I mean not only the drop size distribution, but the effect of spray cone angle, tip penetration length, and other parameters.

Yu: They will definitely depend on the injector and interactions with flow structures.

Chapter 17

ACTIVE PATTERN FACTOR CONTROL ON AN ADVANCED COMBUSTOR

S. C. Creighton and J. A. Lovett

A fuel control system is being assembled to demonstrate active spatial combustor exit temperature control on an advanced aircraft engine combustor. Activity was primarily focused on preparations for upcoming open-loop combustion tests, but tasks were also initiated for subsequent closed-loop experiments. Fuel-injector modifications were made to allow for spatial control of individual fuel-injection sites. Small, lightweight fuel delivery valves compatible with engine requirements were developed, and a flow bench was tested. Optical temperature sensor probes and a traversing gas sampling rake are being integrated into a test rig to quantify the spatial exit temperature distribution for the combustor. Additionally, computational fluid dynamics (CFD) simulations have been created to provide guidance for the open-loop active control experiments. The CFD solutions will also serve as input for model-based control algorithms that are being developed for the closed-loop active pattern factor control tests.

17.1 INTRODUCTION

Active Pattern Factor Control (APFC) is an enabling technology to achieve optimal gas-turbine engine performance and life. The exit temperature distribution of a combustor (pattern factor) has a direct implication to turbine durability and overall engine cycle efficiency. Local high-temperature regions in the combustor may produce undesirable hot spots that significantly reduce engine component life and determine the design margin that must be accommodated in the aerothermal design. Variations in engine operating conditions, transient events, and manufacturing tolerances for engine components can also contribute to adverse temperature variations that again result in increases in design margin. Furthermore, degradation of the engine over its life cycle adds further to the required design margin. An active control system which allows the combustor exit temperature to be precisely controlled can substantially reduce the design margin requirement and enable higher performance levels.

Pattern factor control using fuel control requires a combustion system that incorporates individual fuel-injection site control to allow for fuel modulation to produce the desired temperature pattern. Thus, an advanced combustor with zonal fuel control capability has been selected for demonstrating APFC. Activities associated with the preparation for APFC open-loop experiments with this combustor are described along with results from supporting numerical simulations.

17.2   FUEL DELIVERY SYSTEM

A four-nozzle planar sector rig used to demonstrate an advanced combustor has been selected for the test vehicle for the demonstration of APFC technology. The aerodynamics of this advanced combustion system is well suited to orthogonal spatial control of exit temperature pattern. The combustor fuel injectors were configured to provide individual zonal fuel distribution control capability. A layout of the fuel delivery system is illustrated in Fig. 17.1. Each fuel injector allows for radial fuel flow control to three distinct regions, making it possible for tailoring of the combustor exit temperature profile in the radial dimension. Injection sites at the top of the injector assembly effectively control the temperature at the outside diameter (OD) of the annular combustor, while injection sites at the center and bottom of the injector control the temperature at the center and inside diameter (ID) of the combustor, respectively. Circumferential distribution of temperature is then controlled by controlling fuel delivery to the individual fuel-injector assemblies. Thus, an array of fuel-injection sites is avail-

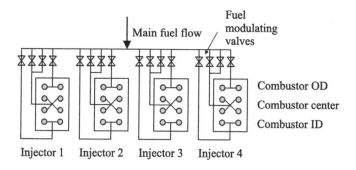

Figure 17.1 Fuel-injector system sketch valves.

able for regulation of fuel mass flow to each injector by individual modulating valves as shown in Fig. 17.1.

17.3   FUEL CONTROL VALVES

The practical application of active control technology on aircraft combustion systems requires the fuel delivery valves to be small and lightweight, so as not to add appreciable weight to the engine system. Proportional solenoid valves were selected as the fuel modulating device to meet these requirements. Prototype solenoid valves were developed by the Lee Co. based on existing on/off solenoid valves. The operation of the proportional solenoid is based on varying the input current to a magnetic coil that creates motion to a plunger which results in a variable orifice area and modulated flow. Standard solenoids snap completely open when the input current reaches a specific level, while proportional solenoids utilize stiff springs that resist the force created by the input current and allow incremental motion. A specific flow rate is maintained as the pressure, magnetic, and spring forces are balanced, accordingly. A cross-section of one of these valves is illustrated in Fig. 17.2. The size of the valve makes it very attractive for engine applications. Figure 17.3 provides a Lee Co. photo of the proportional solenoid valve to illustrate the relative size of the valve. The maximum weight of the valve is 90 g (0.2 pounds), which is also appealing for engine applications.

Flow hysteresis was a major challenge as the prototype proportional solenoid valves were being developed. The earliest version of the proportional solenoid valves produced the flow curve shown in Fig. 17.4a, having a large hysteresis

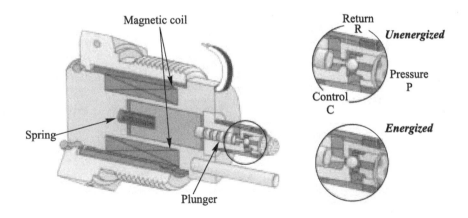

Figure 17.2 Proportional solenoid valve cross-section (courtesy of the Lee Co., Westbrook, CT).

Figure 17.3 Proportional solenoid valve hardware specifications (courtesy of the Lee Co., Westbrook, CT).

depending on the direction of travel of the valve plunger. The flow characteristic of this valve is similar to that of an on/off solenoid because of the relatively low spring stiffness. An increase in the spring stiffness and a change in the magnetic coil material resulted in a valve that has improved flow hysteresis as indicated by the flow curve in Fig. 17.4b. The stiffer spring design produced a more proportional flow curve, while the change in the valve material affected the magnetic characteristics of the coil and thus the hysteresis. A full set of valves was fabricated using the design characterized by flow curves shown in Fig. 17.4b. The valves were configured in a compact manifold assembly for the rig tests as shown in Fig. 17.5. For an engine application, it is envisioned the valves would be integrated into the fuel-injector assembly. The valves shown on each face of the manifold block in Fig. 17.5 provide radial control for one injector assembly. Two valves control the center zone, while the inner and outer zones are controlled by one valve each, respectively.

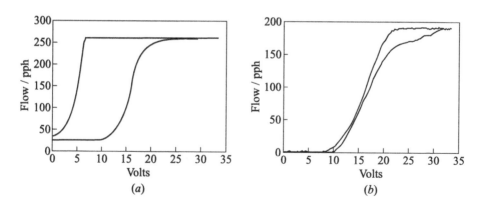

Figure 17.4 Proportional solenoid valve flow curves: (a) initial valve design flow curves and (b) developed valve flow curves.

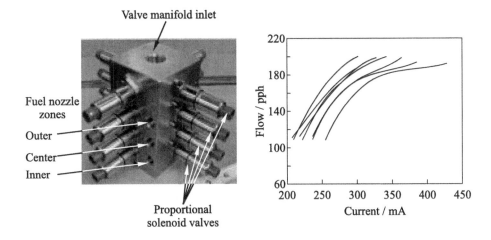

Figure 17.5 Fuel delivery valve hardware photo (courtesy of the Lee Co., Westbrook, CT).

Figure 17.6 Flow curves for proportional solenoid valves.

A PC-based controller was developed to operate the valves for fuel flow modulation in the various zones. The valve hardware and controller were bench tested in a jet-fuel wet-bench test facility to verify valve performance and controller effectiveness. Valve flow tests indicate stable and repeatable valve operation over the range of flow conditions tested. The valve flow curves demonstrated acceptable proportionality at lower mass flows, but became slightly nonlinear at higher mass flow conditions as illustrated by Fig. 17.6.

Some variability in the flow curves of the different valves is also noted in Fig. 17.6. A level of discrepancy is expected for the first generation, prototype valve, and the variability would be reduced with changes in the manufacturing process for production valves. For rig testing, the variability can be accommodated with the computer-based control system.

17.4   OPTICAL SENSORS

Temperature sensors for providing closed-loop feedback for APFC are being developed by Stanford University for the measurement of combustion species and temperature. Laser-diode sensors offer nonintrusive measurements of the combustion exit plane temperature pattern. A detailed description of the sensor measurement methodology can be found in the literature [1, 2]. There is

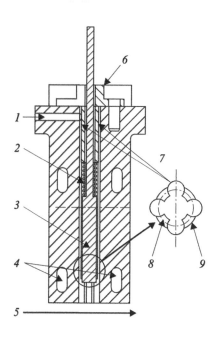

Figure 17.7 Optical sensor mount and cooling configuration: *1* — N$_2$ inlet; *2* — compression mount spring; *3* — optical probe assembly with lens; *4* — water channels; *5* — hot gas flow direction; *6* — two-piece mount plug; *7* — N$_2$ channels; *8* — optical probe diameter; and *9* — 45-degree chamfer.

concern about the durability of optical sensors in aircraft engine combustor applications because of the exposure to high temperature and pressure conditions. Soot deposition on the optical lens is also of concern because it can scatter laser energy and thus significantly reduce the measurement sensitivity and accuracy of the sensor. Therefore, a sensor mount design capable of operation in high temperature and pressure conditions that included a purging mechanism to keep soot away from the optical lens was designed and fabricated (as illustrated in Fig. 17.7).

The optical probe is spring mounted to ensure that it stays seated in the mount cavity with variations in temperature. The probe sits on a chamfered surface that has a minimum diameter that is smaller than that of the sensor cable. Nitrogen is supplied around the sensor cable for the purpose of cooling the optical components and also functions as a mechanism for lens purging at the end of the sensor cable to keep soot off the lens surface. Water channels are also provided in the sensor mount structure to cool the surfaces that are exposed to high-temperature exhaust gases at the combustor exit. A set of four sensors will be utilized in the initial open-loop combustion tests for verification of sensor performance in the intended environment. A retroreflector scheme will be investigated, where the laser beam is reflected back through the transmission probe for a simplified mechanical system and possibly increased measurement sensitivity. A standard transmitter/receiver design will also be examined to determine the relative effectiveness of each methodology. The optical sensor measurements will be validated against data acquired with a traversing gas sample system. The traversing gas sample probe will provide detailed temperature distribution data for the combustor exit plane. Both the gas sample rake and the optical sensor data will be used for validation of CFD simulations of the combustor exit temperature.

17.5   COMPUTATIONAL RESULTS

Computational fluid dynamics analysis is being used to investigate the fundamental processes which govern combustor exit pattern factor and to determine the limits of control authority by spatial modulation of fuel injection. The CFD solutions will also be used to define basis functions that are a central part of the control methodology. The CFD models have been created, and a preliminary set of solutions has been obtained. The latest CFD models incorporate many geometric features of the combustor rig for better exit temperature predictions. A new spray model that accounts for secondary droplet breakup and grid refinement of the CFD geometry was also incorporated for better computational accuracy. Improved agreement with prior experimental data is expected with these new CFD model features.

The previous rig test was run at a relatively low combustor pressure with a small pressure differential across the fuel nozzles. The fuel nozzle spray pattern at lower pressure differentials is not expected to be fully developed, and thus, accurate modeling of the fuel spray pattern in CFD would be difficult. Fuel modulation at the lower differential pressures could cause large changes in the spray characteristics and would not be a good case to demonstrate APFC. In Figs. 17.8a and 17.8b, CFD comparison results are considered reasonable for this case, however, given the spray uncertainties. The figures show a plot of temperature factor vs. span for two cases. The temperature factor is the difference between the local radial temperature and the average combustor exit temperature normalized by the average combustor temperature rise. The "in-line" curves in Fig. 17.8a are data taken from a "circumferential" location in-line with the combustor fuel injectors, while the "in-between" curves in Fig. 17.8b are representative of a location exactly between the fuel injectors.

The APFC tests will be run at higher combustion pressures for better characterization of the spray pattern. Consequently, the new CFD simulations will be run at higher pressure conditions. Preliminary results in Fig. 17.9 show good ability to control pattern factor. Figure 17.9 shows a plot of pattern and profile factor vs. fuel flow rate of the injector center zone. Pattern factor is the difference between the maximum temperature in the CFD sector and the average combustor exit temperature normalized by the average combustor temperature rise. The profile factor is the difference between the maximum circumferentially averaged temperature and the average combustor exit temperature normalized by the average combustor temperature rise. The points on the continuous curves represent temperature profiles from CFD cases with changing center zone fuel flow and an equal amount of ID and OD fuel flow. For example, the nominal fuel flow distribution is 50% fuel in the center zone, 25% fuel in the ID zone, and 25% fuel in the OD zone. The 45% point on the curves would be representative of 45% fuel in the center, 27.5% fuel in the ID, and 27.5% fuel in the OD. On the other hand, the individual points in Fig. 17.9 indicate the associated fuel

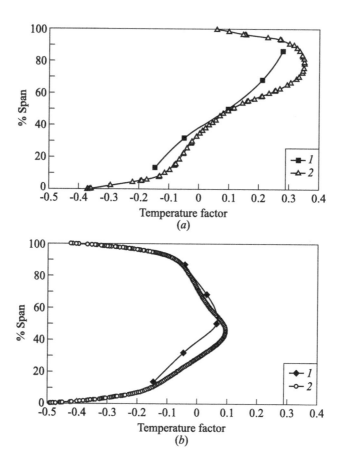

Figure 17.8 Experimental and computational radial temperature profiles at low pressures: (*a*) temperature profiles in-line with fuel injector: *1* — in-line rig, and *2* — in-line CFD base; and (*b*) temperature profiles in-between fuel injectors: *1* — in-between rig, and *2* — in-between CFD base.

flow distributions with "c," "i," and "o" representing the center, ID, and OD fuel flow percentages, respectively.

The complexity of managing the exit temperature distribution can be observed from Fig. 17.9. The profile factor or maximum integrated circumferential average did not change much as fuel was moved from the center zone toward the ID and OD, but the pattern factor or local hot spot peak changed strongly. The complex behavior is still not clearly understood. The 45c30i25o point in Fig. 17.9 demonstrates how with APFC one can attempt to minimize both profile and pattern factors. For this case, fuel was removed from the center zone

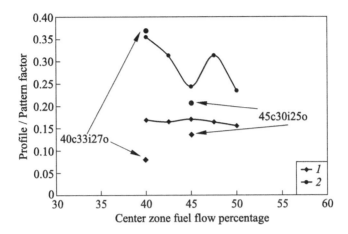

Figure 17.9 CFD combustor exit temperature profile and pattern factor vs. center zone fuel flow: *1* — profile factor and *2* — pattern factor.

and sent only to the ID where temperatures are generally lower. Point 40c33i27o shows that profile factor can be significantly reduced. The pattern factor was not reduced, however, but the peak was changed from OD to ID. These preliminary results show that one can control the spatial exit temperature distribution of an advanced combustor with APFC by modulating fuel in local fuel-injector sites.

17.6 CONCLUDING REMARKS

Modifications of the fuel injectors of an advanced combustor were made to permit spatial control of individual fuel-injection sites. Miniature fuel delivery valves were developed and flow bench was tested to characterize the valve flow characteristics. Optical sensors and a translating gas sampling rake are being integrated in an experimental rig to evaluate the level of pattern factor control at the combustor exit. The CFD simulations were run, and the results will be validated with data that will be acquired from the forthcoming APFC experiments. The numerical solutions will then be used to develop control algorithms that will be used for future closed-loop APFC experiments.

ACKNOWLEDGMENTS

This work was performed under ONR contract N00014-00-3-0021.

REFERENCES

1. Baer, D. S., R. K. Hanson, M. E. Newfield, and N. K. J. M. Gopaul. 1994. Multi-species diode-laser sensor system for H_2O and O_2 measurements. AIAA Paper No. 94-2643.
2. Furlong, E. R., D. S. Baer, and R. K. Hanson. 1996. Combustion control and monitoring using a multiplexed diode-laser sensor system. AIAA Paper No. 96-2763.

Comments

Witton: Does the cone angle on the spill atomizer change with the pulse width and/or load conditions?

Lovett: Yes, the angle changes, but we haven't looked at this parameter in detail.

Gutmark: What is the maximum frequency of the valves used in the study?

Lovett: Approximately 200 Hz.

Chapter 18

SYSTEM DESIGN METHODS FOR SIMULTANEOUS OPTIMAL CONTROL OF COMBUSTION INSTABILITIES AND EFFICIENCY

W. T. Baumann, W. R. Saunders, and U. Vandsburger

This chapter reports experimental and analytical results for pulsed control of combustion instabilities at both fundamental and subharmonic frequencies. Two suites of control algorithms have been developed: one based on least-mean-square (LMS) techniques that is suitable for inner-loop stabilization of combustion instabilities, and one based on direct optimization that can be used either for stabilization or outer-loop optimization of combustion process objectives, such as flame compactness or emissions.

18.1 INTRODUCTION

Control and optimization of combustion processes will require controllers capable of reacting to fast processes, such as combustion instabilities, as well as controllers capable of optimizing the combustion parameters on a slower scale to achieve desired performance objectives, such as flame compactness or minimal emissions. The focus of research has been on refining two suites of control tools: one based on LMS techniques and suited to control of instabilities, and another based on direct descent techniques and suited for either instability control or combustion process optimization. In addition, investigation of the performance consequences of using pulsed (on–off) as opposed to proportional actuators has been completed, and this chapter presents experimental results of the effect of varying the subharmonic order of the control pulses on instability suppression.

191

18.2   PULSED AND SUBHARMONIC CONTROL

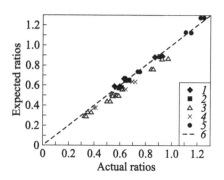

Figure 18.1 Theoretical gain ratios vs. measured gain ratios for proportional, pulsed, and subharmonic control: *1* — subharmonic–subharmonic forcing; *2* — fundamental–fundamental; *3* — subharmonic–fundamental; *4* — subharmonic–linear; *5* — fundamental–linear; and *6* — baseline.

Since actual control systems may be required to use on–off actuators, the analysis of the effects of using pulsed and subharmonic control was continued. The main conclusion of the study is that pulsed control of an instability at a given frequency is achieved through the harmonic of the pulsed control signal at the instability frequency. The amplitude of this harmonic, for a pulse height of X, an on-time of w, and a subharmonic order of M ($M = 1$ corresponds to forcing at the fundamental frequency, $M = 2$ corresponds to forcing at the $1/2$ subharmonic, etc.), is given by $2X \sin(\pi w/T)/(M\pi)$. This hypothesis has been verified on a Rjike-tube combustor using acoustic actuation. By considering the gain at which a linear phase-shift controller stabilized the system, it is possible to predict the gains for stabilization that would be required for pulsed control using various duty cycles and subharmonic orders. Figure 18.1 plots the predicted gain ratios as a function of the observed ratios. The near equality demonstrated by this figure supports the hypothesis that it is the harmonic component at the instability frequency that is important.

Clearly, a pulsed control system can never actually drive an unstable system to a zero level. If no pulses are applied, the unstable oscillation will grow, and if fixed-height pulses are applied, a fixed amount of energy will be introduced into the system with each pulse, resulting in some level of oscillation. To quantify this level of oscillation, note that the effective gain of the pulsed controller is given by $2X/(AM\pi)$, where A is the amplitude of the input oscillation. Thus, as the amplitude of the oscillation decays towards zero, the effective gain increases towards infinity. For any realistic controller, there is a limit to its gain before the closed-loop system is driven unstable. In the stabilization problem considered here, if the pulse height X is large enough, the initial gain of the controller will be sufficient to start to stabilize the system. As the oscillation begins to decrease, the controller gain will increase, eventually reaching a point where the system is driven unstable. At this point, the oscillation begins to increase, decreasing the effective gain. The result will be a stable limit cycle where the effective gain is equal to the gain at which the linear controller will drive the system unstable, k_{ult}. From this equality, the amplitude of the ultimate limit cycle is given by

$$A = \frac{2X}{\pi k_{\text{ult}} M}$$

This formula was verified in the tube combustor by experimentally determining the ultimate gain of a linear phase-shift controller to be 4.2 and then using a pulsed controller with the same phase shift and various pulse heights. By observing the amplitude of the ultimate limit cycle, the ultimate gain k_{ult} can be computed at each pulse height. The plot in Fig. 18.2 shows that this computation yields the expected value of 4.2.

From the expression for the effective gain of the controller, it is clear that for a given amplitude limit cycle, a minimum amplitude of pulse height will be necessary for stabilization. So for stabilization over a large range of instabilities, a large pulse height would be desirable. In addition, the pulse height directly affects the ultimate amplitude that will be achieved, and so a small pulse height is desirable. To get around this tradeoff, a variable subharmonic controller is proposed. For initial instability suppression, $M = 1$ is used, and then M is increased to reduce the ultimate limit cycle amplitude and the number of actuator cycles used to maintain control. The details of this work are reported in [1, 2].

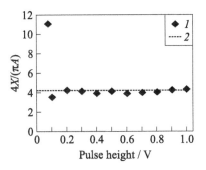

Figure 18.2 Verification of the ultimate gain hypothesis for various pulse heights: *1* — stable gain and *2* — gain = 4.2.

18.3   LEAST-MEAN-SQUARE-BASED ALGORITHMS

In this study, the work in adaptive algorithms for realistic controllers is divided between LMS-based approaches and direct optimization approaches. The difference is in the amount of *a priori* information required for the controller. In the LMS case, a model of the control to error path over the relevant bandwidth is required [3]. In current implementation, this model is identified in real time when the controller is first turned on.

Figure 18.3 shows the layout for a Filtered-E controller, with G denoting the control to error path and $\hat{G}_c$ and $\hat{G}_e$ representing the estimates of this path used to cancel the control signal and to estimate the gradient, respectively. As reported, it was possible to find situations where small errors in the estimate of the transfer function of G could cause feedback-loop instabilities or

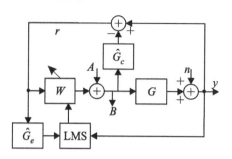

Figure 18.3 Filtered-E algorithm block diagram.

algorithm divergence [4]. Further analysis of the situation shows that this behavior is linked to a transfer function

$$F = \frac{1}{1 + W(\widehat{G}_c - G)}$$

which modifies both the control path and the gradient filter. This transfer function is equal to 1, hence causing no problems, if either there is no estimation error or W is zero. As W adapts, this transfer function can go unstable or generate excessive phase error in the gradient filter. Further investigation has shown that F is the transfer function between points A and B in Fig. 18.3. Thus, this transfer function can be identified during operation of the algorithm, and if F deviates significantly from unity, $\widehat{G}_c$ can be adjusted to force F back towards unity, greatly improving the robustness of the controller.

Although LMS controllers assume a linear control structure, and therefore assume proportional actuation will be used, the algorithm with fixed-pulse-height and on–off actuation has been run and significant suppression of the instability

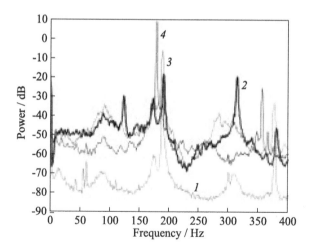

Figure 18.4 Results of pulsed control of a tube combustor using the LMS algorithm with two different height pulses ($\Phi = 0.6$): *1* — proportional; *2* — $H = 0.1$; *3* — $H = 0.2$; and *4* — uncontrolled.

has been achieved. Figure 18.4 shows the results of the adaptive algorithm using two different pulse heights and forcing at the fundamental frequency. The results using proportional actuation are also shown.

As usual, proportional control provides 35 dB of peak suppression and drops the noise level over the entire bandwidth. Using a pulse height of 0.1 V provides almost 30 dB of suppression and raises the noise level over most of the band. There is a competing instability at about 315 Hz that appears to limit the achievable suppression. Doubling the pulse height and adding a bandpass filter (second order, 100–250 Hz) to avoid exciting the 315-hertz instability resulted in about 15 dB of suppression and a similar increase in noise level. In light of the analysis of pulsed control above, it is not surprising that the ultimate suppression is less for the larger pulses.

The LMS algorithm was also used to achieve control using subharmonic pulsing. Figure 18.5 shows the results of starting the controller off with fundamental forcing ($M = 1$) and then transitioning to 1/2 and 1/3 subharmonic forcing. Fundamental forcing yields about 26 dB of peak suppression, and the emerging instability at 315 Hz is evident. Transitioning to the 1/2 subharmonic provides an additional 4 dB of peak suppression and vastly reduces the 315-hertz peak while pumping up energy near the 1/2 subharmonic frequency, as expected. Transitioning to the 1/3 subharmonic results in approximately 50 dB of suppression. At this level the input to the controller will not be dominated by a single sinusoid, and so the control signal is probably fairly erratic. It should be pointed

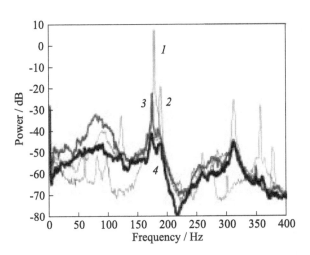

Figure 18.5 Results of subharmonically pulsed control of a tube combustor using the LMS algorithm ($\Phi = 0.56$): *1* — no control; *2* — $M = 1$; *3* — $M = 2$; and *4* — $M = 3$.

out that if the controller is started with $M = 2$, it cannot achieve significant suppression. It is necessary to start with $M = 1$ to generate significant suppression and then to transition to higher orders of subharmonic pulsing.

18.4 DIRECT OPTIMIZATION ALGORITHMS

In an effort to reduce the amount of *a priori* information needed for control, a time-averaged-gradient (TAG) approach has been previously investigated. Not only is this approach conceptually simple, but there is no restriction on the type of controllers that can be adapted. The tradeoff for using no *a priori* information is that this approach may converge more slowly than an approach that relies on additional information. Time-averaged gradient is just one of many potential direct optimization algorithms that can be applied to this problem and has been extended in this work to produce a fairly general control tool that implements several direct optimization algorithms: TAG; Gradient; Conjugate gradient; Hooke and Jeeves; and Rosenbrock.

The control tool runs on a commercially available, digital control prototyping system manufactured by the DSPACE Corporation. These algorithms can be divided into two basic categories: gradient descent and pattern searches. In general, the gradient descent methods perturb controller parameters, compute the corresponding gradient, and move the parameters along the gradient direction. The pattern searches perturb the parameters along search directions, compute the mean-squared error (MSE) of the resulting signal, and make decisions based on these calculations.

Time-averaged gradient [5] is the most basic of the gradient descent algorithms and is based on the method of steepest descent. The gradient descent with line search (Gradient) algorithm is an extension of TAG. The gradient is calculated as in TAG, but instead of taking a fixed step along the gradient, the algorithm searches along the gradient direction for the optimal step size. This is done by taking successively larger step sizes and evaluating the corresponding MSE. When an increase in step size no longer results in increased performance, a new gradient is calculated, and the procedure is repeated. The final gradient descent algorithm is the Fletcher and Reeves Conjugate Gradient Method (Conjugate). The Conjugate algorithm creates search directions that are a linear combination of the steepest descent direction and the previous search directions so that the resulting directions are conjugate (orthogonal). The concept of conjugacy is very important in unconstrained optimization problems. If the function to be minimized is quadratic, conjugate search directions guarantee that convergence will occur in at most n steps, where n is the number of parameters being adapted.

The pattern searches implement the method of Hooke and Jeeves using a line search and the method of Rosenbrock [6]. The Hooke and Jeeves algorithm that was used in this work can be thought of in two separate phases: an exploratory search and a pattern search. The exploratory search successively perturbs each parameter and tests the resulting performance. The pattern search steps along the $(x_{k+1} - x_k)$ direction, which is the direction between the last two points selected by the exploratory search. When both positive and negative perturbations of the parameters do not result in enhanced performance, the perturbation size is decreased. When the perturbation size is less than an arbitrary termination factor, ε, the algorithm stops.

The Rosenbrock algorithm searches along orthogonal parameter directions and is closely related to the conjugate direction methods. If a perturbation results in a lower MSE, an expansion term is used to increase the step size in that direction. If a higher MSE results, a contraction term reverses the sign of the perturbation and decreases the magnitude. This is repeated until a failure occurs along all directions, which leads to the development of new directions through the Gram–Schmidt procedure. The Gram–Schmidt procedure takes the mutually orthogonal, linearly independent weight directions, $d_1, \ldots, d_n$, and forms new directions, $\bar{d}_1, \ldots, \bar{d}_n$. The result is a new set of linearly independent, orthogonal search directions.

The algorithms described above require the *a priori* selection of several parameters that guide the operation of the algorithm. Since good values for these parameters will depend upon the system to be controlled, it is important that these parameters be chosen automatically if the overall control algorithm is to be robust. Two parameters that are involved in each algorithm are the integration length, N, and the perturbation size, δ. The integration length is the number of samples over which the MSE is computed. The perturbation size is the value by which the parameters are changed to perform the gradient calculation or MSE evaluation.

The value of the MSE approaches the true value as the integration length approaches infinity. To minimize convergence time, the goal is to find a minimum integration length while maintaining an accurate estimate of the MSE. This procedure involved filling an array with error signal samples and computing the MSE after each sample. The past n MSE values were then evaluated, where n was some specified evaluation range. When the maximum and minimum MSE values within this range fell within a specified percentage of each other, the integration length was chosen as the total number of samples taken to this point. If this condition was not met, another sample was taken and the next MSE value was computed.

The perturbation size needs to be large enough to cause a measurable change in the signal, yet not so large as to produce a large change in the signal. The initial perturbation size was chosen to be relatively small. The first controller parameter was perturbed, and the change in the MSE was measured. If the

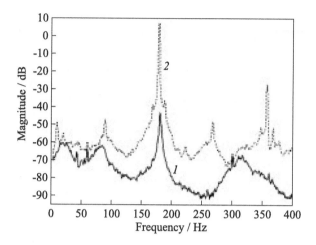

Figure 18.6 Control of tube combustor using the Hooke and Jeeves algorithm (*1*) as compared to no control case (*2*).

percent change was below a minimum value, the perturbation size was increased. If the change exceeded a maximum value, it was decreased. Otherwise, when the change fell into the specified range, it was deemed acceptable.

The fixed step-size parameter, μ, used in the TAG algorithm was also calculated during the tuning phase. From the theory of the LMS algorithm it can be shown that for stability the step size must satisfy the condition [7]

$$0 < \mu < \frac{1}{\lambda_{\max}}$$

where $\lambda_{\max}$ is the largest eigenvalue of the input correlation matrix, R. Since the largest eigenvalue cannot be greater than the trace of R, and for a transversal adaptive filter the trace of R is $(L+1)E[x_k^2]$, the bounds on the step size can be restated as

$$0 < \mu < \frac{1}{(L+1)(\mathrm{MSE})}$$

In practice, the convergence parameter was chosen to be 25% of the upper bound. The above derivation assumes the use of a transversal filter and so is applicable to the adaptation of the finite impulse response (FIR) filter used in this work.

Each of these algorithms was used to adaptively update a two-weight FIR filter and stabilize a Rijke-tube combustor through acoustic actuation. Both the gradient descent and pattern search methods proved quite effective and produced 40 to 50 dB of attenuation of the instability peak. For example, Fig. 18.6 shows the power spectral density of the uncontrolled tube and the system controlled with the Hooke and Jeeves algorithm. The steady-state results of all the

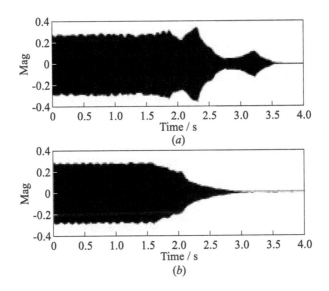

Figure 18.7 Convergence of the Hooke and Jeeves (a) and TAG (b) algorithms.

algorithms were similar. The reason for implementing a variety of algorithms is not so much that one expects significantly different steady-state results on this particular combustor, but that different algorithms might prove faster or more effective in different situations. For example, smooth performance surfaces will probably be best handled by gradient searches, whereas rough performance surfaces might be better handled by pattern searches.

By calculating a gradient, the gradient descent algorithms gain knowledge of the performance surface and thus move towards an optimal solution. The pattern searches, however, are systematically searching the weight space, which may result in steps in nonoptimal directions. Because of this, one might expect the gradient descent algorithms to converge in less time than the pattern searches. This is supported by the results of tests of these algorithms on the tube combustor. While each algorithm was successful in gaining control of the instabilities present in the tube, the gradient descent algorithms were quicker to converge. An example of this can be seen in Fig. 18.7, which shows the decay envelope for the TAG and Hooke and Jeeves algorithms. The initial 1.5 s of each plot is the limit cycling of the tube, at which point the algorithms were started.

18.5 CONCLUDING REMARKS

Future work will involve implementation of a probing function, to eliminate the likelihood of ending at a local minimum, and a monitor function, to detect when the system characteristics have changed significantly and to allow for reinitialization of the controller.

ACKNOWLEDGMENTS

The work was performed under ONR contract N00014-99-1-0752.

REFERENCES

1. Carson, J. M. 2001. Subharmonic and non-subharmonic pulsed control of thermoacoustic instabilities: Analysis and experiment. M.S. Thesis. Blacksburg, VA: Virginia Polytechnic Institute and State University.
2. Baumann, W. T., W. R. Saunders, and J. M. Carson. 2005 (in press). Pulsed control of thermo-acoustic instabilities: Analysis and experiment. *ASME Transactions on Dynamic Systems, Measurement and Control.*
3. Vaudrey, M. A., W. T. Baumann, and W. R. Saunders. 2003. Applying adaptive LMS to feedback control of thermoacoustic instabilities. In: *Combustion and noise control.* Ed. G. D. Roy. Cranfield,U.K.: Cranfield University Press. 232–34.
4. Vaudrey, M. A., W. T. Baumann, and W. R. Saunders. 2003. Stability and operating constraints of adaptive LMS-based feedback control. *Automatica* 39:595–605.
5. Vaudrey, M. A., W. T. Baumann, and W. R. Saunders. 2003. Time-averaged gradient control of thermoacoustic instabilities. *J. Power Propulsion* 19(5):830–36.
6. Bazaraa, M. S. and C. M. Shetty. 1979. *Nonlinear programming: Theory and algorithms.* New York, NY: John Wiley and Sons, Inc.
7. Widrow, B., and S. D. Stearns. 1985. *Adaptive signal processing.* Englewood Cliffs, NJ: Prentice Hall, Inc.

Comments

Ghoniem: Subharmonic frequency forcing must be system dependent. The shear layer might become more unstable.

Baumann: I agree. One needs to know *a priori* the type of system you are trying to control.

Lovett: Could you comment on the pulsed data? There appears to be a drop in the peak, but always some residual remains.

Baumann: As the oscillation decreases and becomes corrupted by noise, the pulsing becomes somewhat erratic, raising the noise floor.

Annaswamy: Your system seems to be too simple to apply an adaptive technique, doesn't it?

Baumann: Yes, it does. We need to move to a more realistic combustor configuration.

Chapter 19

MODEL-BASED OPTIMAL ACTIVE CONTROL OF LIQUID-FUELED COMBUSTION SYSTEMS

D. Wee, S. Park, T. Yi, A. M. Annaswamy, and A. F. Ghoniem

Significant progress has been made during the past year in (i) modeling of combustion dynamics and (ii) model-based active control of combustion systems. The role of hydrodynamic modes in combustion systems was investigated using a reduced-order model. The shear-flow mode frequencies thus predicted using mean-velocity profiles in separating flows and numerical simulations, and the resulting pressure amplitudes were shown to match those in experimental and numerical investigations. These were combined with a new tool based on the Proper Orthogonal Decomposition (POD) method to yield reduced-order models of combustion systems. A Recursive POD (RePOD) algorithm was developed to facilitate a stable on-line control design by providing suitable updates of the POD modes. Finally, model-based posi-cast controllers developed earlier were validated using a 4-megawatt combustor model where it was shown to result in a 20-decibel pressure power-spectrum reduction without generating secondary peaks.

19.1 INTRODUCTION

Progress has been made in both reduced-order modeling and model-based control of combustion dynamics. Advances in modeling were obtained by investigating shear-flow driven combustion instability. The authors of this chapter determined that shear-layer instability occurs when an absolutely unstable mode is present. This mode can be predicted and matched to experimental data. Also, it is shown that the temperature profile determines a transition to absolutely unstable operation. Parameters of the shear-flow modes together with a POD-based approach led to the derivation of a new reduced-order model that sheds light on the interactions between hydrodynamics, acoustics, and heat release. A RePOD algorithm allows the use of this model for control design and guarantees a reliable and stable on-line operation. Finally, it is shown that model-based posi-cast controllers produce favorable results as compared to phase-shift controllers using simulation studies on a large-scale combustor model.

19.2   SHEAR-FLOW DRIVEN COMBUSTION INSTABILITY

Vortical structures are reported to cause flame/heat release oscillations in reacting flows or pressure oscillations in nonreacting flows [1–8]. In these cases, hydrodynamic instabilities leading to the formation of the vortical structures define the mode frequency, perturb the heat release, and feed energy into the acoustic field, which, under the circumstances, acts as an amplifier. During the past years, the authors have developed an approach to capture flame–acoustic shear-layer coupling.

A theoretical framework for the investigation of shear-flow dynamics is found in modern linear stability analysis in which unstable modes have been characterized as:

(1) *Convectively unstable* (CU) modes which grow while moving downstream away from their point of inception, leaving behind decaying oscillations and stable flow (unless the perturbation is continuously applied); and

(2) *Absolutely unstable* (AU) modes which grow in place while spreading spatially, hence by contaminating the entire domain and "regenerating" the instability.

The latter evolves into a self-sustained resonator that, once perturbed, continues to oscillate indefinitely due to the transfer of energy from the mean flow to the fluctuation. Details of how to conduct this analysis can be found in [9].

To understand what flow characteristics give rise to this mode, the study is extended to a family of velocity profiles defined as

$$U(y) = \frac{\beta - 1}{2} + \frac{\beta + 1}{2} \tanh\left(\frac{y}{\delta}\right) - \left(\beta \tanh\left(\frac{y+1}{0.1}\right) + \tanh\left(\frac{y-1}{0.1}\right)\right) \quad (19.1)$$

where β represents the amount of backflow in the recirculation zone, while δ is the shear-layer thickness. The impact of these parameters on the frequency and growth rate of the instability is shown in Fig. 19.1. Switching from a CU to an AU mode occurs at large backflows and small shear-layer thicknesses. Because the shear-layer thickness mostly scales with the step height at the middle of the recirculation zone, the magnitude of backflow within the recirculation zone is the main parameter determining whether or not the flow can sustain an AU mode.

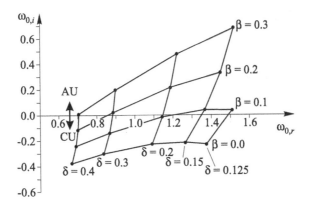

Figure 19.1 Absolute mode frequencies for the family of velocity profiles shown in Eq. (19.1) for different values of the shear-layer thickness δ and the backflow β.

In premixed reacting flows, the recirculation zone is made up of hot products separated from cold reactants by the flame front. To capture some of the impact of combustion on the unstable mode, the stability analysis was repeated while assuming the existence of a density or temperature profile superimposed on the velocity profile.

The temperature distribution within the shear layer is taken as

$$\theta(y) = \frac{1+\gamma}{2} + \frac{1-\gamma}{2} \tanh\left(\frac{y-\Delta}{\delta}\right) \tag{19.2}$$

where γ is the temperature ratio across the flame, and Δ is the local mean offset between the shear and temperature layers. The offset is the distance between the point of maximum shear and the flame front. Note that (i) this offset increases downstream as the flame front slopes upwards while the shear layer slopes downwards [2, 10, 11], and (ii) as ϕ increases, both Δ and γ increase, with the flame moving further away from the shear layer. Results in Fig. 19.2 show that increasing γ improves the flow stability by reducing the growth rate of the AU mode, with higher temperature ratios leading to switching from an AU mode to a CU mode, which does not support self-sustained oscillations. Thus, low ϕ mixtures, with small Δ and γ, are more likely to support AU modes, i.e., leaning out/decreasing ϕ may lead to the onset of an AU mode and hence combustion instability at its frequency, consistent with the experimental results in [12] where the nonacoustic mode around 96 Hz was excited as ϕ was lowered [13].

The existence of this eddy shedding is also reported by the authors' numerical experiment of a separating flow downstream of a rearward-facing step using a vortex code. The eddy shedding and pairing frequencies are determined using frequency domain analysis of local time traces to obtain the local dominant Strouhal

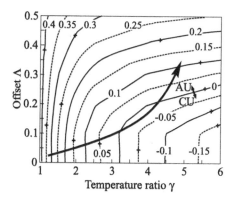

Figure 19.2 The impact of the temperature distribution on the properties of the absolute instability, shown in terms of the ratio of absolute growth rate to absolute frequency $(\omega_{0,i}/\omega_{0,r})$ for various values of temperature ratio and offset for the case with $\delta = 0.3$ and $\beta = 0.4$. The thick arrowed line shows a possible path for increasing the equivalence ratio.

numbers, defined as the frequency of the mode possessing maximum local amplitude. This value drops from 0.3 to close to 0.08 within 2–3 step heights downstream of the step. A similar trend was observed experimentally in [12]. The drop is consistent with an eddy pairing mechanism where two and sometimes three eddies pair before leaving the recirculation zone. The smallest Strouhal number appears first in the middle of the recirculation zone. The amplitude of the corresponding frequency shows a continuous increase of the dominant mode amplitude until the end of the recirculation zone, beyond which it begins to decay. Towards the end of the recirculation zone, the mode with the highest signature has a normalized frequency of 0.08.

Preliminary results were obtained also for a reduced-order model for the nonreacting shear layer described above, as a component of a more comprehensive combustion instability model that includes the acoustic field and heat release dynamics. Model reduction is accomplished using POD, in which numerical results are used to construct a space of optimal basis functions that describe the different modes of the flow, and applying these functions to construct time-dependent ordinary differential equations (ODEs) that determine the amplitudes of the corresponding modes under different conditions. The ODEs for the reduced-order model are obtained from either a Galerkin expansion of the dependent variables in these basis functions [13] or a system identification technique. The current focus is on extending the recent results to more general type of reacting flows.

19.3 A RECURSIVE PROPER ORTHOGONAL DECOMPOSITION ALGORITHM FOR FLOW CONTROL PROBLEMS

The POD method is an efficient tool for extracting the most dominant features of a dynamical system. It has been used in several flow control problems to develop a low-order finite dimensional model, either from a numerical or an experimental database of the plant, for potential use in designing active control strategies for manipulating the flow. Quite often, the underlying POD modes in a dynamical system change with control actuation, thereby introducing a possible divergent behavior [14]. An algorithm for computing the POD modes recursively so as to facilitate a stable on-line control design has been developed recently [15]. If $u(x, t)$ is a zero-mean flow variable, then the POD method seeks to generate an approximation for u by using separation of variables as

$$\hat{u}(x, t) = \sum_{i=1}^{l} a_i(t)\phi_i(x) \tag{19.3}$$

where $\phi_i(x)$ is the spatial mode, $a_i(t)$ is the amplitude, and the subscript represents a mode number. The POD method consists of finding $\phi_i(x)$ such that the error $u(x, t) - \hat{u}(x, t)$ is minimized. The optimization problem can be stated as follows:

$$\min_{\phi} J_m(\phi_1, \ldots, \phi_l) = \sum_{j=1}^{m} \left\| Y_j - \sum_{k=1}^{l} \left(Y_j^T \phi_k \right) \phi_k \right\|^2 \tag{19.4}$$

subject to

$$\phi_i^T \phi_j = \delta_y, \quad 1 \leq i, j \leq l$$

where Y_j is the vector of flow data u at time $t = t_j$ [16]. To determine a recursive procedure that is capable of updating the modes of the system on-line using very few computations, a gradient technique shown below is used:

$$\phi_i^{(m+1)} = \phi_i^{(m)} - s \left. \frac{\partial \Delta J^{(i)}}{\partial \phi_i} \right|_{\phi_i = \phi_i^{(m)}} \tag{19.5}$$

where

$$\Delta J^{(i)} = \left\| Y_{m+1} - \sum_{k=1}^{l} \left(Y_{m+1}^T \phi_k \right) \phi_k \right\|^2$$

Equation (19.5) is the RePOD algorithm, which is the main contribution in this section.

The recursion is performed on new flow data using gradient techniques, which are then followed by the determination of control input. The RePOD method can also be used in isolation to compute the modal basis, potentially resulting in substantial savings in computational resources and time. The algorithm was applied to two classes of problems [15]. In the first, the problem of heat conduction through a one-dimensional rod with a sinusoidal heat perturbation source located inside the rod was considered. The objective is to reduce the temperature perturbation using another zero-mean sinusoidal heat source/sink. The one-dimensional heat conduction equation with two concentrated heat source/sinks is given by

$$\frac{\partial T}{\partial t} = \alpha \frac{\partial^2 T}{\partial x^2} + q_s \delta(x - x_s) + q_c \delta(x - x_c) \tag{19.6}$$

where α is the thermal diffusivity, and x_s and x_c are the locations of the perturbation and control heat source/sink, respectively. A Dirichlet boundary condition was applied at both ends of the rod. Due to boundary conditions, the mean temperature is zero, and hence, resolving temperature in the spatial and time domain, one gets

$$T(x, t) = T'(x, t) = \sum_{i=1}^{n} a_i(t) \phi_i(x) \tag{19.7}$$

where $\phi_i(x)$ is the POD mode. In order to obtain a suitable control strategy, the reduced-order model is derived that governs the behavior of the modal amplitude $a_i(t)$. Using Eqs. (19.6) and (19.7), one gets the following reduced-order model:

$$\dot{a}_j = \sum_{k=1}^{n} (A_{jk} a_k) + q_s \phi_j(x_s) + q_c \phi_j(x_c) \tag{19.8}$$

where

$$A_{jk} = -\int_0^L \left(\frac{\partial \phi_j}{\partial x_j} \right)^2 dx$$

In a first-order reduced model of Eq. (19.8), since A_{11} is negative, the system becomes stable if the perturbation term is eliminated by the control term. Hence, the control law chosen was

$$q_c(t) = -\frac{\phi_1(x_s)}{\phi_1(x_c)} q_s(t) \tag{19.9}$$

The gain of the controller depends on the POD modes, and it was updated using the RePOD algorithm. Figure 19.3 shows the results using the RePOD algorithm and the control law in Eq. (19.9). As shown in Fig. 19.3b, the controller gain converged to -0.85 with a 30% reduction in fluctuation level, while the theoretical value of optimal gain was found to be -0.58 with 40 percent fluctuation reduction.

The second class of problems involves thermal and fluid flow systems, in which an explicit control action changes the system basis. The RePOD method, embedded in a feedback control algorithm, was able to keep track of these changes and resulted in a stable closed-loop system. These POD or RePOD techniques

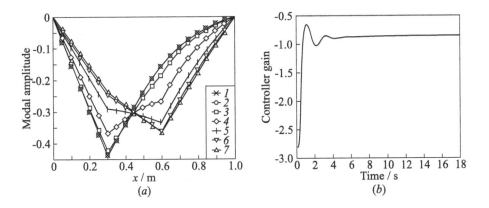

Figure 19.3 First RePOD mode (a): 1 — $t = 0$; 2 — 0.2; 3 — 0.4; 4 — 0.6; 5 — 0.8; 6 — 1.2; and 7 — $t = 1.4$, and controller gain (b).

will be used in combustion processes to develop reduced-order models that capture not only stability characteristics, but also other performance parameters such as emission and mixture homogeneity. Also, a hierarchical multivariable controller will be incorporated in the design with the ultimate goal being a simultaneous optimization.

19.4 ADAPTIVE LOW-ORDER POSI-CAST CONTROL OF A COMBUSTOR TEST-RIG MODEL

An adaptive posi-cast controller has been developed for dynamic systems with large time-delays [17]. Recently, the authors evaluated its performance in the context of a 4-megawatt combustor model that mimics many of the dynamic characteristics of an actual engine including a significant time-delay [18]. Using closed-loop input–output data and system identification (SI), a model of the test-rig was derived. The resulting expression of the SI model is

$$W(s) = \frac{-468s - 1.76 \cdot 10^6}{s^2 + 13s + 2.97 \cdot 10^6} e^{-0.0116s} \tag{19.10}$$

Using this model, adaptive posi-cast controllers were designed, and detailed numerical simulation studies were carried out. These studies consisted of (i) the closed-loop performance of the adaptive controller, (ii) comparison of the adaptive controller with an empirical phase-shift controller, (iii) robustness with respect to parametric uncertainties, (iv) robustness with respect to unmodeled dynamics and uncertain delays, (v) performance in the presence of noise,

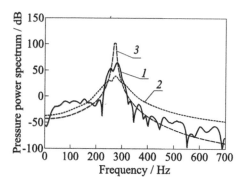

Figure 19.4 Pressure power spectrum of a closed-loop combustor model with a phase-shift controller (*1*) and a posi-cast controller (*2*); *3* — baseline.

and (νi) effect of saturation constraints on the control input amplitude. In Fig. 19.4, the performance of the posi-cast controller is compared with that obtained using the phase-shift controller. As one can see, the posi-cast controller outperforms the phase-shift controller with 20 dB more reduction in the pressure power spectrum. Also, it is worth noting that the posi-cast controller does not generate secondary peaks which limited the performance of the phase-shift controller.

Parameters of the model in Eq. (19.10), such as damping, resonant frequency, and time-delay, can change due to changes in the operating condition. Hence, it is essential to overcome these changes so as to be able to use the controller in a wide range of operating conditions. In the simulation, the resonant frequency was changed by 10% as shown in Fig. 19.5*a*, and the performance of the adaptive low-order posi-cast controller was tested. Figure 19.5*b* shows that regardless of the speed of the frequency change, the controller was able to stabilize the combustor. It was observed that the adaptive controller was capable of handling up to a 20% change in the resonant frequency.

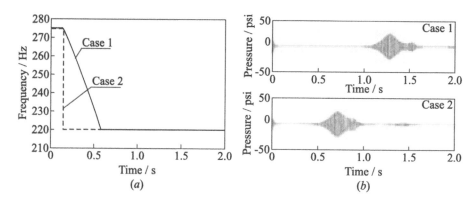

Figure 19.5 Pressure response of the closed-loop system using an adaptive low-order posi-cast controller with changes in frequency of a combustor model: (*a*) changes of a resonant frequency and (*b*) pressure–time responses.

It was also shown that the adaptive low-order control was able to stabilize the combustor model with an order of magnitude change in the damping ratio. The saturation of a control input was also investigated, and the results show that by increasing the saturation level, the performance of the posi-cast controllers can be improved. These studies show that the adaptive posi-cast controllers decrease pressure oscillations much faster than the phase-shift controller without generating peak splitting. It also showed that the adaptive controller was capable of stabilizing the plant in the presence of a 20 percent change in the resonant frequency, an order of magnitude change in the damping ratio, and unmodeled dynamics. The studies showed that the performance of the adaptive controller improved as the magnitude of the saturation constraints on the control input increased.

19.5   CONCLUDING REMARKS

The idea that shear-layer instability produces nonacoustic modes in shear flow and that these modes can be predicted and correlated with experimental results has been investigated. A novel method of recursively calculating POD basis functions to use in on-line control in a stable manner was developed. Finally, it has been demonstrated that a model-based posi-cast controller is able to successfully control a large-scale combustor model with satisfactory results.

ACKNOWLEDGMENTS

The work was performed under ONR contract N00014-99-1-0448.

REFERENCES

1. Keller, J. O., L. Vaneveld, D. Korschelt, G. L. Hubbard, A. F. Ghoniem, J. W. Daily, and A. K. Oppenheim. 1982. Mechanisms of instabilities in turbulent combustion leading flashback. *AIAA J.* 20:254–62.

2. Jou, W. H., and S. Menon. 1986. Numerical simulation of the vortex–acoustic wave interaction in a dump combustor. *AIAA 24th Aerospace Sciences Meeting Proceedings*. Reno, NV.

3. Shadow, K. C., E. Gutmark, T. P. Parr, D. M. Parr, K. J. Wilson, and J. E. Crump. 1989. Large scale coherent structures as drivers of combustion instability. *Combustion Science Technology* 64:167–86.

4. McManus, K. R., U. Vandsburger, and C. T. Bowman. 1990. Combustion performance enhancement through direct shear layer excitation. *Combustion Flame* 82:75–92.

5. Yu, K., A. Trouve, and J. W. Daily. 1991. Low-frequency pressure oscillations in a model ramjet combustor. *J. Fluid Mechanics* 232:47–72.

6. Weller, H. G., G. Tabor, A. D. Gosman, and C. Fureby. 1998. Application of a flame-wrinkling LES combustion mode to a turbulent mixing layer. *27th Combustion Symposium Proceedings.* 899–907.

7. Fureby, C. 2000. A computational study of combustion instabilities due to vortex shedding. *28th Combustion Symposium Proceedings.* 783–91.

8. De Zilwa, S. R. N., J. H. Uhm, and J. H. Whitelaw. 2000. Combustion oscillations close to the lean flammability limit. *Combustion Science Technology* 160:231–58.

9. Huerre, P., and P. A. Monkewitz. 1990. Local and global instabilities in spatially developing flows. *Annual Review Fluid Mechanics* 22:473–537.

10. Ganji, A., and R. Sawyer. 1980. Turbulence, combustion, pollutant and stability characterization of a premixed step combustor. *AIAA J.* 18:817–24.

11. Pitz, R., and J. W. Daily. 1983. Combustion in a turbulent mixing layer formed at a rearward-facing step. *AIAA J.* 21:1565–70.

12. Cohen, J. M., and T. J. Anderson. 1996. Experimental investigation of near blow out instabilities in a lean premixed step combustor. AIAA Paper No. 96-0819.

13. Wee, D., S. Park, T. Yi, A. M. Annaswamy, and A. F. Ghoniem. 2002. Reduced order modeling of reacting shear flows. AIAA Paper No. 2002-0478.

14. Graham, W. R., J. P. Peraire, and K. Y. Tang. 1999. Optimal control of shedding using low-order models, part *ii*-model based control. *Int. J. Numerical Methods Engineering* 44:973–90.

15. Sahoo, D., S. Park, D. Wee, A. M. Annaswamy, and A. F. Ghoniem. 2002. A recursive proper orthogonal decomposition algorithm for flow control problems. Adaptive Control Laboratory, MIT. Technical Report 0208.

16. Holmes, P., J. Lumley, and G. Berkooz. 1996. *Turbulence, coherent structures, dynamical systems and symmetry.* Cambridge, UK: Cambridge University Press.

17. Evesque, S., A. P. Dowling, and A. M. Annaswamy. 2000. Adaptive algorithms for control of combustion. *NATO/RTO Active Control Symposium Proceedings.* Braunschweig, Germany.

18. Park, S., D. Wee, A. M. Annaswamy, and A. F. Ghoniem. 2002. Adaptive low-order posi-cast control of a combustor test-rig model. *IEEE Conference on Decision and Control.* Las Vegas, NV.

Comments

Seiner: In your stability analysis, how did you handle the problem of elliptic vs. hyperbolic equations while predicting the stability?

Ghoniem: We used the dispersion relation to calculate and compare the growth rate.

Seiner: It is not possible to obtain a solution to this elliptic subsonic problem.

Ghoniem: The solution is from linear stability, and, hence, a solution is possible.

SECTION TWO

HIGH-SPEED JET NOISE

Chapter 1

AEROACOUSTICS AND EMISSIONS STUDIES OF SWIRLING COMBUSTOR FLOWS

S. H. Frankel, J. P. Gore, and L. Mongeau

A combined experimental and computational research program has been initiated to investigate the role of partial premixing and swirl on pollutant and noise emissions from aeropropulsion gas-turbine engines. Planned measurements include three-dimensional flow velocity fields using Particle Image Velocimetry (PIV), sound pressure using microphone arrays, and pollutant emissions using gas analyzers. Computational studies will employ computational aeroacoustics approaches based on large-eddy simulation (LES) to predict flow, flame, and acoustics features. Multiple-swirl combustors and trapped-vortex combustors will be studied with geometries relevant to naval and industrial applications. This chapter summarizes previous work, current objectives, and the technical approach. Preliminary results from the current project are presented along with future plans.

1.1 INTRODUCTION

Increasingly stringent global laws require that the design of next-generation navy aircraft and marine engines addresses both improved performance as well as reduced environmental emissions of chemical pollutants and noise. Many strategies have been proposed to reduce gas-turbine pollutant emissions, including unburned hydrocarbons, particulates, and oxides of nitrogen. To maintain performance and stability, these strategies often include some degree of premixing of fuel and oxidizer prior to combustion together with swirl for flame stabilization.

Effectively, premixing facilitates better control of temperatures and reactive species concentrations and hence the kinetic control of pollutant processes, such as NO_x production. Lean, premixed, prevaporized combustion and lean-direct injection (partial premixing) are two important examples of this approach. Future combustion concepts may involve triple swirlers or trapped-vortex cavities.

Noise emissions from gas-turbine engines, resulting from unsteady flow interactions with rigid and moving surfaces (compressor and diffuser noise) and from combustion (core noise), must also be reduced without sacrificing performance. Studies addressing simultaneous pollutant and noise emission reductions are rare. The current study may lead to a better understanding of the coupled acoustic and pollutant behavior of advanced combustor concepts.

The present work involves detailed measurements of flow, pollutant emissions, and acoustics in model multiple-swirl, partially-premixed flames and combustors. Complementary computations are also planned, combining computational aeroacoustics approaches with combustion modeling. The focus is on how pollutant control and flame stabilization strategies, such as partial premixing and swirl, respectively, influence combustor noise sources and potential instabilities. Also of interest are flow and acoustics associated with diffuser–combustor interactions and the utility of the trapped-vortex combustor design. A better understanding of pollutant and acoustic sources and how to modify them will aid in the control of emissions, noise, and instabilities in modern swirl combustors.

1.2   PREVIOUS WORK

Flame Sound

Self-excited combustion instabilities are associated with the propagation and reflection of heat-release-induced acoustic waves and their interactions. Hence, flame sound represents the main source of these acoustic waves. Therefore, sound pressure level (SPL) data for turbulent nonpremixed jet flames have been obtained for two Turbulent Nonpremixed Flame (TNF) workshop flames, DLR-A and DLR-B [1]. The exit Reynolds numbers (Re) for the two flames based on injected gas properties at room temperature were 15,200 (DLR-A) and 22,800 (DLR-B). Air was used for studying the sound emission from equivalent nonreacting jets. The flow in each case had very low exit Mach numbers (M = 0.04–0.18) based on the sound speed at room temperature. The A-weighted SPL was calculated based on recorded sound pressure data acquired using a National Instrument Analog-to-Digital Converter (ADC) board and a PC.

Figure 1.1 shows the axial variation in sound pressure and level for flames (top) and air jets (bottom) as a function of axial distance from burner exit, normalized by the burner diameter ($D = 0.8$ cm). The nonpremixed turbulent flame levels are between 20 and 30 dB greater than those for the cold jets. A careful observation of the changes in directivity shows that the sound

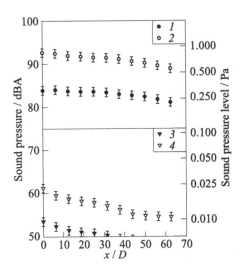

Figure 1.1 Axial variation of sound pressure and sound pressure level for flames (top) and jets (bottom). Top: 33.2% H_2 + 22.1% CH_4 + 44.7% N_2 – air flame, $D = 0.8$ cm, $r/D = 50$, _1_ — Re = 15,200, M = 0.12; and _2_ — Re = 22,800, M = 0.18. Bottom: air jet: _3_ — Re = 15,200, M = 0.08; and _4_ — Re = 22,800, M = 0.12.

generated by the flames remains almost constant at a constant radial location, for x/D values between 10 and 30, whereas the nonreacting jet sound exhibits a monotonous decrease in the same region. The scalar data from the TNF web site* show that significant chemical reactions occur in this region of the flames. These results confirm that the contribution of combustion processes on the radiated sound is significant for the flow regime considered. The contribution from combustion processes is often termed "entropy noise" and is likely the result of local density fluctuations within the combusting flame. The trends in levels and directivity are in agreement with the observations in the literature for premixed turbulent flames and co-flow nonpremixed turbulent flames.

Additional results showing radial variation and Re number dependence allow one to conclude that the combustion noise source is distributed over a wider spatial extent than for the air jet and the difference between reacting and nonreacting SPL decreases with increasing Re number. Further spectral analysis of these results is underway, together with similar measurements in swirling and partially premixed flames and combustors.

---

*http://www.ca.sandia.gov/tdf/workshop.html.

Swirl Combustor Flowfields

A cross-correlation PIV system was used to measure the multipoint instantaneous velocity fields in strongly swirling flows with a theoretical swirl number of 2.4 and an Re number of 72,000 [2]. The system is shown in Fig. 1.2. The mean flow structure obtained was consistent with the literature. The multipoint instantaneous flow structures show significantly different characteristics compared to the mean flow. Many smaller-scale vortices with both directions of vorticity appear compared to a single large vortex observed in the mean flow. Heat release due to lean premixed methane–air combustion seems to break down the multiple vortices into a more homogeneous field and increases the mean and instantaneous vorticity magnitudes. Typical results can be seen in Fig. 1.3 for cold and reacting swirling flow.

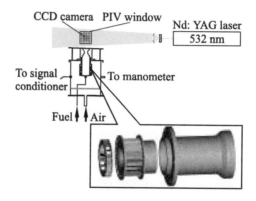

Figure 1.2 Experimental combustor rig and swirler configuration.

Combustor Acoustics

Sound generation by flames within combustors is a problem involving a complex interplay of fluid–sound interactions. Two types of interactions are of interest. First, in cases where there is no strong feedback coupling between the combustion processes and the sound waves within the combustor (or combustion "instability"), the noise generation processes may be construed as a linear source–filter system. Sound is presumably generated by one or a series of low-Mach-number, confined, stationary combusting jets, and the combustor acts as an acoustic enclosure. The acoustic response of the enclosure to the acoustic excitation by the flames may then strongly color the airborne or structure-borne noise emissions

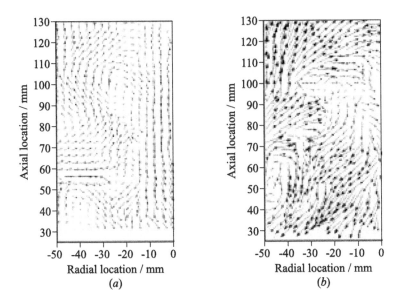

Figure 1.3 Instantaneous velocity vectors in swirl combustor: (*a*) cold flow and (*b*) with flame.

of the combustor. In cases where strong instabilities exist, the system can no longer be modeled using a linear source–filter system. A limit-cycle, closed-loop feedback model may be used instead, as described, for example, in [3] for the problem of a flow-excited cavity.

To investigate the influence of weak (i.e., linear) acoustic loading on aerodynamic noise sources, preliminary experiments were performed using a quiet flow supply, muffler-terminated rigid uniform tubes, and acrylic orifice-plates [4]. The emissions of cold circular jets discharging in a circular tube were investigated. A spectral decomposition method based on a linear source–filter model [5] was used to decompose radiated, nondimensional, sound pressure spectra measured for various gas mixtures and mean flow velocities into the product of (*i*) a source spectral distribution function, (*ii*) a function accounting for near-field effects and radiation efficiency, and (*iii*) an acoustic frequency response function. The acoustic frequency response function agreed, as expected, with the transfer function between the radiated acoustic pressure at one fixed location and the strength of an equivalent velocity source located at the orifice. The radiation efficiency function indicated a radiation efficiency of the order $(kD)^2$ over the planar wave frequency range and $(kD)^4$ at higher frequencies, where k is the wavenumber and D is the tube cross-sectional dimension. This is consistent with theoretical predictions for the planar-wave radiation efficiency of quadrupole sources in uniform rigid anechoic tubes. The effects of the Re number on the source spectral distribution function were found to be insignificant over the range $2,000 < Re < 20,000$. The source spectral distribution function obeyed a St^{-3} power law for Strouhal number values $St < 0.9$ and a St^{-5} power law for $St > 2.5$. The influence of a reflective open tube termination on the source function spectral distribution was found to be insignificant, confirming the absence of a feedback mechanism in this case.

The use of such a dimensional analysis and the spectral decomposition method is useful to isolate spectral features of the radiated sound that are associated with sound generation processes from other features related to near-field effects or the acoustic response of the combustor cavity. It can be used, for instance, to identify the presence of feedback coupling or instability in a combustor system by systematically varying the flow rate of the combusting mixture, the acoustic properties of the working gas, or the physical dimensions of the combustor housing or cavity. The application of this method for the case of the Purdue University research combustor is planned and should provide some fundamental insight into sound generation phenomena in such systems.

Computational Aeroacoustics

A computational aeroacoustics (CAA) code based on the large-eddy simulation (LES) approach has been developed and tested for a variety of nonreacting and reacting axisymmetric and turbulent unconfined and confined jet and combustor flows [6–12]. The code solves the compressible Navier–Stokes equations using fourth-order Runge–Kutta time stepping and sixth-order compact finite-differencing for spatial discretization on fine spatial grids in a generalized curvilinear coordinate system to produce numerical solutions with minimal dispersion and dissipation errors. Far-field characteristic-based boundary conditions with an exit buffer zone are used. Several versions of the dynamic Smagorinsky subgrid-scale turbulence model have been used. Inlet harmonic or stochastic forcing has also been used. As an example, LES-based near-field flow and far-field sound predictions were in good agreement with measured data for a subsonic turbulent jet (see typical result in Fig. 1.4). Acoustic analogies based on Lighthill, Kirchhoff, and Ffwocs Williams–Hawking equations have been used for jets and confined flows to predict far-field sound based on the near-field flow simulation results. Excellent agreement with the directly simulated sound has allowed the use of these analogies to identify specific sound sources with the various flows studied. To complement the available LES research code, it will also make use of the commercial computational fluid dynamics (CFD) code FLUENT in both Reynolds-Averaged Navier–Stokes (RANS) and LES modes.

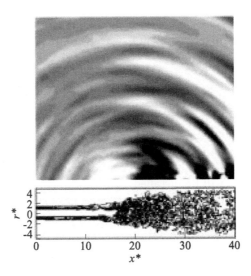

Figure 1.4 Contour plot showing instantaneous vorticity in the near-field and dilatation in the far-field from LES of sound radiation from a subsonic turbulent jet.

1.3 PRELIMINARY WORK

Measurements for swirling jets depicted in Fig. 1.5 show that the sound pressure levels increase with the number of swirlers. This is logical since multiple swirlers introduce swirling shear layers, which are sources of vorticity and noise. The results indicate that the nature of sound produced from swirling jets may be quite different from simple jets, including the sound pressure level and the decay away from the source. The expertise gained from this ongoing preliminary work will be employed for this project. It will measure mean species and temperature profiles together with planar velocity statistics including three-dimensional (3D) and turbulence quantities to help identify the acoustic sources within partially-premixed swirling jet

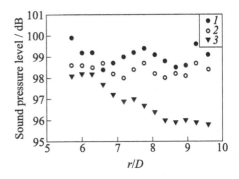

Figure 1.5 Preliminary measurements of sound radiation from different swirling jets: *1* — counter-flow swirlers, *2* — co-flow swirlers, and *3* — single swirler.

flames. Pollutant emissions from these flames will also be measured to quantify the effect of partial premixing and swirl on the relationship between sound and pollutant emissions.

The Purdue University high-pressure combustion facility will be modified to allow detailed acoustic pressure measurements. Flow supply noise sources will be minimized using a muffler with flexible couplings. The inlet plenum and discharge cylinder will be instrumented with an array of 16 flush-mounted, high-temperature microphones. Each sensor will be mounted on a water-cooled adapter with a power supply. The signals will be acquired and processed using an available multichannel high-speed data acquisition system. Circumferential arrays of microphones will be installed with two arrays at two streamwise locations upstream of the combustor, in the plenum, and two downstream. The two-microphone method will be used to compute net acoustic intensity and acoustic impedance on both sides of the combustor within the frequency range where sound waves are planar. Assuming reverberant sound in the main combustor pipe, statistical energy analysis (or room acoustics) methods will be used to determine flame sound power. Acoustic measurements will be made for each type of flame stabilization device (different number and degree of swirler) at high and low mean pressures. Detailed frequency spectral analysis will be used to develop semiempirical laws to quantify combustion noise.

Swirling flows are inherently three dimensional. The authors' current PIV system measures only two velocity components. A newly acquired stereoscopic PIV system will be used to characterize the 3D flowfield. In addition, pollutant characteristics of high-pressure partially-premixed swirling flames will also be obtained. Together with the PIV velocity and acoustic measurements, a full characterization of both pollutant and noise emissions from such flames will be achieved. It is believed that this will help improve the fundamental understanding of these types of emissions from modern gas-turbine combustors and will provide a valuable database for validating LES models.

Preliminary numerical work has involved benchmarking of the LES research code and the FLUENT–LES code. The LES research code is currently being used to simulate the flow in a lean-premixed dump combustor for which benchmark data is available. Simulation results with the FLUENT–LES code for cold fully-turbulent jets, diffusers, and swirl combustors have been compared to experimental data from the open literature.

1.4   FUTURE PLAN

Current experimental efforts are focused on measurement and analysis of far-field sound from turbulent flames to investigate swirl and partial-premixing effects. PIV efforts are focused on obtaining 3D velocity measurements in industrial swirl combustors. Instrumentation of such combustors with appropriate acoustic sensors is underway.

Current numerical efforts are focused on LES benchmarking. Nonreacting jets and jet flames from the Sandia database library will be studied. Combustion models will initially be based on flamelet concepts, but models which address finite-rate chemistry will also be considered [13, 14]. Flame, emissions, and noise predictions will be compared to experimental data. Predictions of swirl combustor flowfields with comparison to data from the literature and the authors' own experiments are also underway. Far-field sound predictions of diffuser flow are currently underway. Coupled diffuser–combustor studies addressing unsteady inlet flow effects and trapped-vortex combustor studies focusing on cavity acoustic resonance effects are planned.

1.5   CONCLUDING REMARKS

Experimental and computational aeroacoustics and emissions of modern swirl combustor flows are underway. Preliminary measurements of turbulent non-premixed flame sound highlight the influence of combustion as a sound source. Particle Image Velocimetry measurements in swirl combustors reveal the influence of heat release and its effect on the complex spatial structures that are present. Acoustic measurements in confined turbulent jets are used to better understand sound sources in such flows. Computational aeroacoustics studies of unconfined and confined flows and flames have allowed acoustic source identification. Preliminary LES of diffuser and swirl combustor flowfields serve as benchmarks for future combustion simulations.

ACKNOWLEDGMENTS

This research project is supported by ONR contract N00014-02-1-0769. Funding from the Indiana 21st Century Research and Technology fund and Rolls Royce, Indianapolis, IN also contributed to this project, and is greatly appreciated. The contributions of Dr. Jun Ji, Kapil Singh, Cheng Zhang, Nagendra Dittakavi, Zhaoyan Zhang, Dr. Wei Zhao, and Jungsoo Suh, postdoctoral or graduate students who have provided results for this chapter, are gratefully acknowledged.

REFERENCES

1. Singh, K., S. H. Frankel, and J. P. Gore. 2003. Effects of combustion on the sound pressure generated by circular jet flows. *AIAA J.* 41(2):319–21.

2. Ji, J., and J. P. Gore. 2002. Flow structure in lean premixed swirling combustion. *29th Symposium (International) on Combustion Proceedings* 29.

3. Kook, H., and L. Mongeau. 2002. Analysis of the pressure fluctuations induced by flow over a cavity. *J. Sound Vibration* 251(5):823–46.

4. Zhang, C., W. Zhao, T. Ye, S. H. Frankel, and J. P. Gore. 2002. Parametric effects on combustion instability in a lean premixed dump combustor. AIAA Paper No. 02-4014.

5. Mongeau, L., D. E. Thompson, and D. K. McLaughlin. 1995. A method for characterizing aerodynamic sound sources in turbomachines. *J. Sound Vibration* 181(3):369–89.

6. Zhao, W., S. H. Frankel, and L. Mongeau. 2000. Effect of spatial filtering on sound radiation from a subsonic axisymmetric jet. *AIAA J.* 38(11):2032–39.

7. Zhao, W., S. H. Frankel, and L. Mongeau. 2001. Large eddy simulation of sound radiation from subsonic turbulent jets. *AIAA J.* 39(8):1469–77.

8. Zhao, W., and S. H. Frankel. 2001. Numerical simulations of sound radiation from axisymmetric premixed flames. *Physics Fluids* 13(9):2671–81.

9. Zhao, W., S. H. Frankel, and L. Mongeau. 2001. Numerical simulations of sound from confined pulsating axisymmetric jets. *AIAA J.* 39(10):1868–74.

10. Zhao, W., S. H. Frankel, and J. P. Gore. 2001. A numerical study of combustion instability in a dump combustor. AIAA Paper No. 01-2720.

11. Zhao, W., S. H. Frankel, and L. Mongeau. 2001. Computational aeroacoustics of an axisymmetric jet in a variable area duct. AIAA Paper No. 01-2788.

12. Zhang, Z., L. Mongeau, and S. H. Frankel. 2002. Experimental verification of the quasi-steady assumption for aerodynamic sound generation by pulsating jets in tubes. *J. Acoustical Society America* 112(4):1652–63.

13. DesJardin, P. E., and S. H. Frankel. 1998. Large eddy simulation of a nonpremixed reacting jet: Application and assessment of subgrid-scale combustion models. *Physics Fluids* 10(9):2298–314.

14. DesJardin, P. E., and S. H. Frankel. 1999. Two-dimensional large eddy simulation of soot formation in the near-field of a strongly radiating nonpremixed acetylene–air turbulent jet flame. *Combustion Flame* 119(1/2):121–32.

Comments

Santoro: What did you match for comparison of cold and hot flows?

Gore (for Frankel): Exit flow velocity was matched.

Ghoniem: Using acoustic analogy, I expected a difference.

Chapter 2

CONSIDERATIONS FOR THE MEASUREMENT OF VERY-HIGH-AMPLITUDE NOISE FIELDS

A. A. Atchley and T. B. Gabrielson

The research discussed here addresses long-range issues in measurement technology in anticipation of advances in computational predictions of radiation from full-scale jets. The focus of this research is a detailed investigation of sensors and techniques for making high-fidelity measurements of very-high-amplitude noise fields, such as might be found in the near-vicinity of high-performance, military jet aircraft. Specifically, this research will (*i*) investigate the performance of standard measurement microphones in very-high-amplitude sound fields with emphasis on potential degradation of performance due to nonlinear mechanical and acoustical processes and (*ii*) develop alternate sensors and sensing techniques against which to compare the performance of standard microphones. The results of this research will be applicable to efforts to experimentally validate representations of unsteady flows used as acoustic sources in computational models of jet noise. They can also provide empirical representations of the sources. Such measurements will contribute to a better understanding of fundamental jet noise generation mechanisms and to better prediction models for radiated noise and its impact on the environment.

2.1 INTRODUCTION

The focus of the research discussed here is the measurement of very-high-amplitude acoustic noise, primarily jet noise. Undoubtedly, research directed at predicting and reducing jet noise has been ongoing since the invention of the jet engine. It is reasonable to ask then, what new information can be gained from pursuing this research and what is its importance? This background is intended to address these questions. Because society is increasingly less tolerant of noise pollution, what passed as an acceptably quiet jet engine before, no longer does so. Prior research has not provided a solution to the contemporary problem, much less the future problem. It is not unreasonable, then, to venture that unless something else is learned, an acceptable solution is not likely.

As explained below, it is suggested that accurate measurements of acoustic noise very near the jet plume may provide critical data which are required to validate computational jet-noise models. Further, it appears that only data acquired very near the plume can provide such information. Although there are several different specific sources of jet-engine noise [1], they are all linked to unsteady flows associated with the mean flow of the engine exhaust.

In general, computational approaches to jet-noise predictions involve three steps. First, the mean flow is determined from analytical or computational models. Given the mean flow, the second step is to predict the statistics of the unsteady flow using analytical or empirical relationships between the steady and unsteady flows. The final step is to use the unsteady flow as the source in an acoustic radiation and propagation model. Consequently, accurate prediction of jet noise hinges on the realistic representation of the unsteady flow.

This research is aimed at facilitating the experimental validation of such representations. To understand how the research advances this goal, it is necessary to distinguish high-amplitude sound generated by periodic sources from that generated by random sources. At any point in the medium, all the properties of the acoustic signal radiated by a periodic acoustic source (in a stationary medium) can be determined from a single realization (in a measurement or calculation) of the waveform at that point. However, if the source is random, many realizations are needed to determine the necessary statistical properties of the field (e.g., probability density functions, cross-correlation functions, the average pressure, etc.). Jet noise is a random process. As such, statistical nonlinear acoustics techniques are needed to describe propagation of high-amplitude jet noise [2–4]. The evolution of high-amplitude statistical signals differs from that of high-amplitude periodic signals in several ways that point to the need for accurate measurements of the pressure fields radiated by jets. References [5] and [6] provide good introductions to statistical nonlinear acoustics.

One of the relevant consequences of nonlinear propagation is that waveforms steepen as they propagate. This steepening manifests itself in the generation of harmonics of the signal frequency. Therefore, as a high-amplitude signal propagates, its spectrum changes, transferring energy from low frequencies to higher frequencies. Waveforms steepen more rapidly with propagation distance for higher-amplitude waves than for lowest amplitude waves. This means that the highest amplitude components of a noise field influence the evolution of a field's spectrum more than the lowest amplitude components. For example, at a given propagation distance, assumed small in comparison to the shock formation distance, the intensity of the nth harmonic of a quasi-harmonic signal is $n!$ times that of an initially-pure tone having the same initial intensity as the quasi-harmonic signal [5–7]. (A quasi-harmonic signal has the form $A(t) \sin[\omega_0 t + \phi(t)]$, where A and ϕ are slowly-varying random functions of time; ω_0 is the constant angular frequency; and t is time. For the result cited, A and ϕ are assumed to be Gaussian distributed.) Therefore, it is important to be able to accurately measure the highest amplitude components of a noise field. This is one of the primary focuses of the proposed research.

A second relevant consequence of nonlinear propagation is that waveform steepening eventually leads to shock formation. Once shocks form, the statistical nature of a signal and the evolution of the spectrum change. Amplitudes of a noise field initially having a Gaussian probability distribution evolve toward a uniform distribution once shocks form [8]. This change implies that the probability distribution measured far enough from the source for shocks to have formed bears little or no relation to probability distribution of the source. This disconnection with the source has an analogy in propagation of finite-amplitude periodic waves. Regardless of the initial waveform, all periodic signals evolve into sawtooth waveforms, unless prevented from doing so by dissipation. Conversely, it is not possible to determine the initial waveform from a sawtooth signal [9]. Because the probability distribution of the noise at the source is related to the unsteady flow of the source, then measurements of the probability distribution that accurately reflect the nature of the source can only be made near the source. The evolution of the frequency spectrum of high-intensity acoustic noise is analyzed theoretically by Rudenko and Soluyan [10] and Scott [11]. It is shown that the high-frequency part of the spectrum obeys a universal law once shocks form. This prediction was verified experimentally by Bjørnø and Gurbatov [12]. This universal behavior further supports the idea that once shocks form, the details of the noise source are lost.

The main points can be summarized as follows. If measurements of noise fields are to be used to validate predictions of the unsteady flow statistics, they must be made at distances close enough to the source so that shocks have not formed. In addition, they must be able to faithfully record the large-amplitude outliers in the probability distribution. This last statement implies that the probability distribution must be extracted from the measurements. These criteria require that the measurement system responds linearly at the highest amplitudes, yet has a sensitivity and electronic noise level sufficient to provide the dynamic range needed to capture the low-amplitude outliers as well. In addition, the frequency response must extend from frequencies low enough to capture large-scale flow instabilities to frequencies high enough to identify shock formation. One can believe that given the complex, nonlinear nature of the source and the propagation, an essential key to developing a better understanding of jet-noise generation and radiation lies in high-fidelity measurements of the acoustic process. A starting point for determining an appropriate technical approach is to estimate the sound levels involved. Measurements made with a sound level meter during military aircraft engine run-up tests with afterburners at 150% indicate that the sound pressure levels (SPLs) at 35 and 50 ft from the engine nozzle are

153 and 150 dB, respectively [13]. The SPL of a broadband signal represents the time average of the squared, frequency-weighted sound pressure [14]. The SPL values given above are reported as being SPL-flat. The filter characteristics of flat weighting, also known as C-weighting, are flat from 100 to 2500 Hz with a slow roll off above and below [14]. The averaging time for the cited measurements is not specified, but it is likely 125 ms (fast response). Therefore, while the SPL gives some indication of the noise levels involved, it offers no information about the noise waveform or the instantaneous noise levels. Further, because statistical nonlinear acoustic processes favor the large-amplitude outliers in a noise field, average levels are not useful for predictive purposes. Finally, no mention is made of whether shocks had formed at the measurement locations. Shocked waveforms have significant spectral components in high frequencies, which are attenuated by sound-level-meter weightings.

In light of this background discussion, one estimates that to accurately record noise signals near the jet plume, microphones need to faithfully reproduce noise levels in excess of 180 dB (re 20 μPa) over frequency ranges up to 150 kHz. Learning the actual requirements is part of the project.

2.2   TECHNICAL APPROACH

The focus of the research discussed here is a detailed investigation of sensors and techniques for making high-fidelity measurements of very-high-amplitude noise fields. The technical approach for this effort is (i) to investigate the performance of standard measurement microphones in very-high-amplitude sound fields with emphasis on potential degradation of performance due to nonlinear mechanical and acoustical processes and (ii) to develop alternate sensors and sensing techniques against which to compare the performance of standard microphones. This work will entail developing a testing apparatus for generating very-high-amplitude, very spectrally-pure harmonic signals; an apparatus for generating very-high-amplitude, very broadband signals; and a shock tube. In addition to focusing on the linearity, dynamic range, and bandwidth of the sensors, attention will also be paid to nonlinear surface interactions between the sensor and the fluid, and how the presence of the sensor perturbs both the acoustic and nonacoustic fluid flow. The initial focus of this work will be on individual isolated sensors. However, the feasibility of using sensor arrays will also be considered for source localization and measuring power flow from the source.

Evaluating Sensors for Use in Very-High-Amplitude Noise Fields

The measurements must accurately represent the acoustic field without contamination from the measurement system. In particular, any nonlinear distortion introduced by the reference transducers (and electronics) must be negligible compared to the nonlinear generation by propagation of the high-amplitude jet noise. The transducers must accommodate very high acoustic levels and must have a large useable frequency range. If these conditions are not satisfied, it is not possible to separate distortion introduced by the receiving system from "distortion" introduced by high-level signal generation and propagation.

For a simple sine wave, nonlinearity in propagation produces harmonics. If the sensing system introduces distortion, then the measured harmonic strength will be inaccurate. In principle, a careful calibration of the distortion of the measuring system would permit separation of the propagation nonlinearity from the transducer nonlinearity, but only if the original signal were predominantly sinusoidal and the nonlinearity were weak. Jet-noise fields are broadband and intense, so compensation is impractical. The distortion introduced by the measurement system must be small compared to the propagation nonlinearity across the frequency band of interest and for the expected range of SPLs.

Establishing transducer linearity over a very large dynamic range is problematic. The advent of 24-bit analog-to-digital converters has led to a need for measurement systems that have sufficient linearity to justify that level of digitization (i.e., 144 dB for 24 bits). The most vexing problem in establishing linearity over such a large range is generation of a reference signal with the required signal-to-noise ratio and distortion. A digital waveform can be generated with the required purity, but it is difficult to produce an acoustic field from that waveform that preserves the purity.

Two approaches are proposed to resolve this difficulty. The first is based on generation of a high-purity acoustic field in a spherical, unharmonic resonator, and the second is based on comparison of a set of transducers with limited but overlapping dynamic ranges. In the second approach, simultaneous measurement of a broadband signal produces an accurate characterization over the entire range. Candidates for sensor fabrication include solid dielectric capacitive sensors, piezoelectric transducers, and high-temperature lithium niobate or quartz transducers.

Sensor/Fluid Interactions

The discussion up to this point has neglected interactions between the sensor and the medium, other than that required to sense the sound, of course. It is anticipated that several problems not apparent in linear-acoustics sensor development will crop up at sound levels relevant to jet noise. There are two areas of initial concern: (i) boundary layer interactions with the sensor surface and (ii) perturbation of the acoustic and nonacoustic flow fields due to the presence of the sensor. The boundary conditions imposed on a fluid by the presence of a solid surface cause both viscous and thermal boundary layers to form in the fluid in contact with the surface. These boundary layers lead not only to dissipation of acoustic energy, but also to a nonlinear flow phenomenon known as acoustic streaming. These effects are exacerbated by the presence of porous structures such as might be used for microphone windscreens, nose cones, probe tubes, or protective grids. Rather than improve performance, these structures might actually degrade it. The dynamics of acoustic boundary layers are reasonably well understood in harmonic sound fields under moderate Reynolds number conditions. In fact, these dynamics are the foundation of thermoacoustic heat transport [15]. However, the dynamics of oscillatory boundary layers in high-Reynolds number compressible random flows (i.e., high-amplitude acoustic noise) are not well understood. Acoustic boundary layers are frequency-domain concepts. In other words, a boundary layer thickness is only well defined for a monofrequency sound field. It is not readily apparent how best to treat boundary layers in high-amplitude noise. This area appears to be ripe for research.

While calculating the boundary layer dissipation in high-amplitude noise fields may be challenging, the consequences are mostly straightforward — surface heating. Deposition of even small amounts of heat onto small, low-heat-capacity sensor elements can result in significant changes in temperature and, hence, in their performance characteristics. This concern should guide sensor design.

There is a second, less well-known, boundary layer effect that also has the potential to lead to performance degradation. Acoustic streaming is a time-averaged net flow associated with the dissipation of acoustic energy. It is a non-linear phenomenon that scales in magnitude with pressure amplitude in roughly the same way as nonlinear harmonic distortion. Although generated by the boundary layer, streaming can result in large-scale fluid motion. Consequences of streaming are enhanced convective heat transfer, vortex shedding, generation of turbulence, and fluid pumping.

It is important to stress that streaming is a consequence of the presence of a surface in a high-amplitude sound field. Therefore, streaming is going to occur whenever a sensor is placed in such a field. The resulting degradation of sensor performance needs to be understood. While the authors are aware of a great deal of prior research on streaming generated by harmonic sound fields, they are not aware of similar work pertaining to noise fields. This area also appears to be ripe for research.

Perturbation of the Acoustic and Nonacoustic Flow Fields by the Sensor

As discussed above, boundary layer interactions can perturb the flow field in the vicinity of the sensor. However, there are other sources of perturbation that need to be investigated as well. One type is diffraction of sound by the sensor. The other is perturbations of nonacoustic flow (e.g., wind or jet wash) present at the measurement location. Even in linear acoustic applications, there is no single calibration factor. The calibration depends upon the direction from which the sound is incident, the frequency of the sound, and the type of protective grid used. The different calibrations are due to the differences in how sound diffracts under the various conditions. Jet-engine noise is caused by an extended, broadband source. Therefore, an individual sensor is exposed to sound coming from many different directions with random frequency content. It is important to understand the consequences of this for high-fidelity measurements. If one envisions using arrays of sensors, then the variation from sensor to sensor needs to be understood as well.

Finally, the presence of the sensor can perturb the nonacoustic flow, which, in turn, can result in vortex shedding and turbulence. The net effect of this can be nonacoustic contributions to the spectrum and an increased noise floor. It will be important to be able to separate this noise source from the one intended to be measured.

2.3 CONCLUDING REMARKS

The research discussed here addresses long-range issues in measurement technology in anticipation of advances in computational predictions of radiation from full-scale jets. In measurements of high-amplitude sound fields, distortion introduced by a measuring apparatus can be difficult to detect and can have considerable impact on the conclusions drawn regarding those fields. If a phenomenon is attributed to the noise source or propagation when it is really an artifact of the sensor, then considerable time and money can be wasted in developing systems based on that phenomenon. Consequently, the focus of this work is on developing and understanding high-fidelity measurement systems for high-amplitude sound fields in the near field of jet engines. The intent is to develop well-controlled and repeatable sound fields and to measure the perturbation of these fields and the distortion introduced by a number of sensor types and configurations. Both high-purity, single-frequency fields and controlled, high-amplitude, broadband fields will be produced in the laboratory to identify and characterize sensor problems. From these investigations, a set of sensors suitable for field measurement will be developed.

ACKNOWLEDGMENTS

The work was performed under ONR contract N00014-02-1-0313.

REFERENCES

1. Howe, M. S. 1998. In: *Acoustics of fluid-structure interactions*. Cambridge: Cambridge University Press. 49–156.

2. Morfey, C. L., and G. P. Howell. 1981. Nonlinear propagation of aircraft noise in the atmosphere. *AIAA J.* 19:986–92.

3. Howell, G. P. 1983. Effects of nonlinear propagation on long-range noise attenuation. AIAA Paper No. 83-0700.

4. Howell, G. P., and C. L. Morfey. 1987. Nonlinear propagation of broadband noise signals. *J. Sound Vibration* 114:189–201.

5. Gurbatov, S. N., and O. V. Rudenko. 1977. *Nonlinear acoustics*. Ch. 13: Statistical phenomena. Eds. M. F. Hamilton and D. T. Blackstock. San Diego: Academic Press. 377–97.

6. Naugol'nykh, K., and L. Ostrovsky. 1998. In: *Nonlinear wave processes in acoustics*. Cambridge: Cambridge University Press. 61–72.

7. Pernet, D. F., and R. C. Payne. 1971. Nonlinear propagation of signals in air. *J. Sound Vibration* 17:383–96.

8. Sakagami, K., S. Akio, I. M. Chou, T. Kamakura, and K. Ikegaya. 1982. Statistical characteristics of finite amplitude acoustic noise propagating in a tube. *J. Acoustical Society Japan* 3:43–45.

9. Naugol'nykh, K., and L. Ostrovsky. 1998. In: *Nonlinear wave processes in acoustics*. Cambridge: Cambridge University Press. 45–46.

10. Rudenko, O. V., and S. I. Soluyan. 1977. *Theoretical foundations of nonlinear acoustics*. New York: Plenum Press.

11. Scott, J. F. 1982. The nonlinear propagation of acoustic noise. *Proc. Royal Society London A.* 383:55-70.

12. Bjørnø, L., and S. N. Gurbatov. 1985. Evolution of universal high-frequency asymptotic forms of the spectrum in the propagation of high-intensity acoustic noise. *Sov. Physics Acoustics* 31:179–81.

13. McKinley, R. 2001. Jet noise models and issues. *Penn State Jet Noise Meeting*. University Park, PA.

14. Beranek, L. L. 1988. In: *Acoustical measurements*. Rev. ed. Acoustical Society of America. New York: American Institute of Physics Press. 800–9.

15. Swift, G. W. 1988. Thermoacoustic engines. *J. Acoustical Society America* 84: 1145–80.

Chapter 3

HIGH-SPEED JET NOISE REDUCTION
USING MICROJETS

A. Krothapalli, B. Greska, and V. Arakeri

This chapter deals with an experimental investigation on the suppression of high-speed jet noise using microjets. The far-field acoustic measurements from a high-temperature (1033 K), supersonic ($M_j = 1.38$), axisymmetric jet issuing from a 50.8-millimeter converging nozzle show the suppression of screech tones, Mach-wave radiation/crackle, and large-scale mixing noise due to the use of microjets. The overall sound pressure level (OASPL) reduction of 4.5 dB in the peak radiation direction is achieved. In the present configuration, eight 400-micrometer diameter microjets generated by a high-pressure air source (500 psia) are used; the total microjet mass flow rate is about 1% of the primary jet mass flux. The A-weighted spectrum, with appropriate scaling to reflect the full-scale nozzle, shows a 6-dBA reduction in the peak noise-radiation direction. Experiments using a nozzle designed for active noise control are also discussed.

3.1 INTRODUCTION

The far-field noise of a supersonic jet is comprised of four major noise components [1]. The first is a high-frequency, short wavelength field that is coherent in phase, commonly referred to as Mach waves. They have plane phase fronts and are confined to a definite wedge sector and emanate from a region within the first few diameters downstream of the nozzle exit, as can be seen in Fig. 3.1. These are generated by small-scale disturbances (or eddies) that are being convected at supersonic speeds so that they emit Mach waves in the direction defined by a disturbance convection velocity and the atmospheric speed of sound [2–4]. Surrounding the jet with a gas stream that has a higher speed of sound eliminates these waves as demonstrated by Oertel and Patz [5] and more recently by Papamoschou [6].

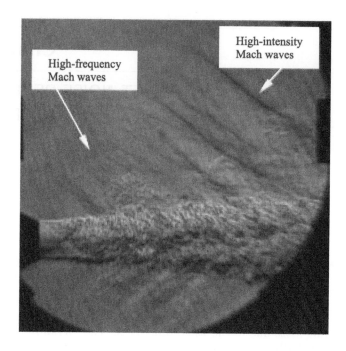

Figure 3.1 Schlieren picture of a Mach 2 round jet: $T_0 = 1250$ K and $U_j = 1050$ m/s.

The second noise field is highly directional, peaking at smaller angles to the jet axis (or larger angles to the inlet axis). This noise field is generated from large-scale instabilities reaching peak amplitudes in the region somewhat upstream of the end of the potential core. These sources of noise are associated with the unsteady flow on a scale that is comparable with the local shear-layer width. The spectral intensity of this sound field consists of two distinct peaks [7]. One is associated with the highly directional Mach waves characterized by high positive pressure peaks in the far-field microphone signal [8]. These Mach waves are of significant strength as compared to those that originate very close to the jet exit as discussed above. This intense radiation is observed to emanate from a region between $5 \sim 10$ nozzle diameters and is associated with supersonically traveling large-scale coherent regions of vorticity [9] (Fig. 3.1). It is found that the far-field intensity contribution of this source is about 30% of the measured total intensity [7, 10]. The sources for the second peak appear to be located much further downstream ($10 \sim 20$ nozzle diameters) and are associated with the unsteady flow generated by the large structures quite similar to those in subsonic jets.

The third noise field is at all angles to the jet axis and at higher frequencies. This sound is generated in precisely the same manner as in subsonic flow by the conventional chaotic turbulence. Recent analysis of experimental data by Tam *et al.* [11] vividly shows the contributions of the distinct components to the total far-field spectrum — that associated with large-scale motions (inclusive of Mach-wave radiation if it exists) and a component to small-scale turbulence.

The fourth noise field is commonly referred to as shock-associated noise and it occurs in nonideally expanded jets. The far-field noise spectrum associated with this noise source typically consists of discrete peaks, representing the screech tones, and a broad peak associated with the shock-associated broadband noise [12].

This chapter presents a possible approach for the suppression of the dominant large-scale mixing noise sources in a supersonic jet. High-pressure microjets are injected into the primary jet at the nozzle exit to manipulate the dominant source region, which extends typically from 5 to 20 diameters from the nozzle exit. Recent results of experiments on an $M_j = 0.9$ round jet suggest that the interaction of the microjets with the jet shear layer reduces the turbulence levels in the noise-producing region of the jet [13]. It appears that the microjets influence the mean velocity profiles such that the peak normalized vorticity in the shear layer is significantly reduced, thus inducing an overall stabilizing effect. Therefore, it is suggested that an alteration in the instability characteristics of the initial shear layer can influence the whole jet exhaust, including its noise field.

The present experiments were conducted using a convergent axisymmetric nozzle operating at a nozzle pressure ratio of 3, which resulted in a jet Mach number of 1.38. The nozzle temperature ratio (stagnation temperature/ambient temperature) was kept nominally at 3. The microjet mass flow rate was kept at less than 2% of the main jet mass flow rate.

3.2   EXPERIMENTAL SETUP AND PROCEDURES

Experiments were conducted in the newly built High Temperature Supersonic Jet Facility at the Fluid Mechanics Research Laboratory of the Florida State University in Tallahassee. A schematic of the facility can be seen in Fig. 3.2. In the present experiments, a converging axisymmetric nozzle having an exit diameter of 50.8 mm was used. The nozzle profile was designed using a fifth-order polynomial with a contraction ratio of approximately 2.25. The stagnation pressure and temperature were held constant to within 0.5% of its nominal value during the experiment.

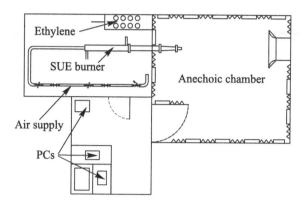

Figure 3.2 A schematic of the high-temperature supersonic jet facility.

A set of ten microphones was set up in an arc that had a radial distance of 50 diameters from the nozzle exit. The arc covered the polar angle, θ, range from 90° to 150° relative to the upstream jet axis. Each of the microphones had a relatively flat frequency response up to 100 kHz and was subsequently sampled at 250 kHz. The dataset for each microphone contained 409,600 samples (1.6 s). This allowed for a fast Fourier transform (FFT) of 4096 points over 100 subsets. Averaging the results for the 100 subsets reduced the random error in the calculation to within 0.1%. The resulting narrowband spectrum had a spectral resolution of 61 Hz.

The OASPL can be found through the use of two methods. The first method involves integrating under the power spectral density curve to obtain the squared pressure value and then applying the following well-known formula:

$$\text{OASPL} = 10 \log \frac{p^2}{p_{\text{ref}}^2} \qquad (3.1)$$

where p_{ref} is 20 μPa. The second method uses the pressure data that is obtained from the microphones to compute the root-mean-square pressure. This value is squared and then Eq. (3.1) can be used. In both cases, the OASPL value was found to be the same.

A number of corrections must be applied in order to obtain accurate data. The sound pressure level (SPL) at each frequency needs to be determined through the use of the raw data. The corrections for the actuator response as well as the free-field response are then applied at each frequency. Lastly, the effect of atmospheric absorption at each frequency needs to be determined and applied. This was done through the use of formulas provided by Blackstock [14]. The corrected SPL values are then converted back into pressure values and integration

Figure 3.3 The microjet arrangement with respect to the primary nozzle. Primary nozzle diameter = 50.8 mm; and microjet nozzle diameter = 400 μm.

is then performed over the corrected spectrum. The resulting squared pressure value can then be used to obtain the OASPL.

To illustrate the suppression technique on a full-scale engine, the frequencies are divided by a factor of 12 (assuming the full-scale nozzle diameter = 0.6 m) and then the frequency spectrum is converted into a discrete one-third-octave spectrum. In order to do this, one must first determine the center frequency of each octave band and then its lower and upper limits. The corrected pressure spectrum is then integrated over these limits to determine the SPL value at the respective center frequency. A-weighting is then applied to the entire spectrum to properly reflect subjective judgments of the noise as commonly used in the literature.

For the laboratory-scale experiments reported here, a converging axisymmetric nozzle with an exit diameter of 50.8 mm was used. The micronozzles were made of 400-micrometer stainless steel tubing. The underexpanded microjets impinge on the shear layer at 6 mm downstream of the nozzle exit. The angle of the microjets with respect to the upstream jet centerline was 60°. Typically, eight microjets are used, but experiments were also carried out with four and sixteen microjets. The microjet arrangement is shown in Fig. 3.3. The stagnation pressure of the microjets was varied from 300 to 700 psia. At 500 psia, the fully expanded Mach number of the microjet is about 2.91 and the corresponding velocity and Reynolds number of the microjet are 618 m/s and about 8.7×10^3, respectively.

The jet exhausted into a quiet surrounding at ambient conditions. The stagnation temperature of the primary jet was kept at 1033 K. The fully expanded jet velocity for the condition of $M_j = 1.38$ was 760 m/s. The corresponding Reynolds number, based on the nozzle exit diameter, was 4.8×10^5.

3.3   RESULTS AND DISCUSSION

Figure 3.4 shows shadowgraph pictures of the primary jet, with and without the microjet injection, taken using a high-speed digital camera operating at 1000 frames per second with an exposure time of 28 μs. The fully expanded Mach number of the primary jet is 1.38. As expected, a bow shock in front of the microjet appears prominently in the picture on the right. A faint but noticeable trace of the microjet is also visible in the picture. The penetration distance of the microjet is known to be primarily a function of the momentum ratio [15]. For the conditions of the present experiment with $M_j = 1.38$, the correlation [14] gives a value of about 5 mm, which is close ($\sim$ 6 mm) to the measurement from the shadowgraph visualization. The total mass flow rate of the microjets at 500 psia is about 1.1% of the primary jet mass flux.

Figure 3.5 shows pictures of the $M_j = 1.38$ jet, with and without the microjets, taken using an infrared camera. Due to the underexpanded nature of the jet, several shock cells are seen in the picture. The observed effect of the microjets is to reduce the shock-cell length and the corresponding temperature distribution as shown in Fig. 3.6.

A prominent characteristic of the signal at the peak radiation angle is the random occurrence of distinct bursts of strong narrow positive pressure transients as shown in Fig. 3.7. Such a signal is representative of a crackling jet, first recognized and described in [8]. Crackle levels tend to peak near the eddy Mach angle given by the following equation [5]:

$$\sin \alpha = \frac{1}{M_c} ; \quad M_c = \frac{M_j + 1}{a_\infty / a_j + 1} \tag{3.2}$$

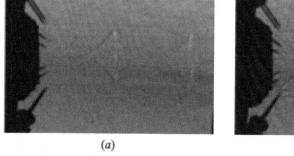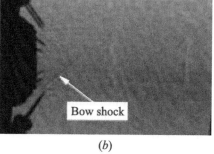

(a)   (b)

Figure 3.4 Shadowgraphs of the jet with and without microjet injection: (a) normal jet and (b) air injection at 500 psia.

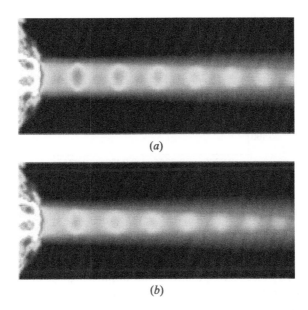

Figure 3.5 Infrared pictures of the $M_j = 1.38$ jet: (a) normal jet and (b) air injection at 500 psia.

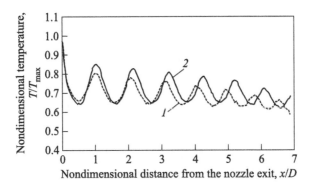

Figure 3.6 Nondimensional jet centerline temperature with (1 — air injection at 500 psia) and without (2 — normal jet) microjet injection.

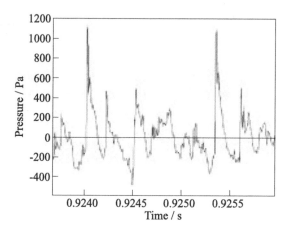

Figure 3.7 Typical far-field microphone crackle signal.

where M_j and a_j are the fully expanded Mach number and speed of sound of the primary jet, respectively; a_∞ is the ambient speed of sound; and α is measured clockwise from the jet axis. The estimated peak radiation angle, θ (the complement of α), using Eq. (3.2) is 139°, which is in agreement with the present results within the experimental uncertainty.

As suggested in [8], the skewness factor of the recorded signal is an effective direct measure of crackle. Keeping this in mind, the skewness of the far-field microphone signal at different angles with and without microjet injection is shown in Fig. 3.8. The skewness for jets peaks near the peak radiation angle when the eddy Mach number, M_c, is greater than 1.2. Away from the peak radiation angle, the skewness drops rapidly, indicating the absence of crackle. The skewness values, and their dependence on the angular position found in the present experiment, are in general agreement with those of [8]. In support of the earlier observation, with respect to the pressure time signals, the crackle suppression in the aft quadrant of the jet exhaust is quite evident by skewness reductions. The energy contained in the spikes responsible for crackle is about 30% of the total energy [7, 10], as such it is expected that the OASPL in the peak radiation angle direction be reduced by at least 1.5 dB.

Figure 3.9 shows the variation of the OASPL with θ for the normal jet and with microjet injection. Significant reductions in the aft quadrant are observed. At the peak radiation angle, the reduction amounts to about 4.5 dB. Since crackle is known to contribute only about 1.5 dB, the additional reductions in the OASPL can be attributed to suppression of part of the unsteady flow generated by the large structures, commonly referred to as large-scale mixing noise. Measurements taken at $M_j = 0.9$, discussed later, suggest that the

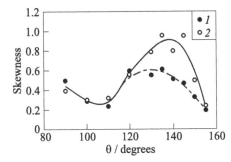 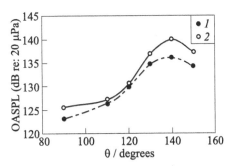

Figure 3.8 The directional distribution of the skewness factor with (1 — air injector at 500 psia) and without (2 — normal jet) microjet injection; $M_j = 1.38$.

Figure 3.9 Far-field directivity of $M_j = 1.38$ with (1 — air injection at 500 psia) and without (2 — normal jet) microjet injection. Nozzle temperature ratio = 3.

large-scale mixing noise is also suppressed to some extent by the microjet injection.

Spectral measures of the microphone signals show the frequencies affected by the microjet injection and thereby lead to some observations about the influenced source types. The spectrum of a sharp-edged spike shown in Fig. 3.7 is flat and evenly distributed over a wide range of frequencies. Hence, the crackle suppression should lead to reductions in a wide band of frequencies in the spectrum. Figure 3.10a shows the narrowband ($\Delta f = 61$ Hz) frequency spectra corresponding to the maximum radiation direction. Although, the computed spectrum extends up to 100 kHz, for compactness only the data up to 60 kHz is shown here. Screech tones that are commonly present in underexpanded jets are clearly seen in the spectrum, with a fundamental tone being at 3662 Hz. The corresponding Strouhal number (St $= fD_j/U_j$) of 0.26 is in agreement with previous measurements reported in the literature [16]. The suppression of screech tones due to microjet injection is clearly evident. But more importantly, the SPL reductions are observed in the entire spectrum. From these observations, it is evident that the microjet injection influences the crackle and the large-scale mixing noise. The source of low-amplitude undulations seen in the spectrum is currently being investigated, but they seem to have little influence on the OASPL based on the agreement of the present data with that of Tanna *et al.* [17].

Using the human response to noise and a scaling factor of 12 to account for the full-scale nozzle, the narrowband spectra are recalculated to yield A-weighted spectra. Shown in this fashion in Fig. 3.10b, at the peak radiation angle, the SPL reductions are quite significant (almost 10 dBA) except at very

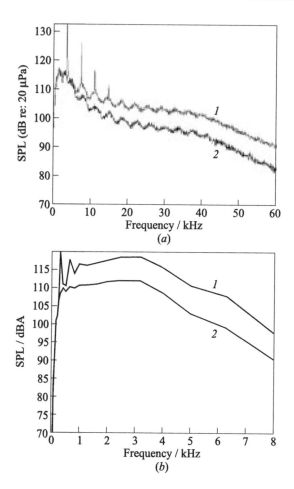

Figure 3.10 (*a*) Narrowband frequency spectra at the maximum radiation direction; and (*b*) A-weighted frequency spectra extrapolated to full scale: *1* — without microjet injection and *2* — with microjet injection.

low frequencies (less than 500 Hz). The corresponding reduction in the OASPL is about 6 dBA.

An experiment was undertaken to determine if jet noise could be affected by influencing the flow inside the nozzle. A special converging axisymmetric sonic nozzle (Fig. 3.11) with an exit diameter of 50.8 mm was built in order to study these effects.

There are four actuator ports located on the nozzle with a pressure port before and after each actuator port. The jet was operated at the same conditions

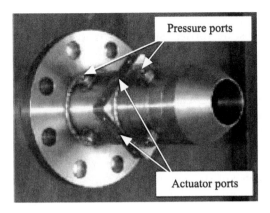

Figure 3.11 Active noise control nozzle.

as the previously mentioned experiments (nozzle pressure ratio NPR $= 3$ and $T_0 = 1033$ K) in order to minimize any variables.

The actuators that were used in this experiment were high-pressure jets with an exit diameter of 2.3 mm. The pressure of the jets was varied from 300 to 700 psia. The jets were operated in two modes: continuous and pulsed. In the pulsed mode, the jets were operated, in phase, at 2000 Hz using a proprietary high-speed valve. A Kulite high-temperature pressure transducer was used in the pressure port following one of the actuators to detect any change in the pressure inside the nozzle. An arc of eight microphones at a distance of 65 nozzle diameters was used to measure the far-field noise. One microphone was placed five diameters downstream of the nozzle exit at a distance of 20 diameters from the jet centerline.

The results from this experiment showed little or no effect on the noise created by the jet. However, data analysis is continuing in an effort to determine if there is any possible coherence between the actuator, the near-field microphone, and the far-field microphones.

3.4   CONCLUDING REMARKS

A novel, high-speed, jet noise suppression technique using high-pressure gas microjet injection at the nozzle exit was developed with promising results using the laboratory-scale jet. The main jet parameters, the nozzle pressure ratio, and the temperature ratio are chosen to correspond with realistic engine operating conditions. Keeping in mind the applicability of the technique to full-scale engines, the microjet mass flow was kept at less than 2% of the primary jet mass flow.

One of the dominant sources of noise from a high-speed jet is intimately related to the large-scale vortical structures that convect at supersonic speeds relative to the ambient medium. The dominant noise from this source consists of two parts: the Mach-wave radiation and a contribution due to the unsteady flow associated with eddying motions whose scale is commensurate with the local shear-layer thickness. The noise radiation is most intense in the direction given by a uniquely defined convective Mach number given by Oertel and Patz [5]. The Mach-wave radiation that is quite distinguishable is characterized by strong narrow pressure transients in the far-field microphone signal, referred to as "crackle," first brought to notice by Ffowcs Williams [8]. It is quantified by the skewness factor of the pressure time signal that does not depend on the scale of the jet. The microjet injection scheme used here significantly reduces its level, resulting in about a 1.5-decibel reduction in the far-field OASPL. Additionally, the microjets interfere with the unsteady vortical flow due to large eddies so as to reduce the mixing noise by about 3 dB. The A-weighted spectra that incorporate the full-scale nozzle diameter show total OASPL reductions of about 6 dBA in the peak radiation angle. The microjets are effective probably because they inhibit the formation of the large eddies, the consequence of basic jet instabilities, which are responsible for the crackle and the low-frequency component of the mixing noise. When the jet is imperfectly expanded, the commonly observed screech tones are completely suppressed by the microjet injection.

Experiments using high-speed actuators affecting the flow inside the nozzle provided little, if any, noise reduction. However, continuing research in this area would be necessary before drawing any definitive conclusion.

ACKNOWLEDGMENTS

The work was performed under ONR contract N00014-01-1-0396. The authors would like to thank the NASA Ames Research Center (Technical Monitor: Dr. James C. Ross) for partially funding this work. Thanks are also due to Mr. Thomas Joseph, Mr. Robert Avant, and Mr. Bobby Depriest for their assistance in operating the facility and the fabrication of the nozzles including the microjets.

REFERENCES

1. Crighton, D. G. 1977. Orderly structure as a source of jet exhaust noise: Survey lecture. In: *Structure and mechanisms of turbulence II.* Ed. H. Fiedler. Lecture notes in physics. Berlin: Springer-Verlag 76:154–70.

2. Phillips, O. M. 1960. On the generation of sound by supersonic turbulent shear layers. *J. Fluid Mechanics* 9:1–28.

3. Ffowcs Williams, J. F., and G. Maidanik. 1965. The Mach wave field radiated by supersonic turbulent shear flows. *J. Fluid Mechanics* 21:641–57.

4. Ffowcs Williams, J. E. 1965. On the development of Mach waves radiated by small disturbances. *J. Fluid Mechanics* 22:49–55.

5. Oertel, H., and G. Patz. 1981. Wirkung von unterschallmanteln auf die Machwellen in der umgebung von Uberschallstrahlen. ISL Report RT 505/81. Institut Franco-Allemand De Recherches, Saint-Louis, France.

6. Papamoschou, D. 1997. Mach wave elimination from supersonic jets. *AIAA J.* 35:1604–11.

7. Laufer, J., R. Schlinker, and R. E. Kaplan. 1976. Experiments on supersonic jet noise. *AIAA J.* 14(4):489–504.

8. Ffowcs Williams, J. E., J. Simson, and V. J. Virchis. 1975. Crackle: An annoying component of jet noise. *J. Fluid Mechanics* 71:251–71.

9. Oertel, H., F. Gatau, and A. George. 1981. Dynamik der Machwellen in der Umgebung von Uberschallstrahlen. ISL Report R 124/81. Institut Franco-Allemand De Recherches, Saint-Louis, France.

10. Krothapalli, A., L. Venkatakrishnan, and L. Lourenco. 2000. Crackle: A dominant component of supersonic jet mixing noise. AIAA Paper No. 2000-2024.

11. Tam, C. K. W., M. Golebiowski, and J. M. Seiner. 1996. On the two components of turbulent mixing noise from supersonic jets. AIAA Paper No. 96-1176.

12. Tam, C. K. W. 1998. Jet noise: Since 1952. *Theoret. Comput. Fluid Dynamics* 10:393–406.

13. Arakeri, V. H., A. Krothapalli, V. Siddavaram, M. Alkislar, and L. Lourenco. 2003. On the use of microjets to suppress turbulence in a Mach 0.9 axisymmetric jet. *J. Fluid Mechanics* 490:75–98.

14. Blackstock, D. T. 2000. *Fundamentals of physical acoustics.* New York: John Wiley & Sons Inc.

15. Papamoschou, D., and D. G. Hubbard. 1993. Visual observations of supersonic transverse jets. *Experiments Fluids* 14:468–76.

16. Krothapalli, A., and P. J. Strykowski. 1996. Revisiting screech tones: Effects of temperature. AIAA Paper No. 96-0644.

17. Tanna, H. K., P. D. Dean, and R. H. Burrin. 1976. The generation and radiation of supersonic jet noise. AFAPL-TR-76-65-Vol III. Dayton, OH: Air Force Aero-Propulsion Laboratory, Wright-Patterson Air Force Base.

Comments

Smirnov: Physically, turbulent kinetic energy should increase first in the near field of the microjet.

Krothapalli: It does.

Mashayek: What is the mechanism for the effect of H_2O injection?

Krothapalli: It is explained in our article in *Journal of Fluid Mechanics*.

Chapter 4

ACOUSTIC TEST FLIGHT RESULTS WITH PREDICTION FOR MILITARY AIRCRAFT DURING FCLP MISSION

J. M. Seiner, L. Ukeiley, and B. J. Jansen

A new program has been initiated, the purpose of which is to develop performance-efficient, engine-noise-suppression devices to reduce community environmental noise impact associated with military aircraft performing Field/Carrier Landing Practice (FCLP). The program involves development of concepts utilizing one-tenth-scale model nozzles operating at realistic engine operating points on the FCLP. In addition, this program will acquire baseline narrowband acoustic data for the unsuppressed aircraft to validate systems noise prediction. After selection of the most promising performance-efficient suppression concept, evaluation will be conducted on an engine test stand with a full-scale engine.

4.1 INTRODUCTION

With the advent of base closings and the ever-expanding civilian communities near military airfields, an urgent need exists to apply low, performance-impact, jet-noise reduction technology to minimize environmental noise exposure. In particular, those aircraft engaged in routine FCLP are liable to produce significant noise exposure during their mission due to the requirement of high landing speed and engine power setting. A successful program requires the coordinated experimental/numerical development of suppression concepts using small-scale laboratory models, evaluation of performance impact on propulsion test stands, and final acoustic suppression evaluation with flight tests. Moreover, to properly address the environmental impact of the candidate noise-suppression system, the use of a system-level code to predict ground contours of a simulated aircraft flyover is required. In addition to the experimental development, numerical simulations are required to project results from model to full scale, especially those associated with forward flight.

A collaborative effort between The University of Mississippi (Seiner), Florida State University (Krothapalli), and Combustion Research and Flow Technology — CRAFT (Dash) has recently been initiated that includes all phases that are required for completion of a successful noise-reduction program. To enable projection of model-scale laboratory acoustic data to full scale, one-tenth-scale models have been constructed. The primary methods being investigated for noise suppression involve the use of micro-air-jet injection, water-drop injection, and modification of nozzle divergent flaps into corrugated shapes with chevrons. All of these concepts are known to produce little impact on aeroperformance.

Since this program is in the early stages of development, this chapter will discuss recent progress associated with a planned flight test of a military aircraft to acquire narrowband acoustics*. The majority of full-scale acoustic studies have been performed to establish the environmental impact on a community by estimating ground contours to set Duration Index Levels. As a result, these studies are generally performed in $^1/_3$-octave bands.

Development of a successful noise-reduction program requires the acquisition of narrowband acoustic data using ground-based microphones. It is only with this type of data that one can identify the jet aeroacoustic sources that drive the particular method to be adopted for noise suppression. In particular, at the engine power settings and engine exhaust temperatures of the FCLP, the predominant jet-noise sources are related to highly efficient processes associated with Mach wave emission and shock noise.

This chapter describes planning for this flight test and pretest predictions obtained using the systems noise prediction code ANOPP [1]. In addition to the flight test, progress has also been made using a one-tenth laboratory-scale model. This chapter discusses preliminary results obtained with this model.

4.2   ACOUSTIC FLIGHT-TEST PREPARATION

A flight test of a military aircraft will be conducted at NAS Patuxent River. The purpose is to obtain narrowband acoustic data to identify jet-noise sources. This same data can be analyzed in $^1/_3$-octave bands and used to evaluate predicted ground-based contours.

The most difficult part of planning the flight test is associated with determining a flight-test profile that is representative of the FCLP mission profile. In order to obtain improved statistical accuracy through ensemble averaging [2], ground-based acoustic measurements are performed with a linear array of microphones, like that shown in Fig. 4.1.

---

*Since the write-up of this chapter, a flight test was conducted at NAS Patuxent River. A total of 27 flyovers were made, and the data collected are being analyzed. (*Editor's remark.*)

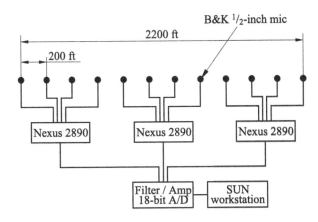

Figure 4.1 Schematic layout of linear microphone array for flight tests.

As can be observed, 12 microphone sensors are utilized that permit acquisition of 12 ensembles for averaging. For the aircraft flight speeds and altitudes associated with the FCLP, the 200-foot microphone spacing is selected to minimize the smear angle.

With this constraint, the array length is 2200 ft. To be able to utilize ensemble averaging, the engine power level is required to be maintained constant, so the jet-noise sources do not change while acquiring acoustic data. In addition, aircraft speed also needs to be held nearly constant due to the effect of flight speed on jet and shock noise. At 200 VKCAS*, the jet noise is reduced by half of that at static conditions. The only way to meet these constraints, particularly at high engine power-level settings, is to allow the aircraft to climb. With a Global Positioning System (GPS), a propagation algorithm, and Rawinsonde (atmospheric condition) data, the acoustic levels that are recorded on the ground can be corrected. In addition, for twin-engine aircraft, the aircraft needs to be flown with small slant angles since the acoustic field is three-dimensional.

The above constraints require that significant deviation occurs between the FCLP mission profile and that adopted for an acoustic study. To understand this, consider the actual flight profile for the FCLP mission shown in Fig. 4.2. At the touchdown point, the engine power is increased to military power (96% N2) and airspeed is 130 knots. Rotation occurs after a 400-foot roll with engine and airspeed maintained. After climbing to approximately 300 ft, engine power is reduced to 90% N2.

At this point, the aircraft is 1000 ft downrange with an indicated airspeed of 145 knots. The aircraft continues to climb to approximately 1000 ft, where the engine power is reduced to 86% N2 with an indicated airspeed of 130 knots.

*VKCAS — velocity in knots calibrated airspeed.

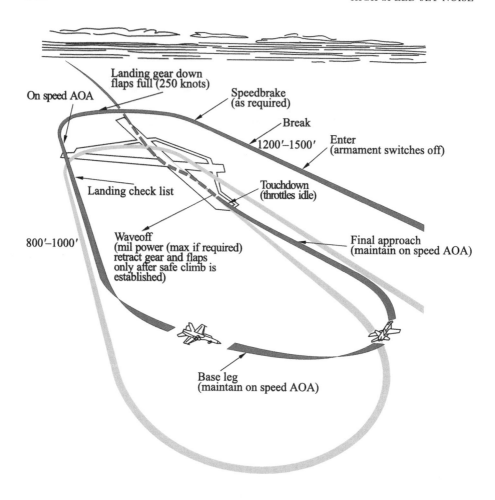

Figure 4.2 Graphic illustration of the FCLP mission.

Holding these conditions, the aircraft proceeds to a downrange distance of 1.3 nautical miles where it initiates a turn to downwind. The aircraft completes its turn, maintaining altitude and airspeed by reducing engine power to 81% N2. Halfway through the final turn, engine power is increased to 84% N2 to maintain the descent rate to 500 ft. Touchdown occurs with engine power at 84% N2 and an airspeed of 130 knots. In addition to these varying engine power settings, pilots almost always take-off from NAS Oceana to NAS Fentress with maximum power (afterburner), although while enroute at 2000 ft reduce power to 81% N2. For acoustics, roll rates are an issue for noise emitted to the ground. The use of

Table 4.1 Mission profile for acoustic study.

Test point	Airspeed (knots)	Altitude (feet)	Engine power (% N2)	Landing gear	Flap position	Profile
1	250	2000	81	Up	Auto	Enroute
2	350	800	82	Up	Auto	Break
3	150	1200	86	Down	Full	Instrument approach
4	150	820	86	Down	Full	Wyle test point
5	135	600	86	Down	Full	FCLP pattern
6	150	200–4000	91	Down	Unknown	Climb out
7	150	200–4000	96	Down	Unknown	Take-off

flaps and landing gear during the mission are not expected to be an issue at this time.

The value of the current acoustic flight study is related to the need to validate the application of a noise systems prediction code, like ANOPP, for selected points of the FCLP mission profile. ANOPP can later be used to predict noise on the ground for the exact FCLP mission profile. Further, ANOPP can be used to evaluate the relative contributions between jet and shock noise with model test data.

With these considerations, the flight-test matrix of Table 4.1 was developed in conjunction with input from test pilots and propulsion engineers from NAS Patuxent River. Seven test points were selected covering all important engine power settings, altitudes, and airspeeds for the FCLP. All test points are flown at constant power and airspeed over the acoustic array of Fig. 4.1. To maintain airspeed at the engine power settings for test points 6 and 7, the aircraft will climb from an initial altitude of 200 ft to approximately 4000 ft. Corrections to received acoustic levels at various altitudes for these test points will be made using standard noise-propagation algorithms.

4.3 SYSTEMS NOISE PREDICTION OF FLIGHT-TEST POINTS

After selection of the mission profile for the acoustic test was determined, NAS Patuxent River's Propulsion Department utilized their flight simulator and cycle deck for the engine to determine both aircraft and engine parameters needed to compute ground-based acoustic signatures. Table 4.2 lists all parameters needed to exercise the NASA LaRC systems noise prediction code ANOPP. In Table 4.2,

Table 4.2 Engine/aircraft parameters for system noise prediction.

A_j, sq ft	T_j, °R	V_j, fps	RHO_j, Slugs/cu ft	A_9, sq ft	M_j	M_d	γ	V_f, fps	AOA, degrees	Altitude, ft
1.4750	1274	1675	1.0909	1.7590	1.064	1.510	1.376	422.3	4	2000
1.4868	1295	1751	1.1407	1.7507	1.111	1.496	1.375	591.2	2	800
1.5896	1461	2034	1.0412	1.7750	1.244	1.458	1.365	253.4	4	1200
1.5868	1462	2032	1.0536	1.7757	1.242	1.459	1.365	253.4	5	820
1.5833	1463	2025	1.0599	1.7764	1.236	1.460	1.365	228.0	11	600
1.8194	1647	2375	1.0163	1.9986	1.411	1.544	1.358	253.4	N/A	200
1.8208	1641	2376	0.9853	2.0028	1.415	1.549	1.358	253.4	N/A	1200
1.9924	1841	2687	0.9573	2.3118	1.552	1.731	1.354	253.4	N/A	200
1.9924	1836	2890	0.9293	2.3153	1.557	1.734	1.354	253.4	N/A	1200

A_j is the fully expanded jet area, T_j is the jet total temperature, V_j is the fully expanded jet velocity, RHO_j is the fully expanded jet density, A_9 is the nozzle exit area, M_j is the fully expanded jet Mach number, M_d is the nozzle design Mach number, γ is the ratio of specific heats at the nozzle exit, V_f is the initial and final aircraft velocity, AOA is the aircraft angle of attack, and Altitude is aircraft height above ground.

In the process of applying ANOPP to the flight-test points, Rawinsonde data from NAS Patuxent River for June 17, 2002, was utilized to exercise the propagation algorithm as a representative atmosphere for pretest predictions. For test points 6 and 7, the flight simulator has not yet determined the aircraft angle of attack. For the present calculations, it was assumed that AOA = 10°. Excess ground attenuation corrections were applied for a hard concrete surface like that to be used during the flight test. Since the tested aircraft has a dual-podded nacelle, the twin-engine option in ANOPP was turned on. For these calculations, the SAE jet-noise prediction module was selected and the TAM Shock module was utilized to predict shock noise. This was necessary, since, as can be observed from Table 4.2, all flight-test points of the FCLP occur with the engines operating significantly overexpanded. The perceived noise level, which is used to determine EPNdB*, was not computed since it does not apply to military aircraft.

Figure 4.3a shows the fly-by predicted A-weighted overall sound pressure level (OASPL) in dBA for all test points of Table 4.2 in terms of observer time, whereas Fig. 4.3b shows the same predicted levels unweighted in dB. The angles of emission cover the range from 12.5° to 167.5°. These pretest predictions are used to determine required microphone gains and required recording time for the fly-by event.

---

*EPNdB — effective perceived noise in decibels.

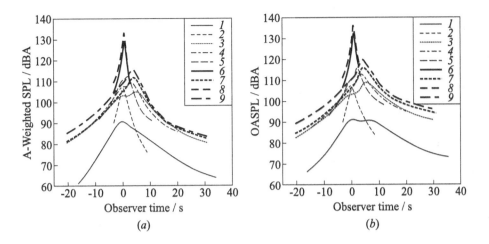

Figure 4.3 ANOPP noise predictions for all flight-test points: (a) predicted A-weighted levels and (b) predicted OASPL. _1_ — test point 1 (TP); _2_ — TP2; _3_ — TP3; _4_ — TP4; _5_ — TP5; _6_ — TP6A; _7_ — TP6B; _8_ — TP7A; and _9_ — TP7B.

The most common engine power settings for the flight-test mission are 81%, 86%, 91%, and 96% N2, which involve test points 1, 4, 6, and 7. One can use these to show representative spectra computed from ANOPP for emission angles near 45°, 90°, and 150° to the engine inlet axis. Figures 4.4a through 4.4d show the computed spectra. For test points 6 and 7, AGL = 200 ft.* was selected as representative. After validation of the systems noise prediction code, it is possible to compute noise contours on the ground for missions that consist of variable engine power settings and flight speeds.

It is clear from the predicted spectra that shock noise is present and makes a significant contribution to the predicted overall sound pressure level. However, near Mil-Power (96% N2), the engine nozzle scheduling is more on-design and the relative shock-noise contribution is less compared to jet noise than can be observed with lower engine power settings.

4.4 MODEL-SCALE DEVELOPMENTS

The model-scale laboratory investigations at the National Center for Physical Acoustics (NCPA) at the University of Mississippi will utilize one-tenth-scale geometry. Figure 4.5 shows a layout of the design presently under construction. Several key design parameters are included in the design that affect noise

---

*AGL — above ground level.

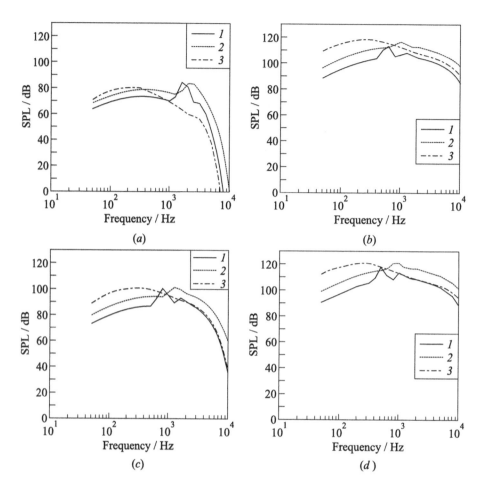

Figure 4.4 Representative predicted spectra for flight test: (*a*) TP1: *1* — $\psi = 45.0°$, *2* — 87.4°, and *3* — 150.2°; (*b*) TP4: *1* — $\psi = 44.7°$, *2* — 89.9°, and *3* — 151.8°; (*c*) TP6A: *1* — $\psi = 44.4°$, *2* — 90.5°, and *3* — 149.1°; and (*d*) TP7A: *1* — $\psi = 44.7°$, *2* — 89.9°, and *3* — 151.8°.

generation. The model geometry will contain a centerbody that geometrically is similar to that in the actual engine. For commercial aircraft engines, the maximum afterturbine Mach number is close to 0.25, whereas for the military engine it is closer to 0.5 due to the centerbody spool design. This means that one expects a greater contribution to occur from internal noise with this engine. The centerbody is a realistic location to inject water droplets in an engine configuration. The Build 2 Jet Rig at the NCPA is being constructed with a centerbody that permits water injection. The internal geometry utilizes exact

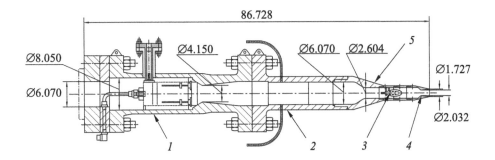

Figure 4.5 NCPA Build 2 Jet Rig with one-tenth-scale aircraft engine model geometry: *1* — propane combustor, *2* — flow straightener and seeder, *3* — transition, *4* — centerbody section, and *5* — nozzle.

replicated distances of the centerbody to nozzle throat and the same internal flap lengths, angles, and throat radius of curvature. The outer divergent flap length and boatail curvature have been reproduced along with the nozzle trailing edge thickness. All of these details are known to affect the radiated noise.

The NCPA Build 2 Jet Rig, when utilized in the NCPA anechoic room, permits operation at full-scale pressures and temperatures associated with the FCLP, has exact-scaled internal and external geometry, and can be used to acquire far-field acoustic data along with flow-field data acquired by stereo Particle

Figure 4.6 Mil-Power nozzle mounted on NCPA Build 1 Jet Rig.

Image Velocimetry (PIV). The exact representation of external nozzle geometry is also important for later when a twin nozzle installation will be constructed. At this time, a Mil-Power model nozzle has been constructed and tested for acoustics using the Build 1 Jet Rig. This Jet Rig contains a propane combustor, but can only operate to jet total temperatures of 960 °R. For the current test sequence, the Mil-Power model was operated unheated. Figure 4.6 shows the model installed in the NCPA Anechoic Jet Facility. Figure 4.7 shows representative spectra obtained at 45° and 150° to the jet inlet axis. The measured acoustic spectra are narrowband, whereas the ANOPP predictions are in $^1/_3$-octave band. In order to compare the model spectra to full-scale predicted spectra, it is required, first, to convert the model data to lossless medium and extrapolate the measured results using spherical spreading to a radius of 1 m. Next, the scale factor needs to be applied to the model data followed by correction for forward motion effects of jet and shock noise. At this point, one then can compute $^1/_3$-octave spectra from the model data and then propagate the sound using atmospheric conditions of the flight test and account for excess ground attenuation typical of the flight test. This exercise will be performed once data are obtained with the correct jet total temperature with the Build 2 Jet Rig. Nevertheless, the current data contains remarkable similarities to the flight-test predictions.

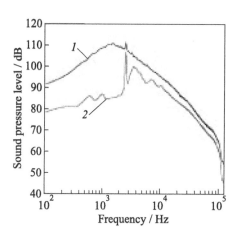

Figure 4.7 Representative model acoustic data for unheated Mil-Power nozzle: *1* — 150° to the jet inlet axis and *2* — 45°.

4.5 BLUEBELL NOZZLE APPLICATION

Several years ago, Seiner and Gilinsky [3] introduced nozzle geometry for noise suppression they termed a Bluebell nozzle. Such examples are shown in Fig. 4.8. These nozzles contain corrugations and trailing edge chevrons. The corrugations produce additional thrust due to an increase in lateral area, and combined with the chevrons they lead to noise suppression as shown in Fig. 4.9.

Application of the Bluebell nozzle geometry to the military aircraft engine requires the research to be accomplished in definite steps. The first embodiment

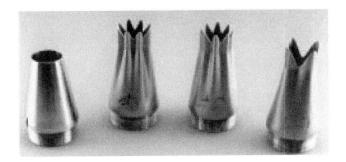

Figure 4.8 Sample Bluebell nozzles compared to reference baseline round convergent-divergent nozzle.

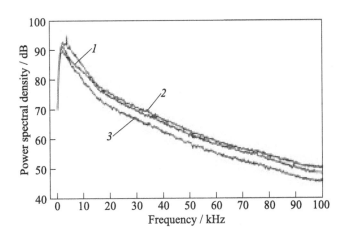

Figure 4.9 Noise suppression of Bluebell nozzle compared to baseline and flexible wire ($M_j = 1.559$; $T_0 = 350$ °F; $\psi = 145.0°$): 1 — Mach 1.5 round with Kelvar material; 2 — Mach 1.5 round; and 3 — 4 chevrons (corrugated).

Figure 4.10 Surface geometry of chevron adaptation for CRAFT CFD application.

of the concept will be to design a corrugated internal divergent flap with an outer divergent flap containing trailing edge chevrons. At this time, a model nozzle has been designed and constructed for study that will enable the evaluation of trailing edge chevrons on the outer divergent flap. Corrugations will not be present for these initial studies. The current design is shown in Fig. 4.10, and Fig. 4.11 shows the fabricated nozzle. Note, in this version, the chevrons can be replaced in any order with inserts that reconstruct the original trailing edge of the Mil-Power nozzle.

4.6 CONCLUDING REMARKS AND FUTURE PLANS

It is expected that the flight test at NAS Patuxent River for baseline narrowband acoustics will take place soon*. Four model nozzles with geometry associated with the 81%, 86%, 91%, and 96% N2 engine-power settings will be available by this time. The Bluebell nozzle with outer divergent flap chevrons will be tested on the Build 1 Jet Rig to determine the most effective chevron shape. The test will be conducted in such a fashion to ensure that a particular configuration will not produce internal nozzle-flow separation. All acoustic data, except for the Bluebell nozzle geometry, will be scaled to see if it matches full-scale data from the flight test using a procedure described in the chapter. Stereo PIV data will be collected for the suppressed and unsuppressed Mil-Power nozzles to assist

*See earlier editor's remark (p. 246).

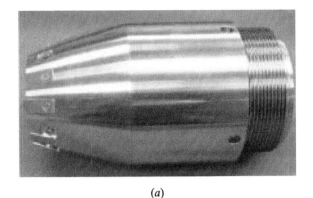

(a)

(b)

Figure 4.11 Fabricated Bluebell Mil-Power nozzle with chevron attachments: (a) profile view and (b) end view.

numerical simulations being performed by CRAFT. The centerbody that enables water-drop injection will be available later with the Build 2 Jet Rig. CRAFT will assist with numerical simulation of the effect on flow with such injection. A Bluebell nozzle with inner divergent flaps will be designed, fabricated, and tested.

ACKNOWLEDGMENTS

The work was performed under ONR contract N00014-02-1-0380.

REFERENCES

1. Zorumski, W. E. 1982. Aircraft noise program theoretical manual. Part I. NASA/TM-83199.
2. Kelly, J. J., M. R. Wilson, J. Rawls, Jr., T. D. Norum, and R. A. Golub. 1999. F-16XL and F18 high speed acoustic flight test databases. NASA/TM-1999-209529.
3. Seiner, J. M., and M. M. Gilinsky. 1995. Nozzle thrust optimization while reducing jet noise. CEAS/AIAA Paper No. 95-149.

Comments

Soto: How is weather affecting the results?

Seiner: We performed two identical runs, and data for seven points was obtained twice. Testing was stopped when the wind exceeded 12 knots.

Chapter 5

COMPUTATIONAL FLUID DYNAMICS SIMULATIONS OF SUPERSONIC JET-NOISE REDUCTION CONCEPTS

S. M. Dash, D. C. Kenzakowski, C. Kannepalli, J. D. Chenoweth, and N. Sinha

A collaborative effort with Florida State University (FSU) and the University of Mississippi (UM) to develop jet-noise suppression technologies for military aircraft is described. In the first phase of this effort, Reynolds-Averaged Navier–Stokes (RANS)-based computational fluid dynamics (CFD) is being used to support laboratory subscale experiments via establishing optimal geometries and conditions. Jet-noise suppression technologies initially being investigated include microjet injection and Bluebell tab/chevron nozzles (with various designs for divergent flap corrugations). Scale-up studies to permit such concepts to work on a military aircraft are also in progress to support tie-down tests. To date, gaseous microjet injection studies have been performed on a round Mach 1.35 imperfectly expanded jet, and baseline studies have been performed on a generic model aircraft exhaust for both static and Mach 0.5 flight conditions. Studies for this engine exhaust with chevrons have just been initiated.

5.1 INTRODUCTION

A collaborative effort (with Krothapalli/FSU and Seiner/UM) was recently initiated to develop jet-noise suppression technologies for military aircraft. The role of Combustion Research and Flow Technology, Inc. (CRAFT) in the initial phases of this program is that of utilizing CFD to support this research effort via:

(1) Pretest RANS simulations which permit selection of optimal noise reduction actuation details (e.g., location/angle of microjet nozzles, chevron/flag corrugation nozzle geometry details, etc);

259

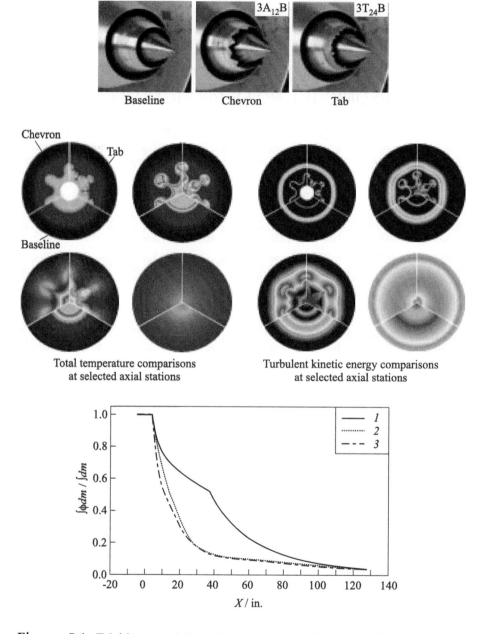

Figure 5.1 Tab/chevron mixing enhancement study for separate flow nozzles: the plot above shows the resulting integrated mixing extent of core stream, $\phi = (T_0 - T_{0,\text{fan}})/(T_{0,\text{core}} - T_{0,\text{fan}})$: *1* — baseline, *2* — chevron, and *3* — tab.

Figure 5.2 Military aircraft plume with shock train.

(2) Post-test RANS simulation to provide an enhanced understanding of flow details, to corroborate experimental findings, and to provide turbulence inputs needed to make noise predictions; and

(3) Scale-up studies which permit transitioning the laboratory isolated jet studies to a full-scale aircraft.

Concepts initially under investigation include those of microjet injection [1] and the use of Bluebell (corrugated) nozzles with tabs/chevrons and/or nozzle lip injection [2]. These concepts embody several types of technologies for jet-noise reduction which include those of conventional vortical mechanisms, turbulent structure alteration, and shock attenuation.

The RANS CFD methodology, such as that contained in the CRAFT CFD structured grid and CRUNCH CFD unstructured grid codes, is well established for the investigation of varied jet-noise reduction concepts. These codes contain specialized, well calibrated turbulence models for jet/plume applications [3–5] and have been used to investigate the effects of tabs and chevrons in NASA's AST program [6], as well as to analyze lobed mixer nozzle concepts in NASA's HST program [7]. Figure 5.1 (from [6]) shows some results of chevron and tab mixing enhancement studies with results corresponding favorably with NASA Glenn data. Much of the current full-scale aircraft plume simulation work has focused on infrared signature prediction for both cargo and fighter aircraft [8]. Aircraft-specific modules with integrated cycle deck data have been developed for the SPIRITS aircraft target signature model for isolated plumes. Such modules have been recently developed for the engine used on the military aircraft with a typical plume prediction shown in Fig. 5.2 [9]. Current work has focused on the development of complete aircraft models using multi-element unstructured numerics [10] with specialized plume grids as depicted in Fig. 5.3.

From a jet-noise perspective, the experience base with the aircraft is invaluable in making the scale-up from laboratory to aircraft (since realistic internal mixing details need to be addressed), and the availability of a complete aircraft capability is also felt to be important since installation effects and plume/plume interactions can have a first-order effect on jet noise [11].

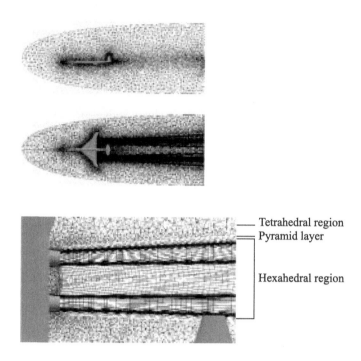

Figure 5.3 Complete aircraft/plume configuration: grid approach.

5.2   MICROJET INJECTION STUDIES

Based on discussions with Krothapalli, work was initiated on analyzing a Mach 1.35 hot laboratory jet (into still air) with eight equally spaced microjet injectors angled at 60° into the main jet. Preliminary studies evaluated the microjets on a stand-alone basis to ascertain a working range of static pressure ratios and Mach numbers. Sonic injection at very high pressure ratios was found to be problematic due to large total pressure losses across Mach discs. Next, studies were performed using the original microjet locations suggested by Krothapalli. These studies indicated that at the lowest microjet momentum conditions there was insufficient penetration, and, thus, the microjet nozzle exit was moved closer to the prime jet. A grid for the configuration analyzed is shown in Fig. 5.4 (symmetry plane through one injector and cross-cut at exit plane of microjet) — this grid provided grid resolved results and required a high concentration of points about the microjet to avoid numerical diffusion errors. Bounding limits for the microjet momentum were established for air injection, yielding minimal penetration and excessive pene-

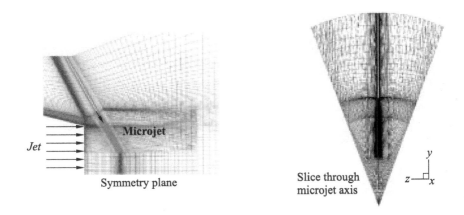

Figure 5.4 Grid and configuration for the microjet simulations.

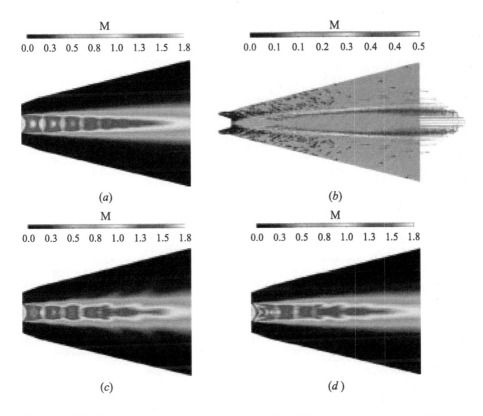

Figure 5.5 Mach number contours: (*a*) baseline, (*b*) velocity vectors, (*c*) LM case, and (*d*) HM case. (Refer color plate, p. XVIII.)

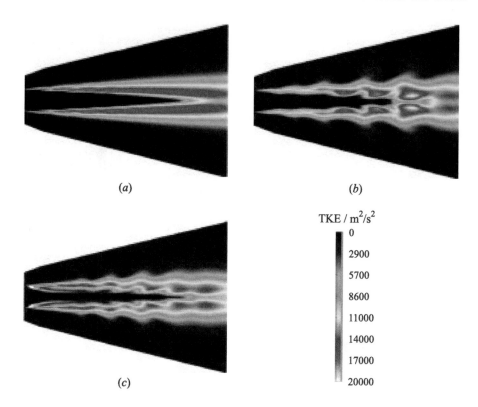

(a) (b)

(c)

TKE / m²/s²

Figure 5.6 TKE contours: (a) baseline, (b) LM case, and (c) HM case. (Refer color plate, p. XIX.)

tration (the work was conducted to establish optimal conditions with geometry fixed). Figure 5.5a shows symmetry plane Mach number contours for the baseline flow with entrainment velocity vectors superimposed in Fig. 5.5b. The corresponding Mach number contours for the low momentum (LM) and high momentum (HM) microjet solutions are shown in Figs. 5.5c and 5.5d, respectively. Turbulent kinetic energy (TKE) contours for the same solutions are shown in Figs. 5.6a–5.6c. From these contours, it is seen that the LM case may not penetrate sufficiently (or may not contain sufficient mass) to produce a significant effect, while the HM case overpenetrates forming an "inner shear layer."

From a jet-noise perspective, variations of pressure along the axis and TKE along the axis and lip line provide some insight. Pressure variations are shown in Fig. 5.7, while TKE variations are shown in Fig. 5.8. Note that microjet 1 is the LM case and microjet 2 is the HM case. Shock strengths are at-

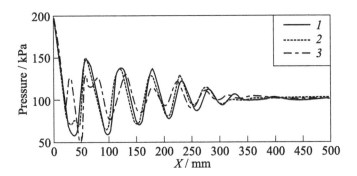

Figure 5.7 Axis pressure variations: *1* — baseline, *2* — microjet 1 (LM case), and *3* — microjet 2 (HM case).

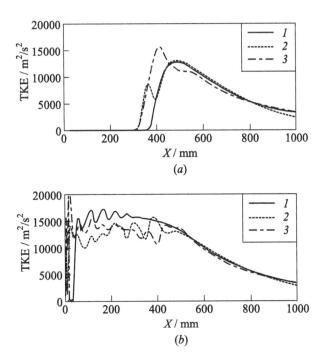

Figure 5.8 Axis (*a*) and nozzle lip line (*b*) TKE variations: *1* — baseline, *2* — microjet 1 (LM case), and *3* — microjet 2 (HM case).

tenuated significantly for the HM case, but minimally for the LM case. Lip line TKE is initially reduced for both cases, but values revert to those of the baseline case. The peak level of TKE at the axis is increased for the HM case, and both the LM and HM solutions have a shorter core length than the baseline case. Work is continuing to find optimal mass and momentum conditions for the air microjet and to examine injection at smaller angles (30° and 45°).

5.3   MILITARY AIRCRAFT MODEL STUDIES

A wind tunnel model has been designed by Seiner that replicates many geometric features of the real engine (plug, nozzle geometry for varied thrust settings, etc.). Operating conditions were representative of 96% MRT (Military Rated Thrust). Figure 5.9 shows a near-field grid of the baseline geometry, while Fig. 5.10 shows contours of Mach number and TKE for static conditions ($M_e = 0$) at sea level. Figure 5.11 shows comparable contours for a flight Mach number of 0.5, showing the effects of flight Mach number on the shock structure and turbulent mixing. Figure 5.12 compares axis variations of pressure and TKE for the static and flight calculations. The well-known effect of flight velocity is to reduce the turbulent mixing rate (velocity ratio effect) with shock strengths somewhat higher since shear layer interactions are reduced (shocks interact with thinner shear layers having lower turbulence levels). Present work is examining modifying the nozzle geometry via the addition of chevrons. Some details of the chevron geometry being analyzed are provided in Fig. 5.13.

5.4   CONCLUDING REMARKS

The studies performed and presented here are very preliminary. In the near term, it will explore the effects of varying microjet conditions (jet size, exit Mach number and pressure, and injection angle) to achieve maximum noise reduction using an overall microjet mass flux that is about 1% of the main jet. This work is being performed in collaboration with experiments at FSU [12]. In addition, it will analyze the military aircraft laboratory model with chevrons to be tested by Seiner at NCPA for static conditions and assess their effectiveness with forward flight. Finally, it will unify chevrons and microjets and obtain a noise reduction design that extracts the maximum effectiveness out of both concepts. This design will then be scaled-up for operation on the real engine (with requisite modifications to accommodate moveable nozzle features),

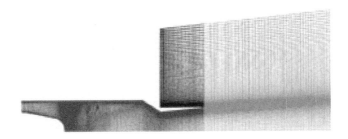

Figure 5.9 Near-field baseline grid of military engine nozzle.

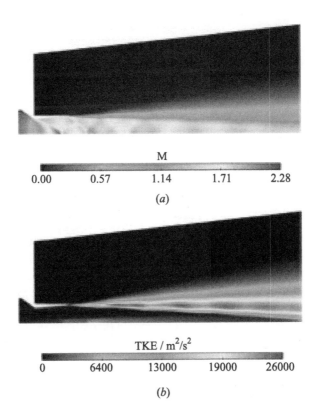

Figure 5.10 Contours for $M_e = 0$: (a) baseline Mach number and (b) baseline TKE. (Refer color plate, p. XX.)

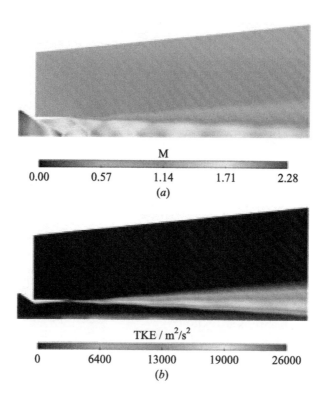

Figure 5.11 Contours for $M_e = 0.5$: (a) baseline Mach number and (b) baseline TKE. (Refer color plate, p. XX.)

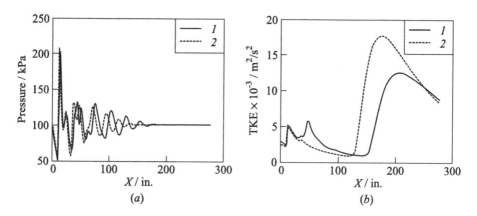

Figure 5.12 Comparison of static and flight variations of pressure and TKE along the jet axis: (a) axis pressure variation and (b) axis TKE variation. 1 — $M_e = 0.5$, and 2 — $M_e = 0.0$.

Figure 5.13 Nozzle geometry with chevrons.

and CFD studies will be performed using actual geometry and cycle information.

ACKNOWLEDGMENTS

Research was supported under ONR Contract N00014-02-1-0380 under subcontract to Florida State University.

REFERENCES

1. Krothapalli, A., B. Greska, and V. Arakeri. 2002. High-speed jet noise reduction using microjets. In: *Energy conversion propulsion: New horizons.* Ed. G. D. Roy. Washington, D.C.: Office of Naval Research. 231–44.

2. Seiner, J. M., L. S. Ukeiley, and B. J. Jansen. 2002. Acoustic test flight results with prediction for the F/A-18 E/F aircraft during FCLP mission. In: *Energy conversion propulsion: New horizons.* Ed. G. D. Roy. Washington, D.C.: Office of Naval Research. 245–58.

3. Papp, J. L., and S. M. Dash. 2001. Turbulence model unification and assessment for high-speed aeropropulsive flows. AIAA Paper No. 2001-0880.

4. Kenzakowski, D. C., J. L. Papp, and S. M. Dash. 2002. Modeling turbulence anisotropy for jet noise prediction. AIAA Paper No. 2002-0076.

5. Dash, S. M., D. C. Kenzakowski, and J. L. Papp. 2002. Progress in jet turbulence modeling for aero-acoustic applications. AIAA Paper No. 2002-2525.

6. Kenzakowski, D. C., and S. M. Dash. 2000. Study of three-stream laboratory jets with passive mixing enhancements for noise reduction. AIAA Paper No. 2000-0219.

7. Kenzakowski, D. C., and S. M. Dash. 1998. Advances in jet aircraft mixer/nozzle and plume simulation. *JANNAF 6th SPIRITS User Group Proceedings*. NASA Kennedy Space Center, FL.

8. Kenzakowski, D. C., J. D. Shipman, S. M. Dash, P. Markarian, M. Borger, and G. Smith. 2001. Increasing fidelity of aircraft plume IR signatures. *The 2001 Electromagnetic Code Consortium (EMCC) Annual Meeting Proceedings*. Kauai, Hawaii.

9. Cavallo, P. A., D. C. Kenzakowski, and S. M. Dash. 1997. Axisymmetric flowfield analysis of the F414-400 turbofan and exhaust plumes. Combustion Research and Flow Technology, Inc. Report No. CRAFTR-10/97.014.

10. Shipman, J. D., P. A. Cavallo, and A. Hosangadi. 2001. Efficient simulation of aircraft exhaust plume flows using a multi-element unstructured methodology. AIAA Paper No. 2001-0598.

11. Seiner, J. M., M. K. Ponton, B. J. Jansen, S. M. Dash, and D. C. Kenzakowski. 1997. Installation effects on high speed plume evolution. *ASME Fluids Engineering Division Summer Meeting Proceedings*. Vancouver, British Columbia, Canada. Paper No. FEDSM97-3227.

12. Krothapalli, A., B. Greska, and V. Arakeri. 2002. High speed jet noise reduction using microjets. AIAA Paper No. 2002-2450.

Comments

Smirnov: What turbulence model and boundary conditions did you use?

Dash: We used the k-ε model with correction on compressibility effects. Boundary conditions were more sophisticated than in other approaches.

SECTION THREE

PULSE DETONATION ENGINES

Chapter 1

INVESTIGATION OF SPRAY DETONATION CHARACTERISTICS USING A CONTROLLED, HOMOGENEOUSLY SEEDED, TWO-PHASE MIXTURE

B. M. Knappe and C. F. Edwards

A fundamental study of the effect of a two-phase mixture state on spray detonation characteristics has been initiated; preliminary results using hexane are presented. Emphasis has been placed on the creation of a known two-phase mixture state that is invariant along the detonation wave's axis of propagation. In so doing, the effects of heterogeneities and axial stratification of the spray on detonation behavior can be separated from those of the droplet size distribution, equivalence ratio, etc. The impact of each variable can thus be studied with all other important parameters held constant. A 5.8-meter detonation tube with a 10-centimeter inside diameter has been fitted with gas manifolds and 80 Siemens DI fuel injectors. One of the tube's key features is its modular design. Each module is 10 cm in length and has access ports for two fuel injectors, an ionization detector, and one additional instrument (e.g., a pressure transducer). The fuel injectors in each module are positioned directly opposite one another and are fired under identical conditions. The resultant momentum cancellation and vortex–vortex interaction both minimize the amount of wall wetting that takes place and promote the homogeneous distribution of droplets throughout the module and into neighboring modules. Using this opposed injection configuration and the modular tube design, a two-phase mixture for which the properties are known and do not vary significantly with position can be created. Results show that for a given liquid-based equivalence ratio, detonation velocity decreases with increasing droplet diameter (D_{32}). There is also a tendency for increased instability as larger drops are used. The two-phase detonation velocity was found to be slower than gas-phase Chapman–Jouguet (CJ) predictions for equivalence ratios greater than 0.2. This difference increased with increasing equivalence ratio and droplet diameter.

1.1   INTRODUCTION

There is currently an interest in the development of a viable Pulse Detonation Engine (PDE). This interest stems from the possibility of increased thermodynamic efficiency when combustion occurs as detonation rather than deflagration. The potential exists that PDEs will emerge as a competitive platform for propulsion in the near future.

Along with this interest in PDEs, there is growing attention being paid to spray detonations. If PDEs are to be truly viable, a high-energy-density liquid fuel (e.g., JP-10) will likely have to be utilized, as systems running on gaseous fuel are simply too large to be effective. For this to happen, spray detonations must be fully understood. The complexity that arises from the interaction between a detonation's three-dimensional structure and a two-phase mixture must be managed and controlled.

Spray detonation is a difficult phenomenon to study. One cannot easily create a two-phase mixture that is practical, well-controlled, well-characterized, and detonable. It is even less straightforward to create such a mixture on a scale large enough to allow a detonation propagating through the mixture to approach steady state. Past studies of spray detonation have exhibited only some of these features. They have typically utilized uniform arrays of impractically large droplets of fuel or uncontrolled clouds of fuel in which little beyond mean droplet size has been determined.

To achieve a solid understanding of spray detonation phenomena, however, one must have the ability to control and characterize all of the pertinent system variables. These variables must be held constant over the length of the detonation's run. Only in this way can one perform a properly designed experiment in which the effect of each variable is investigated and analyzed independent of the others.

This is the philosophy behind the set of experiments currently underway. Fuel properties, equivalence ratio, droplet size distribution, gas-phase composition, liquid–vapor fuel split, prevaporization time, and homogeneity can all be controlled independently. Testing will determine the effect that each of these variables has on detonability and detonation characteristics.

1.2   EXPERIMENTAL SETUP: TUBE SEEDING

The key to controlling two-phase mixture properties throughout an extended volume is modularity of design. A two-phase mixture can be controlled on a small scale far more readily than on a large scale. In this facility, the mixture is controlled in a number of small modules. Several modules can be combined to yield a large-scale two-phase mixture that can be homogeneous or can be tailored to give inhomogeneity in a manner chosen by the experimenter.

A homogeneous mixture is obtained within each module using an opposed injection configuration. A module is fit with two Siemens DI fuel injectors aligned directly opposite one another, as seen in Fig. 1.1.

The opposed injection configuration provides a number of benefits. There is linear momentum cancellation between the two entrained gas jets, with the stagnation plane at the center of the tube. As a result, small droplets are not thrown against the walls, but are left to disperse throughout the tube. Excessive wall wetting is avoided in this

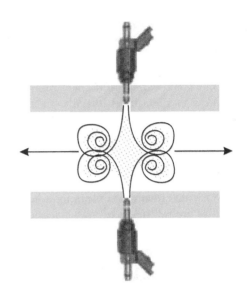

Figure 1.1 Opposed injection configuration.

manner. Ballistic droplets are not significantly affected by the cancellation, however. They penetrate the opposing jet, strike the cold tube wall, and are sequestered there. As most of these ballistics droplets are large-diameter outliers, one can consider it advantageous to have them removed from the system in this manner. Cold walls are further advantageous in that they prevent captured liquid from evaporating and creating an uncontrolled vapor-phase fuel concentration gradient in the radial direction.

Opposed injection also promotes good mixing. There is a strong interaction between the two torroidal vortices that are created in the gas phase upon injection. These vortices expand toward one another until they meet at the center of the tube. At this point, they push each other outward, along the tube's centerline, as shown in Fig. 1.1. When one of the vortex pairs encounters a similar pair from a neighboring module, the vortices are redirected toward the tube walls. Finally, upon encountering the tube walls, the remaining momentum tends to push the vortices back toward the injectors. The vortices carry entrained droplets with them as they move. Their travel throughout the cell leads to a homogeneous distribution of droplets in the gas phase.

This process is completed in a time scale on the order of 100 ms. Homogeneity is therefore achieved before there is significant separation of small and large particles due to settling. With this in mind, detonation is initiated 100 ms after fuel injection. At this time, the system is stationary relative to the time required

for the detonation to travel the length of the tube (approximately 2 ms). Therefore, there is no significant difference in homogeneity between the first module
encountered and the last. This represents a more or less optimal mixing time,
as longer times would result in little, if any, improvement in homogeneity while
leading to greater deposition of fuel droplets on the walls.

One would expect that there might be void spaces near the injectors, where
the linear momentum is initially highest. Initial experiments with this configuration showed that this is indeed the case. In order to counter this effect, the
experimental facility described herein has been built such that the axis connecting the two injectors is rotated 90° in each module. It is expected that this will
permit back-filling from neighboring modules to eliminate these voids. Early
results support this notion.

With this means of obtaining a homogeneous mixture in place, other spray
variables are relatively easy to control. The droplet size distribution is changed
by altering the fuel pressure — lower fuel pressure tends to give larger droplets.
The fuel-based equivalence ratio is controlled by varying the width of the injection pulse (given a known gas-phase composition). The chemical and physical
properties of the liquid phase are altered by changing fuels. So long as both injectors in a module are given the same control parameters, any of these variables
can be investigated without sacrificing mixture homogeneity.

1.3   EXPERIMENTAL SETUP: DETONATION TUBE

The detonation facility used for these experiments is a vertical 5.8-meter-long
stainless-steel tube with an inner diameter of 10.25 cm and a wall thickness
of approximately 1.9 cm. It is made up of 56 modules, 10.16 cm in height.
Each module features four tapped holes separated by 90°. Each hole can be fit
with a plug, an ionization detector, a pressure transducer, a gas inlet, or a fuel
injector.

The top 4 modules in the tube serve as the "initiation section." The bottom
12 modules provide null space that prevents a transition to detonation in the
vicinity of the bottom flange (which can lead to dangerously high local pressures).
These 16 modules contain only plugs and gas inlets. The remaining 40 modules,
in the middle of the tube, are outfitted as described in the previous section
with two Siemens DI fuel injectors (for a total of 80 injectors), a plug, and an
ionization detector. The detonation facility is shown in Fig. 1.2.

Initiation of spray detonation is achieved using an incident gas-phase detonation wave. The top section of the detonation tube (which includes the top
two-phase modules) is filled with a stoichiometric mixture of ethylene and oxygen
immediately prior to firing the fuel injectors. The fuel injectors are then fired.
After a predetermined delay period, a plasma jet in the top flange is fired, which

causes gas-phase combustion that rapidly transitions to detonation. This detonation wave meets and propagates through the two-phase region, where the overdriven velocity quickly decays to a level sustainable by the spray detonation. If a two-phase mixture is not detonable, the detonation wave velocity will continue to decay, and the detonation will fail.

The velocity of the detonation wave is measured along the detonation tube. Detonation velocity is the primary parameter according to which detonation behavior is characterized in this facility. Velocity can be measured using two different instruments — pressure transducers or ionization detectors. For reasons discussed below, ionization detectors are the primary instrument used for this measurement, with pressure transducers able to provide confirmation. Slightly modified Champion N19V surface gap spark plugs are used as ionization detectors. As the combustion wave passes each plug, current flows along the center wire. This current is noted by a logic circuit that then computes the time required for the combustion wave to

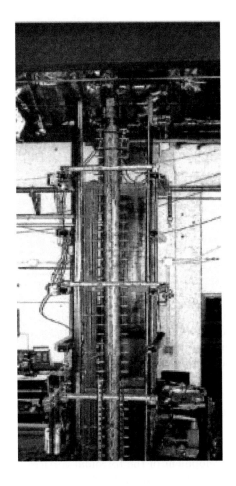

Figure 1.2 Modular detonation tube.

pass from detector to detector. With known detector spacing, this time is readily converted to a velocity. Since there is an ionization detector in each module, the average detonation velocity between each module can be determined.

1.4 RESULTS: TWO-PHASE MIXTURE HOMOGENEITY

A preliminary investigation of the homogeneity achieved using the opposed injection configuration in a tube has been performed. A more in-depth investi-

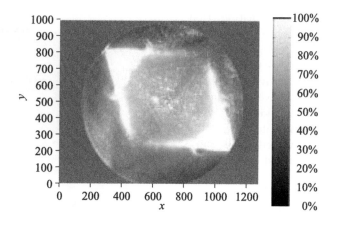

Figure 1.3 Opposed injection configuration at 2 ms.

gation is currently underway. In the original study, injectors were fired into a glass tube geometrically identical to the detonation tube described above. A 10-centimeter laser sheet (with a thickness of about 1 mm) was formed, using an Nd:YAG laser at a frequency of 532 nm, and brought through the tube along the injectors' centerline. The delay between injector firing and laser pulse illumination was varied to permit visualization of the two-phase mixture in various stages of development. Utilizing this system, the two-phase mixture's progress toward homogeneity was observed and recorded. Figure 1.3 shows the opposed injection configuration 2 ms after the injectors have fired, just before the jets intersect.

In this figure, the symmetry of the two cones is apparent, as is the early development of the torroidal vortices that are responsible for the distribution of the mixture's small droplets.

Figure 1.4 shows the opposed injection configuration 70 ms after the injectors have fired. It can be seen that the system is approaching homogeneity. However, the image certainly does not exhibit uniform intensity. There are a few reasons for this. The laser sheet used has a Gaussian intensity distribution, which tends to make the image brighter in the center of the tube. Also, there is extinguishing of the laser intensity as it passes through the spray, which tends to make the image brighter to the side on which the laser was incident. Neither of these problems was corrected for in this image. Finally, a uniform intensity is not the expected result of a homogeneous spray. A spray will always have structure, owing to its random nature. This structure is a consequence of Poisson statistics and will vary from shot to shot in a homogeneous mixture. Current efforts are focused on further characterization of the opposed injection mixture state.

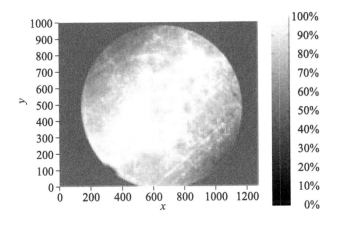

Figure 1.4 Opposed injection configuration at 70 ms.

1.5   RESULTS: TWO-PHASE DETONATION OF HEXANE

A series of two-phase detonations were performed using liquid hexane as the fuel and gaseous oxygen as the oxidizer. The experiments were performed at atmospheric pressure and room temperature. The equivalence ratio and fuel pressure (which controls mean droplet diameter) were varied, and the effect on detonation velocity was observed. The results of this series of experiments are collected in Fig. 1.5 and shown alongside gas-phase CJ velocity predictions.

For experiments in which the equivalence ratio was greater than 0.2, it was found that the measured velocity was lower than that predicted by CJ calculations. This was not unexpected — previous studies have reported velocity deficits for spray detonation. It was found that the velocity deficit increased with equivalence ratio. Liquid mass loading also increases with equivalence ratio. It is therefore possible that this effect is due to gas-phase saturation or to processes that tend to slow the evaporation rate of the droplets (e.g., cooling due to latent heat). It can be further noted that the velocity deficit increases with decreasing fuel pressure (i.e., with increasing droplet diameter). This may also be attributable to a decreased rate of evaporation. Velocity deficit and its causes will be explored in more detail in future experiments.

It should be noted that in the context of these experiments, a stable detonation is defined as one in which the detonation wave completes its run through the two-phase mixture without an obvious failure and asymptotically approaches a stationary state. Significant fluctuation about the mean velocity is to be expected (as it would be for a poorly mixed gas-phase detonation). A detonation

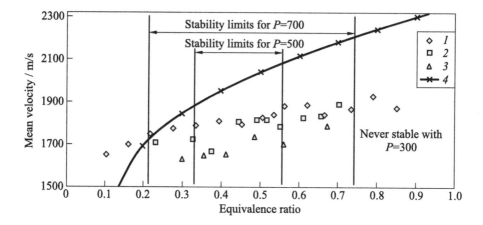

Figure 1.5 Two-phase hexane–oxygen detonation velocity and stability vs. droplet diameter and equivalence ratio: *1* — $P_{\text{fuel}} = 700$ psi; *2* — 500 psi; *3* — 300 psi; and *4* — gas-phase V_{CJ}.

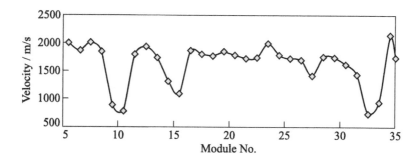

Figure 1.6 Velocity profile for a marginally stable detonation.

whose velocity decays steadily without asymptoting or drops suddenly is considered unstable.

A marginally stable detonation is one in which the detonation fails briefly one or more times, but rapidly reignites in all cases. In Fig. 1.6, one can see the velocity profile of a marginally stable or "stammering" detonation. Marginal instability is best observed using ionization detectors. When the detonation fails, the combustion wave dies rapidly, while the leading shock velocity decays more slowly. If reignition is rapid, there will not be sufficient decay in the shock wave's velocity (measured using pressure transducers) to permit the failure to be distinguished from normal signal fluctuation.

A detonation is said to be unstable if there are long failures with only brief periods of reignition or with no reignition whatsoever. A mixture that yields such detonations is essentially not detonable.

Detonation stability was found to vary with fuel pressure (droplet diameter) and equivalence ratio. As can be seen in Fig. 1.5, the range of equivalence ratios for which stable detonation was achieved grew wider as the droplet diameter decreased. For the experiments with the largest mean droplets (those performed with a fuel pressure of 300 psi = 20.7 bar), stable detonation was never obtained. On the other hand, there was a large range of stable equivalence ratios when using smaller droplets. In all cases, stability degraded as the mixture became either too lean or too rich. This is potentially explained by random droplet clustering. As a mixture becomes richer, the possibility for the formation of an excessively rich pocket increases. If such a pocket is rich enough to cause a momentary detonation failure, instability will result. Likewise in a lean mixture, the possible formation of an excessively lean pocket (with resultant instability) is increased. Mixtures with a larger droplet diameter are expected to be less homogenous (since larger particles are not as prone to be entrained by the gas flow). This would tend to aggravate the clustering problem described above and lead to decreased stability.

1.6   CONCLUDING REMARKS

A homogeneous mixture can be produced using the opposed injection configuration. By combining several modules that employ this configuration, a homogeneous two-phase mixture can be developed in an extended region. Experiments with mixtures of liquid hexane and oxygen at room temperature and atmospheric pressure show that spray detonation velocity is lower than CJ gas-phase predictions and that the deficit increases with increasing equivalence ratio and droplet diameter. Current efforts are focused on further characterization of two-phase mixture homogeneity within an opposed injection module.

ACKNOWLEDGMENTS

The work was performed under ONR contract N00014-99-1-0475.

Comment

Mashayek: From a modeling point of view, do you have phase equilibrium before detonation is initiated in your facility?

Edwards: As initiation starts 100 ms after injection, I believe the mixture will saturate.

Smirnov: I think it is still worthwhile considering nonequilibrium effect. For small droplets, a nonequilibrium model can be used to estimate vaporization.

Edwards: For the time of 100 ms, hexane will saturate independent of the evaporation rate dependence.

Shepherd: Is not the equivalence ratio of 0.2 kind of low for detonation?

Edwards: For an inhomogeneous system, how do you define detonation? If it is thermally chocked behind, I call it detonation. The forward state is not well defined*.

---

*Regarding the question of Shepherd, it is worth it to keep in mind that Edwards studies two-phase hexane–oxygen detonations. At the lean detonation limit, the equivalence ratio of a gas-phase propane–oxygen mixture is about 0.15 and benzene–oxygen is about 0.25 (see Nettleton, M. A. 1987. *Gaseous detonations*. London–New York: Chapman and Hall). In two-phase detonations, a local instantaneous equivalence ratio ahead of the lead shock will be immediately changed behind the shock due to drop motion, breakup, and evaporation. After vapor-phase ignition in proper locations, the energy release behind the detonation front will most probably proceed via diffusion combustion of microdrops. In view of it, the overall equivalence ratio required to support detonation propagation can be sufficiently low. (*Editor's remark.*)

Chapter 2

DEFLAGRATION-TO-DETONATION STUDIES FOR MULTICYCLE PDE APPLICATIONS

R. J. Santoro, S.-Y. Lee, C. Conrad, J. Brumberg, S. Saretto, S. Pal, and R. D. Woodward

A series of studies of the phenomena involved in the transition of a detonation for a geometry in which a significant area change occurs has been investigated. The application of the work is to further the understanding of the appropriate geometry to enhance establishing a detonation in the main thrust tube of a Pulse Detonation Engine (PDE) using a compact predetonator. Results for several geometric configurations have been obtained for a conical transition section in which a transition obstacle may be placed. Studies with and without an obstacle in the transition section have been conducted using both flat disks of varying blockage ratio and a conical obstacle with a 45-degree half-angle. All experiments were conducted with ethylene as the fuel. Successful detonations in the main thrust tube were achieved for most geometries at oxygen-to-nitrogen ratios less than that for air, that is, oxygen-enriched conditions. However, with careful selection of the length of the predetonator, establishment of detonations in the main thrust tube was observed when the region for which localized explosion has been observed in the predetonator is located close to the conical transition section. The use of transition obstacles was generally observed to be helpful in establishing a detonation in the main thrust tube. However, successful detonation events could be achieved without the presence of obstacles for enriched oxygen conditions. A potential explanation for the observed results is that in the region where localized explosions occur, an overdriven detonation forms which, when combined with shock reflections, generated in the conical section or enhanced by the transition obstacles, results in establishing a detonation in the main thrust tube.

2.1 INTRODUCTION

In the past decades, several studies have been devoted to determining the conditions for successful propagation of a Chapman–Jouguet (CJ) detonation wave from a small tube to either an unconfined or a confined environment. It is now commonly accepted that when the predetonator diameter is large enough to accommodate 9 to 13 detonation cells, depending on the geometry and fuel–oxidizer mixture, successful transition of the detonation occurs [1–3]*.This tube diameter is usually referred to as the critical diameter. Unfortunately, meeting this critical diameter requirement would require predetonator tubes for a practical PDE to be of a large diameter, which would, in turn, result in increased difficulty in initiating a detonation. Although this problem could be avoided by using more sensitive mixtures in the predetonator, such as fuel–oxygen mixtures, the added PDE engine complexity discourages pursuing such a solution.

During the same period, another process that leads to the formation of a detonation was the subject of several studies, namely, the deflagration-to-detonation transition (DDT). The first experiments conducted by Urtiew and Oppenheim [4] on this topic demonstrated that at the time immediately preceding the onset of a detonation wave, an "explosion in the explosion" occurs, which accelerates the flame and leads to the formation of a detonation wave. This explosion (referred to here as localized explosions) can either occur between the leading shock and the flame shock, at the flame front, at the shock front, or at the contact discontinuity formed by the coalescence of shock waves that precede the flame [4, 5]. No matter which phenomena cause the onset of a detonation, the detonation wave that emerges from the DDT process presents a high degree of overdrive. Such a highly unsteady state of a detonation wave can be exploited, along with shock-focusing obstacles, to ease the transition of the detonation from the predetonator to the main tube. The same concept can be utilized in multicycle PDE operation. Furthermore, the volume of the predetonator can be significantly reduced by employing obstacles to enhance flame acceleration in the predetonator [6–8].

2.2 EXPERIMENTAL SETUP

The detonation tube used in the current studies is composed of three sections shown in Fig. 2.1a: a 33-millimeter diameter, 1.118- or 2.032-meter-long pre-

---

*As a matter of fact, there exist explosive mixtures that require accommodating of more than 20 detonation cells across the predetonator diameter to ensure detonation transition to unconfined volume (see, e.g., Moen, I. O., A. Sulmistras, G. O. Thomas, D. Bjerketvedt, and P. A. Thibault. 1986. Influence of cellular regularity on the behaviour of gaseous detonations. In: *Dynamics of explosions*. Eds. J. R. Bowen, J.-C. Leyer, and R. I. Soloukhin. Progress in astronautics and aeronautics ser. New York, NY: AIAA Inc. 106:200–43; Lee J. H. S., and C. M. Guirao. 1982. Fuel–air explosions. *Conference (International) on Fuel–Air Explosions Proceedings*. Waterloo, Ontario: University of Waterloo Press. 1005). (*Editor's remark.*)

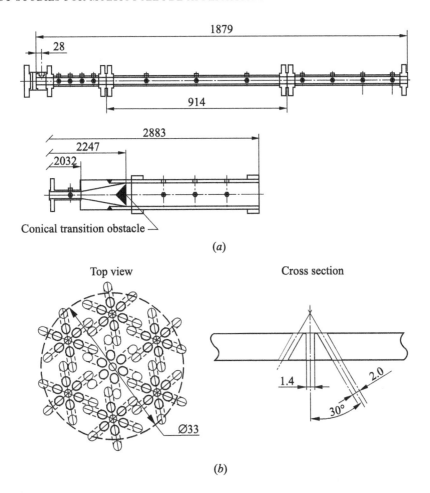

Figure 2.1 (*a*) Detonator tube including predetonator, transition section, and main detonator. (*b*) Impinging injector face. The fuel is injected along the center holes with surrounding oxidizer holes. Dimensions are in mm.

detonator; a 215-millimeter-long conical transition section with a 10-degree divergence angle (with respect to horizontal); and a 109-millimeter diameter, 636-millimeter-long thrust tube. The geometry and angle of divergence of the transition section were designed to minimize diffraction effects at the area change [9]. Ethylene–oxygen–nitrogen mixtures, with a fixed equivalence ratio of 1.1, are injected at the head-end of the predetonator through an impinging jet injector, shown in Fig. 2.1*b*. In order to confirm the target equivalence ratio, fuel concentration was monitored using the 3.39-micrometer He–Ne laser absorption measurement.

The evolution of the detonation wave is monitored by using high-speed piezo-electric pressure transducers and photodiodes mounted along the tube at identical axial positions. Throughout the experiments, the level of nitrogen dilution, i.e., the ratio of moles of nitrogen to moles of oxygen, is varied from 0 to 3.76 by keeping constant the fuel and oxygen mass flow rates and by changing the nitrogen flow rate. In order to explore the effects of shock-focusing obstacles on the developing detonation wave in and around the transition section, shock-focusing obstacle disks of 50% and 78% BR (blockage ratio) and a 78% BR-45-degree (half-angle) conical transition obstacle were used in the experiments.

2.3   RESULTS AND DISCUSSION

A summary of the experiments conducted is given in Table 2.1. Experiments where successful detonation transition was observed are marked with "S," whereas detonation failures are marked with "F." A total of 13 different configurations were investigated in the current study. These configurations can be divided into two groups. For the first group, a 1.118-meter-long predetonator was used, whereas for the second group the predetonator length was extended to 2.032 m. Within each group, several shock-focusing obstacles placed at different locations downstream from the transition exit were used. As seen from the table, detonation transition is easily obtained for the most sensitive mixtures, i.e., mixtures characterized by a low nitrogen dilution. As the nitrogen dilution is increased, the detonation wave fails to transit first for those configurations where a shock-focusing obstacle is not used and eventually for those where a shock-focusing obstacle is used as well. It is apparent that a shock-focusing obstacle enhances the chance that a detonation wave has to survive over the transition from the predetonator to the main thrust tube; however, for the 1.118-meter-long predetonator, none of the configurations explored here allowed a successful detonation transition with a nitrogen-to-oxygen dilution ratio higher than 3.0.

It is also seen from the table that, when a longer predetonator is used, the range of nitrogen dilution for which successful detonation transition occurs is extended to a nitrogen-to-oxygen ratio equivalent to air (3.76). In fact, as nitrogen dilution is increased, the onset of the detonation in the predetonator is moved further downstream toward the transition section. In the final stages of the DDT process, localized explosion occurs, and the resulting detonation wave is characterized by a high degree of overdrive. Thus, the overdriven detonation wave enters the transition section before it decays back to the CJ state. It is believed that within the transition section, further interaction of the decaying detonation with the shock-focusing obstacle generates hot spots where a localized explosion occurs. Again, this phenomenon sustains the detonation and allows it to successfully transit into the thrust tube.

Table 2.1 Summary of configurations tested

	1118-millimeter predetonator tube							2032-millimeter predetonator tube					
	78% BR flat disk						50% BR	No direct	78% BR flat disk			45-degree cone	
	Direct				With dump		Direct		Direct			Backward	Forward
N_2/O_2	I	II	III	IV	II	III	II		II	III	IV	I	I
0.00	S	S						S					S
0.75	S	S						S					S
1.50	S	S	S					S					S
2.25	S	S		S				S					S
2.40		S			F	S							S
2.50							S	F					S
2.60		S	S/F		F	S	S/F	F	S				S
2.70								S					S
2.80		S			F	F	F		S				
3.00	F	S	S/F	S/F				F	S			S/F	S
3.20		F	F	F					S			F	
3.40		F	F	F					F	S/F	F		
3.60		F	F	F					F	F	F		
3.76	F	F	F	F				F	F	F			S

S: Successful detonation; F: detonation failure.

Shock-focusing obstacles are placed at different locations downstream from the transition exit: 0 mm (I), 51 mm (II), 102 mm (III), and 152 mm (IV). Velocity is measured 407 mm downstream of the transition exit (estimated from pressure transducer measurements at the halfway point between the last two pressure transducers).

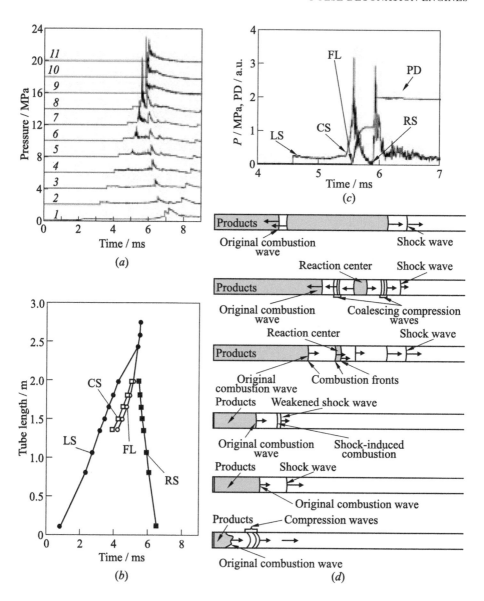

Figure 2.2 (*a*) Pressure traces up to $x = 1981$ mm in predetonator: *1* — $x =$ 115 mm, *2* — 812 mm, *3* — 1066 mm, *4* — 1346 mm, *5* — 1498 mm, *6* — 1650 mm, *7* — 1803 mm, *8* — 1981 mm, *9* — 2426 mm, *10* — 2578 mm, and *11* — $x = 2730$ mm; (*b*) space–time diagram; (*c*) sample of simultaneous pressure and flame traces at $x = 1981$ mm from different run: *LS* — lead shock, *CS* — compression shock, *FL* — flame, *RS* — reflected shock, and *PD* — photodiode; and (*d*) schematic diagram of DDT process and the photodiode trace (PD) with the flame (FL).

An example of pressure traces measured during the experiment along the tube axis is plotted in Fig. 2.2a along with the space–time diagram (Fig. 2.2b). Examples of the pressure and corresponding photodiode traces, at an axial location close to the transition section, are also shown in more detail in Fig. 2.2c. It can be seen from these traces that as the flame accelerates along the tube, compression waves generated by the dilation of the hot gases resulting from the combustion coalesce in front of the flame and generate a leading shock. While the shock propagates along the tube at a constant Mach number (approximately equal to 2), the flame accelerates and tends to catch up with the leading shock. Inside the transition section, the DDT process is finalized, and thus, an over-driven detonation wave and a corresponding retonation wave are generated. Subsequently, the detonation wave propagates as a CJ detonation down the main tube.

Figure 2.3 depicts a schematic representation of the process that is believed to be occurring in the conical transition section. The key phenomena include the propagation of overdriven detonation occurring in the end region of the predetonator which is followed by initiation of significant energy release in the volume of gases located in the transition section. The enhancement observed with the addition of the transition obstacles is postulated to be a result of shock focusing as compression waves are reflected from the obstacle and walls.

Unfortunately, the lack of any diagnostics within the transition section does not allow us to directly identify the process responsible for the successful detonation transition. Nevertheless, since for the current apparatus it was observed that the strength of a reflected shock off the shock-focusing cone is less than 10% of the strength of the incident shock, one can infer from the pressure trace recorded directly upstream from the transition section that a localized explosion has occurred in between this location and the exit of the transition section. Thus, it can be concluded that the overdriven state of the detonation is responsible for the successful transition of the detonation wave into the main thrust tube.

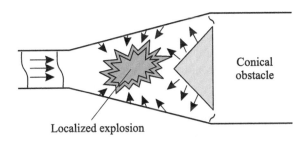

Localized explosion

Conical
obstacle

Figure 2.3 Schematic representation of the initiation of detonation transition from the predetonator to the main thrust tube.

Based on the results presented, several important effects arise for further investigation. First, the present results indicate that the conical obstacle is superior to the flat-disk obstacles of comparable blockage ratio. The explanation presented argues that shock focusing is enhanced by the conical obstacle as compared to the flat-disk obstacles. This hypothesis has yet to be proven using direct visualization of the transition section, which is planned for future investigations. If shock focusing is a key aspect of the process, then specific contouring of the transition obstacles may provide a means to further optimize the obstacles in terms of enhancing the transition from the predetonator to the main thrust tube.

Second, the volume of the transition section may be of importance in terms of the energy-release process that occurs. Only a few experiments have been done to date to investigate this effect, and the results appear to indicate that the transition section volume does have an effect as mixtures having nitrogen-to-oxygen ratios close to air are used. Finally, the extension of this work to fuels other than ethylene is essential for practical PDE operation. Preliminary studies using propane have shown that similar phenomena occur in terms of the transition mechanism. However, as one would expect, it is more difficult to successfully initiate a detonation in the main thrust tube for propane than for ethylene, particularly as the nitrogen-to-oxygen ratio approaches values close to that of air.

2.4   CONCLUDING REMARKS

A series of studies have been completed to investigate transition of a detonation from a small-diameter predetonator tube to a large-diameter main thrust tube using obstacles placed in the transition section connecting the two tubes. These results have shown that transition can occur for a predetonator tube diameter that is less than 9 to 13 detonation cell sizes that is often the criterion observed for unconfined detonation transition. A key result of the current studies is that the conditions in the region just upstream of the transition section are critical in determining whether a successful transition is observed. In particular, achieving a CJ detonation in the predetonator does not assure that a detonation will be initiated in the main thrust tube. In fact, the critical phenomenon appears to be the occurrence of localized explosions just prior to establishment of a CJ detonation. It has been postulated that the resulting overdriven detonation coupled with shock reflections from the transition obstacle that are focused in the volume of combustible gas in the transition section leads to a rapid energy release in a confined region and results in the initiation of the detonation in the main thrust tube.

ACKNOWLEDGMENTS

The work was performed under ONR contract N00014-1-0744.

REFERENCES

1. Knystautas, R., J.H. Lee, and C. Guirao. 1982. The critical tube diameter for detonation failure in hydrocarbon–air mixtures. *Combustion Flame* 46:63–83.
2. Bartlma, F., and K. Schröder. 1986. The diffraction of a plane detonation wave at a convex corner. *Combustion Flame* 66:237–48.
3. Desbordes, D. 1988. Transmission of overdriven plane detonations: Critical diameter as a function of cell regularity and size. *AIAA J.* 114:170–85.
4. Urtiew, P., and A. Oppenheim. 1966. Experimental observations of the transition to detonation in an explosive gas. *Proc. Royal Society London A* 295:13–28.
5. Sileem, A., D. Kassoy, and A. Hayashi. 1991. Thermally initiated detonation through deflagration to detonation transition. *Proc. Royal Society London A* 435:459–82.
6. Santoro, R., J. Broda, C. Conrad, R. Woodward, S. Pal, and S.-Y. Lee. 1999. Multidisciplinary study of pulse detonation engine propulsion. *JANNAF, 36th CS/PSHS/APS Joint Meetings.*
7. Lee, S.-Y., C. Conrad, J. Watts, R. Woodward, S. Pal, and R. J. Santoro. 2000. Deflagration to detonation transition using simultaneous Schlieren and OH PLIF images. AIAA Paper No. 2000-3217.
8. Santoro, R. J., C. Conrad, S.-Y. Lee, and S. Pal. 2001. Fundamental multi-cycle studies of the performance of pulse detonation engines. *15th International Symposium on Air Breathing Engines Proceedings.* Bangalore, India.
9. Nettleton, M. 1973. Shock attenuation in a gradual area expansion. *J. Fluid Mechanics* 60:209–23.

Comments

Smirnov: In the predetonator, you also have "explosion in the explosion" phenomenon and the stage of overdriven detonation propagation prior to establishment of normal detonation that enters the transition chamber. For mixtures of practical interest, the DDT distance is longer than for ethylene–air. Will you need longer predetonators?

Santoro: We use the empirical approach. For ethylene–air, we changed the tube length and found out that a 2-meter-long predetonator was required. Detonation was not observed for shorter predetonator lengths.

Frolov: We must keep in mind that prospective PDEs are supposed to operate on a liquid fuel. In view of it, obstacles inside a tube seem to be avoided as fuel sprays will impinge to their surface that may deteriorate the performance.

Santoro: We will do prevaporization.

Chapter 3

INITIATOR DIFFRACTION LIMITS IN A PULSE DETONATION ENGINE

C. M. Brophy, J. O. Sinibaldi, and D. W. Netzer

A coaxial initiator geometry, currently being used on an integrated pulse detonation engine (PDE) system, is being characterized by both experimental and computational efforts. The goal of the parallel research efforts is to develop the ability to accurately model and predict the detonation diffraction process from a small "initiator" combustor to a larger-diameter main combustor so that a detailed understanding of the mechanisms responsible for successful transmissions can be obtained. Single-shot detonation experiments have been performed on both axisymmetric and two-dimensional (2D) diffraction geometries for both ethylene and propane mixtures involving both oxygen and air as the oxidizer. Homogeneous and heterogeneous mixtures involving the use of fuel–oxygen in the smaller combustor and fuel–air elsewhere have demonstrated the increased benefit of generating an overdriven detonation condition near the diffraction plane for enhanced transmission. Mach reflections have been observed on the outer wall downstream of the diffraction plane for the 2D geometry and appear to be the primary reinitiation mechanisms for the reestablished fuel–air detonations for that geometry. A combination of spontaneous reignition and wall reflection of existing shock waves was observed to be present on the successful axisymmetric cases.

3.1 INTRODUCTION

The interest in PDEs has increased dramatically in recent years due to their high theoretical performance and wide range of potential applications [1]. Practical operation of these systems requires the use of fuels that have already gained acceptance/approval by the military and/or aviation industry, such as kerosene-based Jet-A, JP-5, or JP-10. The use of such fuels has inherent difficulties since such fuel–air mixtures are often difficult to detonate [2, 3], especially in a repetitive and reliable manner. Therefore, various research teams are currently investigating the use of an initiator, which consists of a small tube or auxiliary combustor filled with mixtures highly sensitive to detonation as the means to initiate a detonation in a larger main combustor containing a less sensitive fuel–air mixture. Thus, the importance of detonation diffraction or transmission from the small tube into a larger main combustor tube often arises.

Various initiator concepts exist and can vary from coaxial designs to transverse or splitter plate concepts, just to name a few. Most concepts operate on fuel–oxygen mixtures, while others utilize a blend of oxygen-enriched air as the oxidizer. Although the use of oxygen provides excellent reliability, repeatability, and a very rapid ignition event, the minimization of the oxygen required is of paramount importance since it is treated as "fuel" for specific impulse (I_{sp}) and specific fuel consumption (SFC) calculations and directly reduces the overall system performance. Thus, efficient coupling between an initiator and the larger combustor is of high priority.

The PDE geometry under development at the Naval Postgraduate School in Monterey, CA utilizes a continuous air flow design, which does not possess valves for the air supply to the combustion chamber. The geometry utilizes the initiator approach depicted in Fig. 3.1, where the initiator combustor operates on an oxygen-enriched air–fuel mixture to rapidly and reliably generate a detonation wave which can then be used to initiate the less sensitive fuel–air mixture located in the main combustor. The absence of a large valve on the air flow has permitted a convenient flow path to rapidly fill, detonate, and purge the combustion chamber at rates up to 100 Hz, but has also introduced difficulties into the initiation process due to lowered confinement conditions when compared to conventional PDE concepts which involve some type of valve on the air supply. A critical region of interest is at the initiator exit plane where the exiting detonation wave experiences a diffraction to the main combustor. The concern for this area of the system is the motivation for characterizing the effects of the diffraction condition between the main combustor and the initiator at the diffraction plane. The effects of diameter ratio (D/D_0), mixture variation (overdrive), and varying degrees of confinement are being evaluated so that a more optimized condition can exist.

A large body of historical work has investigated the classical case of detonation transmission from a confined tube to an unconfined volume [4] for homogeneous mixtures. The well-documented critical diameter value of 13 times the cell size (λ) of the mixture for transmission of a detonation wave to an unconfined volume has been verified many times to hold true for most mixtures. The 13λ value has been shown to be specifically valid for mixtures containing more irregular cell spacing, typically fuel–air mixtures with higher activation energies. Mixtures containing highly regular detonation cell structure, such as argon-diluted fuel–oxygen mixtures, have been shown to often require a larger

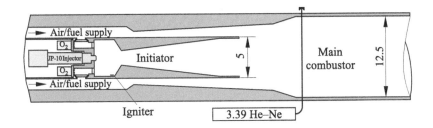

Figure 3.1 Valveless PDE configuration. Dimensions are in cm.

critical diameter than the 13λ rule, thus revealing the increased importance of wavefront structure during the diffraction processes in producing gas-dynamic hot spots for spontaneous reignition to occur. The increased irregularity in the cellular structure for fuel–air mixtures often aids in the adjustment to sudden expansion conditions and can be interpreted as possessing more levels of instability and therefore more modes by which spontaneous reinitiation may occur near a critical diameter value [5].

Teodorczyk et al. [6] and Oran et al. [7, 8] have looked at the reinitiation mechanisms of Mach reflections at a rigid wall from the propagation of a quasi-detonation in an obstacle-laden channel and an imparting spherical blast wave on a rigid wall, respectively. Both studies stress the importance of the rapid reignition sites immediately behind the generated Mach stems at the wall. Murray et al. [9] also demonstrated the importance of shock–shock and shock–wall collisions for different exit conditions at the diffraction plane, including tube bundles, annular orifices, and cylindrical diffraction*. The reinitiation mechanism associated with the Mach reflections observed in those studies is extremely important for the initiator concept utilized in the current engine. It also becomes increasingly important as the combustor diameter approaches the cell size of the mixture and few transverse waves exist to assist with adjusting to the expansion condition occurring at the diffraction plane. The reinitiation process for such conditions appears to be a very local process, and the influence of the wave front structure [10] and reflection cannot be ignored during analysis.

Desbordes [11] and Desbordes and Lannoy [12] have investigated the effects of overdriving a detonation wave during diffraction from a smaller combustion tube to a larger volume. In both studies, it was determined that a definitive benefit existed when a detonation wave was allowed to propagate into a less reactive mixture immediately before diffraction occurred, thus creating an over-driven condition in the less reactive mixture. Recently, Murray et al. [13] investigated the direct benefit of utilizing a fuel–oxygen driver section and an initiator and propagating the generated detonation wave into a fuel–air mixture in or-

---

*See also [13]. (*Editor's remark.*)

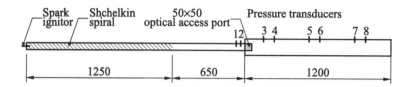

Figure 3.2 Experimental setup used for single-shot tests. Dimensions are in mm.

der to generate the overdriven condition. Overall values of the effectiveness of driver–receptor mixtures and diameter ratios approached 30 for some conditions, indicating a dramatic reduction in the required critical diameter for the receptor mixture. Thus, a combination of Mach reflections and overdriven conditions are the mechanisms which appear to dominate initiator transmissions on the scale of most PDEs and will likely be responsible for the successful application in such systems.

3.2   EXPERIMENTAL SETUP

The single-shot imaging experiments were performed on a test apparatus, schematically shown in Fig. 3.2, which allows for either circular or 2D diffraction into a larger combustor. A 5.08-centimeter inner diameter, 1.5-meter-long combustor was used as the initiating combustor for all tests. A Shchelkin spiral was installed at the head-end of the combustor to promote the rapid formation of the fuel–air detonations. For the axisymmetric case, the detonation wave exiting from the initiating combustor was then allowed to diffract into a larger circular combustor with a maximum diameter of 125 cm. The detonation wave exiting from the initiating combustor was converted into a 5.08 × 5.08 cm square wave over a 25-centimeter axial length for the 2D imaging tests. The test sections used for each series of testing provided 4.95 × 4.95 cm of optical access near the diffraction plane and are shown in Fig. 3.3. Additional diameter and channel ratios were obtained by the choice of an alternate diameter test section and insertion of channel plates.

The mixtures investigated were composed of ethylene–oxygen and ethylene–air at various equivalence ratios and were injected dynamically into the system at the head-end of the test apparatus. A 3.39-micrometer infrared (IR) He–Ne laser was transmitted across the combustor 2.54 cm upstream of the diffraction plane so that the fuel–oxygen to fuel–air interfaces could be located spatially and characterized temporally. The typical mixture transition interface between the fuel–oxygen and fuel–air mixtures was approximately 6 cm in length. All gaseous flow rates were determined from metering orifices. Detonation wave speeds were

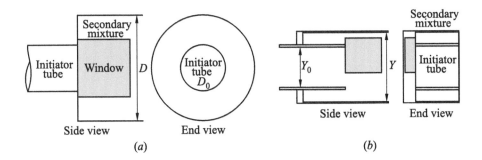

Figure 3.3 Optical access test section configurations used for imaging studies: (*a*) axisymmetric test section and (*b*) 2D test section.

monitored using Kistler 603B1 pressure transducers along the combustor axis to monitor the initial and final velocities in each combustor. If a wave speed indicative of a Chapman–Jouguet detonation was not observed within 50 cm from the diffraction plane, the test condition was considered unsuccessful.

The detonation wave diffraction at the initiator/main detonation tube was observed using high-speed Schlieren and CH^*-chemiluminescence imaging which utilized a 10-nanometer FWHM interference filter centered at 430 nm and both DRS Hadland and Princeton Instruments intensified charge-coupled device (CCD) cameras.

3.3   RESULTS

Axisymmetric Geometry

The first series of tests performed on the axisymmetric geometry ($D/D_0 = 2.5$) were for homogeneous mixtures of either C_2H_4–air or C_2H_4–O_2 at varying equivalence ratios and have been previously reported. All tests were performed at a nominal pressure and temperature of 100 kPa and 283 K, respectively. The recent diffraction tests evaluated the condition of an overdriven condition. The test matrix evaluated the transmission of selected high-energy initiator mixtures into stoichiometric fuel–air mixtures. Two diameter cases were evaluated, and the results are presented in Fig. 3.4. Time sequenced CH^* images are presented in Fig. 3.5 and reveal the relative level of heat release behind the detonation front as the wave experiences the expansion. Figure 3.5c shows a region of spontaneous rapid chemical reaction, likely due to the collision of two laterally propagating shock waves. This appeared to be a very common observation on many of these tests. The reinitiation mechanisms for the near-critical conditions appear to be

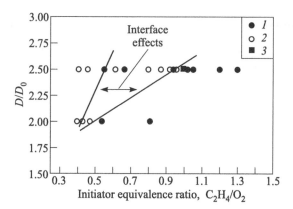

Figure 3.4 Results for axisymmetric geometry with various initiator mixtures driving a stoichiometric main combustor (C_2H_4–air, $D_0 = 5.08$ cm; initial pressure 100 kPa; and initial temperature 283 K): *1* — successful; *2* — unsuccessful; and *3* — computational result. At $\varphi = 0.4$, uncertainty is ± 0.1, and at $\varphi = 1.4$, uncertainty is ± 0.04.

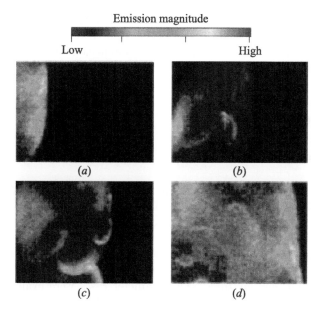

Figure 3.5 Time sequence CH^*-chemiluminescence images of a successful diffraction of a locally overdriven C_2H_4–air detonation ($\varphi = 1.0$) from a 5.08- to 12.7-centimeter circular combustor; $t = 12$ μs (*a*); 19 μs (*b*); 26 μs (*c*); and 33 μs (*d*). (Refer color plate, p. XXI.)

a combination of spontaneous reignition behind the diffracting shock front, seen in the experimental images, as well as reinitiation sites along the wall periphery due to strong Mach reflections. Previous computational results for this test condition from the Naval Research Laboratory (NRL) revealed the strong wall reinitiation sites before evidence of their contribution was observed in the experimental portion of the program, thereby directly aiding in providing a better overall view of the developing flow field. A wider range of initiator conditions is planned for the computational work.

Two-Dimensional Geometry

The 2D geometry evaluated the behavior of an overdriven detonation diffraction at a rapid expansion region, but only on one axis. The head-wall confinement present on the axisymmetric case was also removed so that the geometry was representative of the initiator in use on the current engine design. The result was a 2D test section which provided diffraction on one axis in order to provide

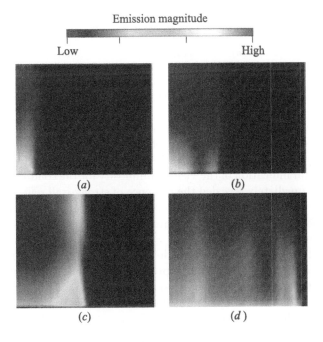

Figure 3.6 Images of the visible emission from a successful 2D diffraction of a locally overdriven C_2H_4–air ($\varphi = 1.0$) detonation in the square detonation-tube geometry ($Y/Y_0 = 1.33$); $t = 15$ μs (a); 22 μs (b); 29 μs (c); and 36 μs (d). (Refer color plate, p. XXII.)

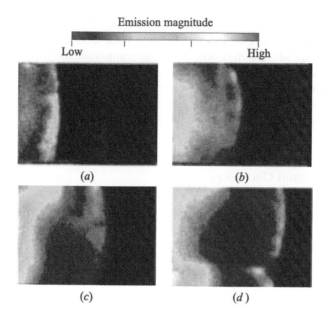

Figure 3.7 C_2H_4–air ($\varphi = 1.1$) detonation transmission in the 2D geometry ($Y/Y_0 = 1.33$); $t = 15$ μs (*a*); 22 μs (*b*); 29 μs (*c*); and 36 μs (*d*). (Refer color plate, p. XXII.)

more appropriate optical access for the Schlieren and CH* imaging. The results are for test conditions similar to the axisymmetric geometry, but with varying vertical diffraction conditions rather than the diameter ratios.

The first series of images taken were of the natural luminosity of a detonation wave as it propagated through the test-section field of view. This was done to provide a general indication of the visible luminosity for the conditions to be evaluated, yet it immediately revealed the presence of strong wall reflections immediately downstream of the diffraction plane, seen in Fig. 3.6c. The interference filter for the CH* images was then implemented, and images were obtained for three expansion conditions. Results for $Y/Y_0 = 1.33$ and 2.00 (where Y_0 is the height of the channel and Y is the height of the main channel) are shown in Figs. 3.7 and 3.8, respectively. They reveal the presence of regions of increased chemical reaction near the walls of the test section as expected. As mentioned earlier, this observation was very repeatable and is due to the strong Mach reflections resulting from the residual shock of the decaying blast wave. A schematic of what is believed to occur is shown in Fig. 3.9 and appears to agree with computational results from Oran and Boris [8] and Kailasanath and Patnaik [14]. Figure 3.7d shows the CH* emission commonly found to correspond with the near-wall Mach stem reflection and the resulting increase in local reaction rate. This behavior was normally observed at distances greater than $1.0Y_0$ from the diffraction plane.

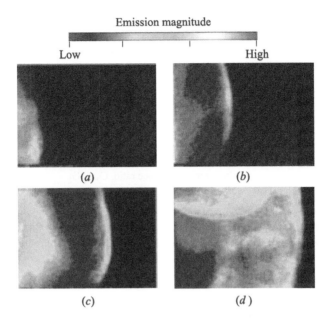

Figure 3.8 CH*-chemiluminescence images of a successful diffraction of a locally overdriven C_2H_4–air ($\varphi = 1.1$) detonation in the 2D geometry ($Y/Y_0 = 2.00$); $t = 15$ μs (a); 22 μs (b); 29 μs (c); and 36 μs (d). (Refer color plate, p. XXII.)

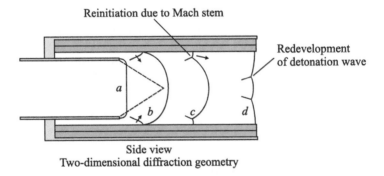

Figure 3.9 Representation of the observed detonation diffraction/reinitiation process.

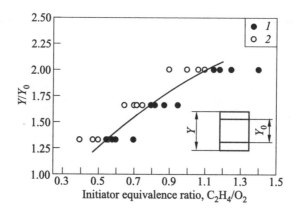

Figure 3.10 Results for the 2D diffraction geometry (main combustor — stoichiometric C_2H_4–air; $Y_0 = 5.08$ cm; initial pressure 100 kPa; and initial temperature 283 K): *1* — successful and *2* — unsuccessful detonation transmission. At $\varphi = 0.4$ uncertainty is ±0.1 and at $\varphi = 1.4$ uncertainty is ±0.04

Figure 3.10 summarizes the successful and unsuccessful test conditions for which detonation transmission occurred for the ethylene–air mixtures.

The required initiator strength (equivalence ratio) was found to be higher than the axisymmetric case. This is likely due in part to the decrease in confinement at the diffraction plane and the fact that the wall reflection from the 2D geometry does not possess the benefits of wall curvature found on the axisymmetric geometry. The effect of wall curvature has been found to be important for the generation of "X"-shaped compression waves and increased local heating.

With the exception of an occasional reignition due to a localized explosion some distance behind the leading shock, it appears that the shock reflection at the confining surface of the outer wall is the primary mechanism for reinitiation in this geometry. The increased heating and associated chemical activity behind the Mach stem provides the rapid energy release required for reinitiation. It is believed that if the exiting detonation wave from the initiator can be tailored to possess a very large Mach number, the transmission across the diffraction will be substantially enhanced.

3.4 CONCLUDING REMARKS

The generation of strong Mach stem reflections downstream of the diffraction plane of an initiator combustor is a reliable reinitiation mechanism for the transmission of a detonation wave into a larger combustor. The reliable generation of Mach stems along the wall downstream of the diffraction plane for coaxial initiators/combustors will be a valuable mechanism to take advantage of in an actual engine since they do not depend directly on a cell size, but on the strength of the exiting shock wave and the physical diffraction condition. Nevertheless, the sensitivity of the mixture to reignition can be correlated to a cell size.

A flowing coaxial region of air, such as on an actual engine, has not yet been investigated on the partially confined initiator geometry, but it is to be evaluated on the next phase of this project. The presence of turbulence and increased kinetic energy should improve the overall transmission properties. Additional computational results will evaluate these effects in conjunction with the experimental test program.

ACKNOWLEDGMENTS

The work was performed under ONR contract N00014-01-2-0153.

REFERENCES

1. Kailasanath, K. 2002. Recent developments in the research on pulse detonation engines. AIAA Paper No. 2002-0470.

2. Lee, J. H. S., and C. M. Guirao. 1982. Fuel–air explosions. *Conference (International) on Fuel–Air Explosions Proceedings*. Waterloo, Ontario: University of Waterloo Press. 1005.

3. Beeson, H. D., R. D. McClenagan, W. J. Pitz, C. K. Westbrook, and J. H. S. Lee. 1991. Detonabilty of hydrocarbon fuels in air. In: *Dynamics of detonations and explosions: Detonations*. Eds. A. L. Kuhl, J.-C. Leyer, A. A. Borisov, and W. A. Sirignano. Progress in astronautics and aeronautics ser. Washington, D.C.: AIAA, Inc. 133:19–36.

4. Knystautas, R., J. H. Lee, and C. M. Guirao. 1982. The critical tube diameter for detonation failure in hydrocarbon–air mixtures. *Combustion Flame* 48:63–83.

5. Moen, I. O., M. Donato, R. Knystautas, and J. H. Lee. 1981. The influence of confinement on the propagation of detonations near the detonability limit. *18th Symposium (International) on Combustion Proceedings*. Pittsburgh, PA: The Combustion Institute. 1615–22.

6. Teodorczyk, A., J. H. Lee, and R. Knystautas. 1991. Photographic study of the structure and propagation mechanisms of quasi-detonations in rough tubes. In: *Dynamics of detonations and explosions: Detonations*. Eds. A. L. Kuhl, J.-C. Leyer, A. A. Borisov, and W. A. Sirignano. Progress in astronautics and aeronautics ser. Washington, D.C.: AIAA, Inc. 133:223–40.

7. Oran, E. S., D. A. Jones, and M. Sichel. 1992. Numerical simulations of detonation transmission. *Proc. Royal Society London A* 436:267–97.

8. Oran, E. S., and J. P. Boris. 1993. Ignition in a complex Mach structure. In: *Dynamic aspects of detonations*. Eds. A. L. Kuhl, J.-C. Leyer, A. A. Borisov, and W. A. Sirignano. Progress in astronautics and aeronautics ser. Washington, D.C.: AIAA, Inc. 153:241–52.

9. Murray, S. B., F. Zhang, and K. B. Gerrard. 2001. The influence of driver power and receptor confinement on pre-detonators for pulse detonation engines. *18th Colloquium (International) on the Dynamics of Explosions and Reactive Systems Proceedings*. Seattle, Washington.

10. Edwards, D. H., G. O. Thomas, and M. A. Nettleton. 1979. The diffraction of a planar detonation wave at an abrupt area change. *J. Fluid Mechanics* 95:79–96.

11. Desbordes, D. 1988. Transmission of overdriven plane detonations: Critical diameter as a function of cell regularity and size. In: *Dynamics of explosions*. Eds. A. L. Kuhl, J. R. Bowen, J.-C. Leyer, and A. Borisov. Progress in astronautics and aeronautics ser. Washington, D.C.: AIAA, Inc. 114:170–85.

12. Desbordes, D., and A. Lannoy. 1991. Effects of a negative step of fuel concentration on critical diameter of diffraction of a detonation. In: *Dynamics of detonations and explosions: Detonations*. Eds. A. L. Kuhl, J.-C. Leyer, A. A. Borisov, and W. A. Sirignano. Progress in astronautics and aeronautics ser. Washington, D.C.: AIAA, Inc. 133:170–86.

13. Murray, S. B., P. A. Thibault, F. Zhang, D. Bjerketvedt, A. Sulmistras, G. O. Thomas, A. Jenssen, and I. O. Moen. 2000. The role of energy distribution on the transmission of detonation. In: *Control of detonation processes*. Eds. G. Roy, S. Frolov, D. Netzer, and A. Borisov. Moscow, Russia: ELEX-KM Publ. 24–26.

14. Kailasanath, K., and G. Patnaik. 2001. Multilevel computational studies of pulse detonation engines. *ISABE 2001-1172 Proceedings*. Bangalore, India.

Chapter 4

THE ROLE OF GEOMETRICAL FACTORS IN DEFLAGRATION-TO-DETONATION TRANSITION

N. N. Smirnov, V. F. Nikitin, V. M. Shevtsova, and J. C. Legros

The chapter contains the results of theoretical and experimental investigations of control of the deflagration-to-detonation transition (DDT) processes in hydrocarbon–air gaseous mixtures relative to propulsion applications. The influence of geometrical characteristics of the ignition chambers and flow turbulization on the onset of detonation and the influence of temperature and fuel concentration in the unburned mixture are discussed.

4.1 INTRODUCTION

Development of modern technology needs more powerful energy converters for propulsion issues. Conversion of fuel chemical energy into propulsion energy of the vehicle is limited by the rates of chemical energy release in combustion. The rate of energy release in the detonation mode of gaseous combustion is three orders of magnitude higher than in the deflagration mode, making the use of the detonation mode more efficient for creating high-power energy converters. The rate of combustible mixture supply is usually lower, thus requiring less fuel for operation at the same power levels.

The operation mode of the pulsed detonation-wave generator has been shown to be closely related to periodical onset and degeneration of a detonation wave. Those unsteady-state regimes should be self-sustained to guarantee reliable operation of such devices. Thus, DDT processes are of major importance. Minimizing the predetonation length and ensuring stability of the onset of detonation enables an increased effectiveness of pulsed detonation devices.

The probable requirement of DDT to create the new generation of engines brings the problem of DDT to the top of current research needs and turns out to be a key factor characterizing the Pulse Detonation Engine (PDE) operating cycle.

Experimental investigations of pulsed detonation initiation in gaseous mixtures of hydrocarbon fuels with air were undertaken. A detailed description of the experimental procedure and apparatus can be found in [1]. Here, a brief summary of the results is presented. Physical experiments were undertaken for comparative studies of the role of different turbulizing elements: Shchelkin spiral, orifice plates, and turbulizing chambers of a wider cross-section. The results showed that the wider turbulizing chambers incorporated into the ignition sections of the tube promoted the onset of detonation by shortening essentially the predetonation length for hydrocarbon fuel–air mixtures.

The mathematical models for simulating turbulent flame acceleration and the onset of detonation in chemically reacting flows were described in detail in [1–3]. The system of equations for the gaseous mixture was obtained by Favre averaging. The standard k–ε model was modified: an equation was added that determined the mean squared deviate of temperature in order to model the temperature fluctuations.

4.2   NUMERICAL STUDIES OF COMBUSTION PROPAGATION REGIMES

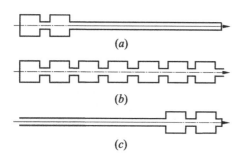

(a)

(b)

(c)

Figure 4.1 Geometry of the computational domain: (a) two chambers in the ignition section; (b) chambers incorporated into the tube along the whole length; and (c) two chambers in the far-end section of the tube.

Experimental investigations of DDT in combustible gases proved that the presence of different turbulizing elements in the initial sections of detonation tubes promotes DDT by shortening the predetonation length and time. Turbulizing chambers of a wider cross-section were found to be one of the most effective geometrical promoters for the DDT. Numerical experiments were undertaken for comparative studies of the role of different turbulizing elements and their location in the tube. In numerical simulations, the test vessel contained a detonation tube with a number of chambers of a wider cross-section incorporated in different places of the tube filled with a combustible gaseous mixture at ambient pressure (Fig. 4.1). Ignition of the mixture was performed by a concentrated energy release in the center of the first chamber or the tube itself in the left-hand side near the closed end. The number of chambers was varied from 1 to 20. Simulations were performed for the following cases:

(1) The initial section had two turbulizing chambers of a wider cross-section;

(2) Turbulizing chambers were distributed over the whole tube; and

(3) The far-end section had two similar turbulizing chambers of a wider cross-section.

4.3 TURBULIZING CHAMBERS AT THE IGNITION SECTION

The influence of turbulizing chambers mounted at the beginning of the tube was investigated earlier [1, 3].

Analysis of the results presented in Fig. 4.2 shows that decreasing the fuel content in the mixture decreases its detonability. The decrease of fuel molar concentration, C_f, below 0.011 brings the formation of galloping combustion regimes.

4.4 TURBULIZING CHAMBERS ALONG THE TUBE

The detonation tube was 2.95 m in length and had 20 turbulizing chambers uniformly distributed along the tube axis (Fig. 4.1b). The tube was 20 mm in diameter; turbulizing chambers were 100 mm in diameter and 100 mm long. The bridge between each of the two chambers was 50 mm long. Thus, the geometry of the test tube was similar to that used in the ignition section to initiate DDT in the previous set of numerical experiments.

The results showed that for the fuel concentration $C_f = 0.012$, the DDT process did not take place at all. The galloping combustion mode was established and was characterized by velocity oscillations within the range from 80 to 300 m/s, with an average flame-front velocity of 156 m/s. The calculated maps of density and velocity for successive times in the section of the tube incorporating chambers number 6 and 7 are shown in Fig. 4.3.

The results of numerical experiments showed that increasing the number of turbulizing chambers did not promote DDT for the present configuration, but did just the opposite — it prevented the onset of detonation and brought to establish the galloping combustion mode. The effect is due to very sharp variations in the cross-section area in the chambers and periodic flame slowing down due to its expansion.

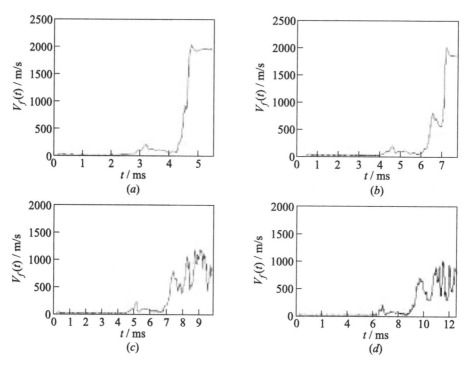

Figure 4.2 Reaction front velocity for different fuel concentrations in a two-chamber device: (*a*) $C_f = 0.015$; (*b*) 0.012; (*c*) 0.011; and (*d*) $C_f = 0.010$.

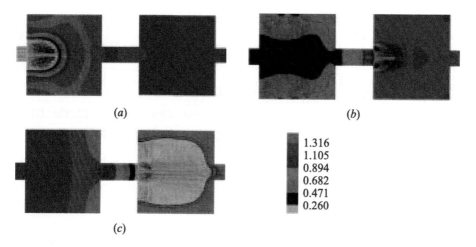

Figure 4.3 Density maps in the 6th and 7th chambers. Expansion ratio $\beta_{\mathrm{ER}} = 96\%$, $C_f = 0.012$: (*a*) $t = 9.01$ ms, (*b*) 9.82 ms; and (*c*) 10.60 ms. (Refer color plate, p. XXIII.)

To characterize the cross-section area variation within the structures of such a type, one could use the expansion-ratio and the volume-ratio parameters:

$$\beta_{ER} = \frac{S_{ch} - S_t}{S_{ch}}$$

$$\alpha_{ER} = \frac{S_{ch} L_{ch} + S_t L_t}{S_{ch}(L_{ch} + L_t)}$$

where S_{ch} is the chamber cross-section area, S_t is the cross-section area of the tube, L_t is the length of a bridge connecting two chambers and L_{ch} is the length of a chamber of a wider cross-section. In the numerical experiment considered, the expansion ratio was very high ($\beta_{ER} = 96\%$), promoting flame acceleration in the beginning of the tube but blocking it in other parts.

To investigate the influence of the expansion ratio on the onset of detonation, a set of numerical experiments was carried out on the DDT in tubes with diameters of 64.3 and 76.7 mm, which had similar chambers 100 mm in diameter and 100 mm long. These two cases correspond to β_{ER} values of 60% and 40%, respectively. The tube had 10 chambers distributed uniformly with 50-millimeter intervals, thus the whole structure was 1.45 m long. Comparison of the results (Fig. 4.4) shows that for high values of the expansion ratio, low-velocity galloping combustion was established with very regular velocity oscillations. For a lower expansion ratio, high-velocity galloping combustion was established characterized by irregular oscillations. For an even smaller expansion ratio, low-velocity galloping detonation was established, characterized by average velocity much less than that of Chapman–Jouguet (CJ) velocity for the given mixture composition.

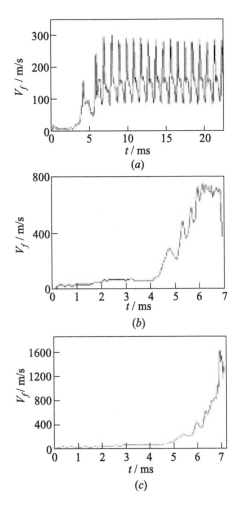

Figure 4.4 Reaction front velocity ($C_f = 0.012$) for different expansion ratios: (a) $\beta_{ER} = 96\%$; (b) 60%; and (c) $\beta_{ER} = 40\%$.

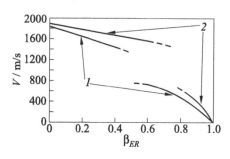

Figure 4.5 Mean reaction front velocities for different expansion ratios: *1* — $C_f = 0.012$ and *2* — $C_f = 0.015$.

The increase of fuel concentration from $C_f = 0.012$ to 0.015 (Fig. 4.4c) shortened the predetonation length (transition took place in the 5th chamber instead of the 8th) and increased the mean value of the galloping detonation velocity from 1450 to 1650 m/s.

Figure 4.5 shows the velocities of self-sustained modes of reaction-zone propagation, V, in tubes with uniformly distributed chambers along the tube axis. The values of V are presented as functions of the expansion ratio and fuel concentration. It is seen that the transition values of the expansion ratio increase with the increase of fuel concentration.

4.5 TURBULIZING CHAMBERS AT THE FAR-END OF THE TUBE

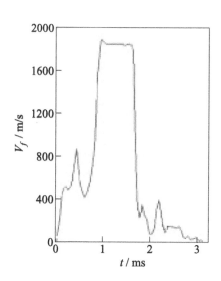

Figure 4.6 Reaction front velocity for the case of two chambers at the end of the tube: $C_f = 0.012$, $\beta_{ER} = 96\%$.

To provide comparative data, the role of two chambers with a wider cross-section mounted at the far-end of the tube was investigated (Fig. 4.1c). Numerical results showed that, after ignition in a narrow tube, acceleration of the flame zone accompanied by a number of oscillations brought the formation of the detonation wave propagating with a mean velocity of 1850 m/s. On entering the first chamber, decoupling of the shock wave and reaction zone took place and the mean velocity of reaction-zone propagation decreased to 200 m/s. Then, in a narrow bridge, the flame accelerated up to 400 m/s and slowed down in the second chamber to 100 m/s. The average velocity in the chambers was 140 m/s. The results show that a normal detonation wave propagating at the CJ velocity degenerates to a galloping combustion mode on entering a set of chambers of a wider cross-section ($\beta_{\mathrm{ER}} = 96\%$) (Fig. 4.6). This galloping combustion mode was characterized by a self-sustained velocity similar to that in a tube with chambers distributed along the axis (Fig. 4.4a) for the same expansion ratio.

4.6 EFFECT OF INITIAL TEMPERATURE

The influence of initial temperature on DDT in gases is an intriguing issue. On one hand, the increase of temperature promotes chemical reactions, thus promoting flame acceleration due to kinetic reasons. On the other hand, transition to the detonation takes place after turbulent-flame-propagation relative velocity exceeds the speed of sound in gas, which increases with temperature, thus inhibiting the transition process. The decrease of density with temperature at constant pressure could also be considered as an inhibiting factor. Probably due to the competition of these opposite effects, the available data on the influence of initial temperature on the DDT process in gases are contradictory. The experiments on DDT in hydrogen–oxygen mixtures in tubes of constant cross-section at constant pressure showed the increase of the predetonation length with the increase of temperature. The DDT process in tubes with chambers of a wider cross-section in the ignition section is essentially promoted by the piston effect of the expanding reaction products, which penetrate the narrow tube from a wide chamber, thus pushing the turbulent flame in the tube assisting it in achieving high velocities exceeding the speed of sound. The use of these turbulizing chambers neutralizes the effect of sound speed increase with temperature. Thus, the effect of reduction of chemical induction time with temperature could have a predominant effect.

Numerical simulations of ignition and flame propagation in lean mixtures (fuel concentration $C_{\mathrm{f}} = 0.011$–0.012), at an elevated initial temperature (350 K), showed that the increase of initial temperature of the combustible mixture brings to shortening the predetonation length and time. Figure 4.7 shows the flame-zone velocities in a two-chamber tube ($C_{\mathrm{f}} = 0.011$) for normal temperature (Fig. 4.7a) and elevated temperature (Fig. 4.7b). It is seen that in the case when the temperature was low (300 K), the onset of detonation did not take place within the length of the test section. High-speed galloping combustion was established. For higher temperature (353 K), DDT took place, and a stable detonation mode was achieved via an overdriven regime (Fig. 4.7b). For higher fuel concentration, $C_{\mathrm{f}} = 0.012$, DDT takes place for both temperatures. Nevertheless, the onset of detonation in the case of an elevated temperature takes place earlier than for the lower temperature.

Summarizing the above, it should be noted that in detonation tubes with chambers in the ignition section filled with lean hydrocarbon–air mixtures, pre-

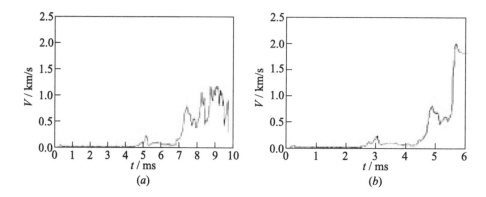

Figure 4.7 Flame-zone velocities in a two-chamber tube ($C_f = 0.011$) at different initial temperatures: (a) $T = 300$ K and (b) 353 K.

heating of the mixture promotes the DDT and shortens the predetonation length. In tubes of constant cross-section (without chambers), the opposite effect can occur. Thus, the role of mixture initial temperature on detonation initiation needs further investigation.

4.7   CONCLUDING REMARKS

Investigations of detonation initiation via DDT showed that:

(1) The presence of one or two turbulizing chambers in the ignition section shortens the predetonation length for gaseous hydrocarbon–air mixtures;

(2) The presence of turbulizing chambers makes the onset of detonation more stable and fixed to a definite place depending on chamber geometry, mixture composition, and initial temperature;

(3) The presence of turbulizing chambers makes the DDT process less sensitive to variation of ignition conditions;

(4) The absence of turbulizing chambers increases the predetonation length for similar initial conditions and makes the transition process very unstable with the onset of detonation having a sporadic character, the predetonation length varying stochastically and making the DDT process very sensitive to ignition conditions;

(5) In narrow tubes with uniformly distributed chambers along the tube axis, onset of detonation did not take place: galloping high-speed combustion modes or low-velocity galloping detonations were established;

(6) At small expansion ratios, low-velocity galloping detonation was established; at large expansion ratios, self-sustained galloping high-speed combustion took place;

(7) Reaction front velocities are lower for fuel-lean mixtures and increase for fuel-enriched mixtures;

(8) The transition values of the expansion ratio, which characterize the transition from low-velocity detonation to a high-speed galloping combustion, increase with fuel concentration within detonability limits;

(9) Chambers of wider cross-section mounted at the end of the detonation tube cause detonation suppression even in case of strong initiation; and

(10) In detonation tubes with chambers mounted in the ignition section and filled with lean hydrocarbon–air mixtures, preheating of the mixture promotes the DDT and shortens the predetonation length; in the absence of chambers, the effect could be the opposite.

Practical Recommendations*

Pulsed detonation devices based on the DDT principles should contain wider chambers in the ignition section (fore-chambers) to guarantee a stable periodical operating mode insensitive to initiating conditions. In case of direct initiation of detonation, such chambers prevent detonation formation and propagation.

ACKNOWLEDGMENTS

This work was performed under ONR contract N68171-01-M-6138.

---

*Computational results presented by Smirnov *et al.* need to be validated both qualitatively and quantitatively. To our best knowledge, there are serious fundamental problems in numerical simulation of a DDT process that encounters a wide spectrum of spatial and temporal scales associated with turbulence, convection, chemistry, acoustics, etc. Simplified approaches applying single-scale turbulence modeling and several-step kinetics, while being useful to some extent, cannot be the basis for practical recommendations without further validation. (*Editor's remark.*)

REFERENCES

1. Smirnov, N. N., V. F. Nikitin, A. P. Boichenko, M. V. Tyurnikov, and V. V. Baskakov. 1999. Deflagration to detonation transition in gases and its application to pulsed detonation devices. In: *Gaseous and heterogeneous detonations: Science to applications.* Eds. G. Roy, S. Frolov, K. Kailasanath, and N. Smirnov. Moscow, Russia: ENAS Publ. 65–94.

2. Smirnov, N. N., V. F. Nikitin, and J. C. Legros. 2000. Ignition and combustion of turbulized dust–air mixtures. *Combustion Flame* 23(1/2):46–67.

3. Smirnov, N. N., V. F. Nikitin, M. V. Tyurnikov, A. P. Boichenko, J. C. Legros, and V. M. Shevtsova. 2001. Control of detonation onset in combustible gases. In: *High-speed deflagration and detonation: Fundamentals and control.* Eds. G. Roy, S. Frolov, D. Netzer, and A. Borisov. Moscow, Russia: ELEX-KM Publ. 3–30.

Chapter 5

PSEUDOSPARK-BASED PULSE GENERATOR FOR CORONA-ASSISTED COMBUSTION EXPERIMENTS

A. Kuthi, J. Liu, C. Young, L.-C. Lee, and M. Gundersen

This chapter describes the design and operation of a pulse generator using an advanced pseudospark device for corona- or transient plasma-assisted flame ignition and combustion. This is an extrapolation of transient plasma work at the University of Southern California (USC) in Los Angeles, CA, wherein various fuels have been studied for ignition enhancement. The pulse generator is comprised of a single-stage lumped element Blumlein and 1:3 pulse transformer designed to produce 50-nanosecond-wide pulses with 70-kilovolt peak amplitude. Details of the pseudospark operation are presented. This apparatus is under development for experiments in pulse detonation engine (PDE) operation in collaboration with researchers participating in the U.S. Office of Naval Research (ONR) PDE program.

5.1 INTRODUCTION

Corona-discharge-assisted flame ignition has interesting properties. Pulsed corona discharge usually consists of many streamers, efficiently filling the gas volume [1]. The plasma produced by the streamers contains ~ 10 eV electrons, while sparks only produce ~ 1 eV average electron energies [2]. Compared to conventional spark ignition, flames ignited by a short-pulse corona discharge have a shorter delay time, faster pressure-rise time, higher maximum pressure, and more complete combustion [3, 4].

The desired pulse amplitude depends on the exact geometry of the combustion chamber and corona electrode. In the case considered here, at least 50 kV is needed for efficient energy deposition. The pulse length must be longer than the streamer formation time, but shorter than the arc formation time. In this experiment, it is estimated that the optimum pulse length is between 30 and 70 ns. The pseudospark is a glow-discharge switch that is an extrapolation of thyratron technology, using a different emission process than a traditional externally heated cathode. For reviews, see [5–8].

5.2 DESIGN

The pulse generator is expected to deliver pulses of $V_P = 70$ kV peak amplitude
and $\tau_P = 50$ ns duration. In order to extinguish occasional arcs, the output
impedance of the generator should be matched to the load, so peak currents are
limited to twice the operating current. The pulse shape is relatively unimportant,
as long as the rise time is short and there are no significant oscillations or ringing.
Bursts of 100 pulses with a pulse repetition rate of 1–100 Hz are needed for the
experiment. The rise time, repetition rate, and reliability requirements can be
met using a pseudospark switch. The final pulse amplitude is achieved by using
a pulse transformer.

Electrical Load

The load is a coaxial combustion chamber. The center corona electrode is positive, and the chamber wall is at ground potential. Previous experiments indicate
that the corona discharge impedance is approximately 250 Ohm, sharply drop-
ping to < 100 Ohm when streamers reach the opposite electrode and an arc
forms [3, 4].

Pulse Forming Network

The known load characteristics and the desirability of the lowest possible turn ratio for the pulse transformer suggest the lumped element Blumlein-pulse forming
network (PFN) configuration (Fig. 5.1). Ordinarily, a Blumlein-pulse forming

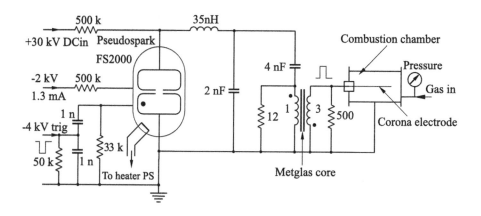

Figure 5.1 The corona-assisted combustion experiment.

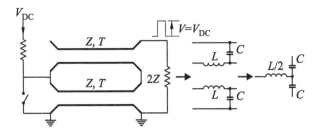

Figure 5.2 Synthesis of the lumped element Blumlein.

network consists of two identical series connected transmission lines charged to a common voltage. Each individual line has a characteristic impedance half that of the load. Placing the two lines above each other and recognizing that the lower half of the top line is at the same potential as the upper half of the bottom line, one gets the simplified diagram shown on Fig. 5.2.

Unfortunately, LC-circuits do not combine as cleanly as transmission lines [7, 8]. The ideal output pulse from the transmission lines is a square pulse with length twice the single line delay. On the other hand, the ideal, i.e., shortest possible nonringing, output pulse from a single LC-circuit is a critically damped pulse. The peak amplitude of such a pulse is $\sim 64\%$ of the charging voltage. The combination of the two LC-circuits will not produce a nonringing pulse unless the capacitor values are adjusted. The actual values of the capacitors shown in Fig. 5.1 give a good pulse (Fig. 5.3) with the shortest rise and fall times into a resistive load of 12 Ohm.

The inductor in the circuit is formed by the combined inductance of the pseudospark switch arc channel and the connecting 16 awg

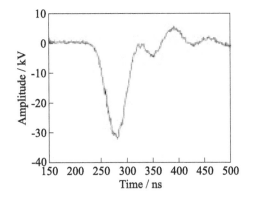

Figure 5.3 Output pulse of the Blumlein only, charged to 30 kV. There is a small amount of ringing caused by a small inductance in series with the load.

wire to the capacitor midpoint. This arrangement results in the shortest possible pulse. The capacitors can be reduced further and the pulse shortened if the load impedance can increase. Unfortunately, the pulse transformer leakage inductance introduces severe ringing in the absence of the 12-ohm primary-load resistance, so one could not take this approach. An optimized pulse transformer may allow further reduction in the pulse length.

Pulse Transformer

The output pulse transformer raises the Blumlein output to the required final voltage. The transformer has a single turn primary and three turns secondary windings. Both windings are made of 60-kilovolt-rated, 18 awg, silicon high-voltage (HV) wire. The transformer output voltage is consistent with the turn ratio in the absence of a load. The voltage ratio drops to 2.4 with the $\sim$ 500 Ohm resistor across the secondary.

The transformer core is a metglas-polyethylene-foil toroid, with an inner diameter of 2.5 cm, outer diameter of 7.4 cm, and length of 5.5 cm. The effective cross-section is $A_C = 8.6$ cm^2. At a constant 30-kilovolt primary voltage, the core would saturate in 90 ns. In the case under study, the primary waveform is approximately triangular. Therefore, saturation is not a problem. The transformer core was available in the laboratory from an earlier project and was not optimized to this application. Optimization of the core would result in a factor-two smaller cross-section and significantly reduced leakage inductance. The smaller leakage inductance, in turn, would reduce the transformer output impedance and enable the full factor-three voltage multiplication into the load.

Pseudospark Switch

The switching device is a commercial Pseudospark, model FS 2000 (Alstom). The rated maximum anode voltage and current are 32 kV and 30 kA. In the present application, the anode voltage is 30 kV, but the peak anode current is only 4 kA. The switch operates at the transition between the hollow cathode and superemissive modes. The switch-current rise-time is $\sim$ 15 ns, limited by the current channel and connection inductances.

Housekeeping requirements are modest. The heater requires 4.8 V at 4 A. The preionization — or keep alive — supply used by the trigger circuit needs −2 kV at 1.3 mA, and the trigger pulse amplitude is −4 kV and > 1 μs long.

The switch lifetime is specified as > 200 kC, so in the present application, at a continuous 100 Hz repetition rate, it would last longer than 3000 h.

5.3   OPERATION

The present system uses a resistive charging supply. The anode voltage of a two-pulse burst of 100-hertz repetition rate is shown in Fig. 5.4.

A resonant charging supply enabling 10-kilowatt peak burst power for 100 pulses is under development. Typical output into the corona chamber is shown in Fig. 5.5. The pulse is the second one of the two pulse bursts. The working gas in this case was air at atmospheric pressure. Noise on the current trace is typical of corona discharges and is a consequence of the large number of streamers carrying varying amount of current.

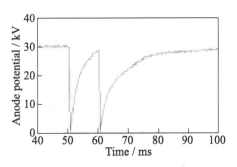

Figure 5.4 Anode voltage of a two-pulse burst at 100 Hz repetition rate.

Output voltage is sensed by a commercial 1000:1 HV probe. Load current is measured by a Pearson 411 current transformer. Due to the relatively slow rise-time of this transformer (20 ns), the current signal is only qualitative. A fast current transformer is being designed and will be used for precise energy deposition measurements.

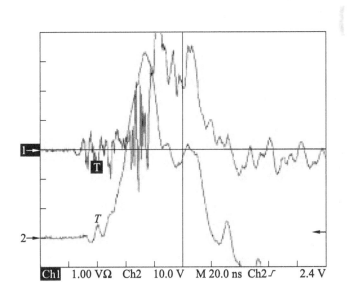

Figure 5.5 Typical load voltage and current. Channel 1 is current (100 A/div). Channel 2 is voltage (10 kV/div). PFN is charged to 25 kV.

5.4   CONCLUDING REMARKS

The design, construction, and operation of a pseudospark-based pulse generator has been described. Reliable long-life operation is made possible by the pseudospark switch. This pulse generator is being developed to support robust and reliable ignition for PDEs.

ACKNOWLEDGMENTS

This work was supported by the Air Force Office of Scientific Research, the Army Research Office, and the Office of Naval Research under contract N00014-02-1-0736.

REFERENCES

1. Liu, J., L. C. Lee, P. Ronney, and M. Gundersen. 2005. Premixed flame ignition by transient plasma discharge. Under preparation.

2. Bazeluan, E. M., and Yu. P. Raizer. 1997. *Spark discharge.* Boca Raton, FL: CRC Press.

3. Puchkarev, V., and M. Gundersen. 1997. Energy efficient plasma processing of gaseous emission using short pulses. *Appl. Phys. Lett.* 71(23):3364.

4. Liu, J. B., P. D. Ronney, and M. A. Gundersen. 2002. Premixed flame ignition by pulsed corona discharges. *Western States Meeting of The Combustion Institute Proceedings.* San Diego, CA.

5. Frank, K., E. Boggasch, J. Christiansen, A. Goertler, W. Hartmann, C. Kozlik, G. Kirkman, C. G. Braun, V. Dominic, M. A. Gundersen, H. Riege, and G. Mechtersheimer. 1988. High power pseudospark and BLT switches. *IEEE Trans. Plasma Science* 16(2):317.

6. Gundersen, M., and G. Schafer, eds. 1990. The physics and applications of pseudosparks. NATO ASI Series B 219. New York: Plenum Press.

7. Kirkman-Amemiya, G., H. Bauer, R. L. Liou, T. Y. Hsu, H. Figueroa, and M. A. Gundersen. 1990. A study of the high-current back-lighted thyratron and pseudospark switch. *19th Power Modulator Symposium Proceedings.* San Diego, CA. 254.

8. Gundersen, M., and G. Roth. 1999. *High power switches.* Eds. A. Chao and Maury Tigner. Singapore: World Scientific Publishing Co.

9. Puchkarev, V., and M. Gundersen. 1998. Power modulators for control of transient plasmas for environmental applications. *23rd International Power Modulator Symposium Proceedings.* Rancho Mirage, CA.

10. Gundersen, M., V. Puchkarev, and G. Roth. 1998. Transient plasma for environment applications with low energy cost. *IEEE International Conference on Plasma Science Proceedings.* Raleigh, NC.

Chapter 6

BREAKUP OF DROPLETS UNDER SHOCK IMPACT

C. Segal, A. Chandy, and D. Mikolaitis

The studies represented here evaluate shock–droplet interactions for mixtures of JP-10 with ethyl-hexyl nitrate, a compound used in the automotive industry to accelerate the reactions in diesel engines. The interest derives from this and other high-energy-density (HED) fuels' capability to accelerate the ignition processes in practical devices, such as, for example, a Pulse Detonation Engine (PDE). Initial studies involve Mach 2 shock interaction with a stream of individual, 0.8-millimeter diameter drops, to be followed by 0.4-millimeter drops impacted by stronger shocks, up to Mach 5. Previous studies indicated that addition of HED to JP-10 did not accelerate ignition-delay times; however, the preignition processes including droplet dispersion, vaporization, and mixing, along with the higher exothermicity of the HED formulations, can contribute to reduce the ignition energy required for direct detonation initiation. Furthermore, the inclusion of HED additives can contribute to the sustainability of detonation waves once they are formed. In the results presented here, the more rapid drop breakup of mixtures of JP-10 with 5% and 10% ethyl-hexyl nitrate was observed.

6.1 INTRODUCTION

Droplet–shock interactions have been studied extensively, and several phenomena have been described, including vaporization, aerodynamic material stripping, and heating [1–3]. It has been shown that, depending on the jet geometry and the aerodynamic regime, the liquid jet column and, subsequently, the formed drops undergo several different breakup mechanisms. Early experimental studies indicated the presence of two critical Weber numbers, We_{c1} for primary and We_{c2} for secondary breakup modes. The We_c variation as a function of the droplet diameter has been identified in these early works, and some influence of liquid viscosity on the critical Weber numbers has been found. Later studies determined that the critical Weber numbers depended on flow regime; Reynolds number, Re; and Ohnesorge number, Oh [4].

The mechanism to which the mass stripping is attributed was found dependent on the flow regime, with surface deformation and rupture being characteristic for large convective flows in a crosswise direction relative to the jet column, whereas under quiescent conditions, the column breakup is due to surface instabilities [5]. These observations, obtained for water injection, were consistent with the observations of Ranger and Nicholls [2] on single droplets. It was determined that even in the case of strong air–liquid interaction, such as that caused by the passage of a shock wave across a liquid droplet, the disruption of the droplet is due to the shear formed at the droplet surface causing a stripping of the boundary layer but, otherwise, without any visible effect of the shock–droplet interaction. Some disagreement exists in the regime of strong shocks, as previous reports in the Russian literature indicated that strong shocks might induce an "explosion" of the droplet before aerodynamic stripping takes effect [4].

It has been shown that in an initial stage, the droplet is deformed to an ellipsoid of revolution under the drag effects. Unsteady wave disturbances grow on the surface, and a boundary layer is formed on the upwind side of the droplet. Blowout of boundary layer appears after some time of induction. It takes place until a critical deformation is reached, and the droplet is transformed into a disk with a thickness $\sim 0.1 d_0$ (d_0 is initial diameter). Then the disk breaks, forming several smaller droplets. This induction time is estimated to be $\sim 2t^*$, and the time of breakup is estimated to be $\sim 2.5t^*$, where $t^* = (d_0/\nu)(\rho_l/\rho_g)^{1/2}$ is some characteristic breakup time [4]. Further correlations have been suggested for droplet mass reduction in time depending on various stages of the breakup. These earlier results have been confirmed by more recent data on primary and secondary droplet breakup modes presented by Chou et al. [5] under conditions of low-speed crossflow velocities. In the experiments described below, higher speeds are encountered, and the characteristic times appear to increase due to inertial effects.

A recent analysis of 2- to 3-millimeter diameter, viscous and viscoelastic drop breakup under high accelerations at Mach 2 to 3 offered in [6] reveals the sequences of the breakup events including the deformation, bag and bag-stamen formation, and, finally, mist formation. Breakup at these large accelerations is attributed to the onset of the Rayleigh–Taylor instability.

This study is focused on the analysis of higher accelerations with smaller sized drops, i.e., 0.8–0.4 mm, and, in particular, the breakup of mixtures formed by addition of detonation-sensitizing compounds to existing liquid hydrocarbons to identify the potential of these additives to enhance the relevant processes present in PDEs. Specifically, in this study, ethyl-hexyl nitrate has been added in concentrations of 5% and 10% on a mass fraction basis to JP-10, and shocks of nominally Mach 2 have been generated to study the extent to which the non-homogeneity of the mixture accelerates the drop breakup process. In an initial study of ignition times of these mixtures in air [7], the 5% ethyl-hexyl nitrate mixtures did not show faster ignition times when compared with pure JP-10. Furthermore, none of the other HED compounds included in that study indicated accelerated ignition times within the 1000–2500 K temperature interval.

Theoretical analyses [8] have shown that JP-10 has a substantial affinity to destroy free radicals, which may explain the inability of HED to accelerate JP-10 ignition. However, the droplet breakup and vaporization processes are enhanced by the nonhomogeneity of the mixture, and the net result is the acceleration of the processes that precede ignition. Indeed, during the experiments described earlier, mixtures of JP-10 with ethyl-hexyl nitrate ignited after the passage of the reflected shock, whereas drops of pure JP-10 did not. Visualization of the breakup sequence shows that drops of JP-10 with additives break up faster than compared to pure JP-10.

6.2   EXPERIMENTAL SETUP

The experimental setup is based on the shock tube shown in Fig. 6.1. The length of the driven section (minus test-section length) is 6.1 m and is made of 4″ Schedule 80 seamless pipe. The driver section is 1.8 m in length. Helium is used as the driver gas and heated, when necessary, until a nominal operating condition of 500 K and 54 atm absolute is attained to generate Mach 5 shocks. At the same time, the driven section/test section is evacuated to a pressure of 0.20 atm absolute and maintained at nominal room temperature. Double diaphragms are employed because with such an arrangement the pressure difference between the

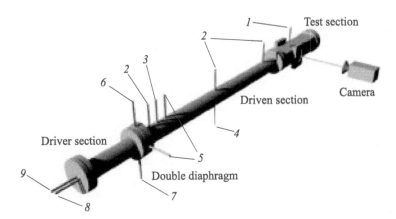

Figure 6.1 Schematic diagram of the shock tube with test section, droplet injection mechanism, and recording devices: *1* — droplet generator; *2* — high-speed pressure transducers; *3* — low-speed pressure transducer; *4* — vacuum port; *5* — thermocouples; *6* — diaphragm chamber bleed; *7* — low-speed pressure transducer; *8* — thermocouple; and *9* — helium input and low-speed pressure transducer.

driver and driven gases can be precisely controlled even if the diaphragm burst pressure is uncertain. In the experiments described below, the Mach number ranged between 2.0 and 2.16 for a nominal Mach 2 condition with a corresponding variation in the Weber number, We, from 4200 to 4300 for pure JP-10 drops.

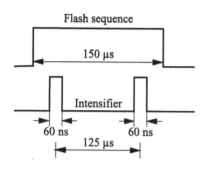

Figure 6.2 Experimental timing. The drop injection takes place during the diaphragm breakup. The flash lasts for 150 μs. During this interval the intensifier is triggered twice for 60-nanosecond duration at an interval of 125 μs, thus recording two images on the same frame.

A droplet injection mechanism has been calibrated for the droplet size and dispersion designed for the experiments. Several drops are captured within the viewing port of the test section. Images are recorded with a Princeton Instruments Intensified Pentamax® Camera using 150-microsecond, 300-joule pulse flash lamp. Within the duration of this flash, the intensifier is triggered twice at an interval of 125 μs. Each intensifier trigger lasts for 60 ns, thereby obtaining, in general, two images of the drops on the same frame, as indicated in Fig. 6.2. The viewing window has been set initially for 15×15 mm for the JP-10 drops and reduced subsequently to 6×6 mm to increase image resolution.

The triggering of the droplet injection, flash, and camera's intensifier is done by detecting the diaphragm rupturing with one of the high-speed pressure transducers and delaying the trigger by a precalculated time interval based on the speed of the generated shock. The pressure transducers have small piezoelectric elements that can detect rapid variations in pressure with resonance response to 1 MHz. The signal from the high-speed pressure transducer is sampled at a rate of 500 kHz. Thus, all the times in the experiment are fixed and known, with the only variation due to material differences from one set of diaphragms to the other.

6.3   RESULTS

Figure 6.3 shows a sequence of images collected in different experiments with a nominal Mach 2 number with pure JP-10 drops. The field of view in these experiments was 6×6 mm. Only a vertical strip of the image is shown. The incident shock Mach number ranged between 2.11 to 2.26 and resulted in a Weber number range from 3890 to 5160; therefore, the drops are in the regime of catastrophic breakup. The shock velocity at Mach 2 is 690 m/s, and the speed of the air behind the incident shock is 435 m/s. The Reynolds numbers were around 11,000, and the Ohnesorge numbers were around 0.015. Since the timing

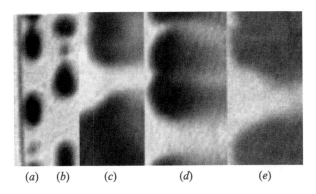

(a)   (b)   (c)   (d)   (e)

Figure 6.3 (a) Shock arrival; (b) at 5 μs after shock impact, the drops begin to flatten (M = 2.2; We = 5060); (c) at 30 μs from the impact, the drops have begun to deform and material is being stripped from the regions of higher shear (M = 2.11; We = 3890); (d) at 46 μs, the drops experience a substantial disruption; the finer mist is carried along by the air faster than the denser, more inert, droplet core (M = 2.26; We = 5150); and (e) at 90 μs, the drops have essentially disintegrated into a fine mist (M = 2.22; We = 4796).

of the shock arrival is measured from the dynamic pressure transducers and the timing of the other events is linked to the dynamic pressure transducers, an exact time of the event shown in the images can be deducted.

Figure 6.3a shows 0.8-millimeter diameter drops at the time of shock arrival. The drops in this image have not acquired a spherical shape yet.

The shock wave does not show deformations despite the cumulative effects of diaphragm rupture, travel through the 6-meter-long tube, and transition from the round cross-section of the shock tube to the square cross-section of the test section. Figure 6.3b is taken at 5 μs after the shock impact, and the drops begin to flatten under the aerodynamic drag. Figure 6.3c is taken at 30 μs from the impact and shows drops that have begun to deform under the impact and material that is being stripped from the regions of higher shear. This is more clearly indicated in the top drops in the figure that have partially coalesced. These effects are shown more clearly in Fig. 6.3d, taken at 46 μs after the impact, when the drops experience a substantial disruption and the finer mist is carried along by the air faster than the denser, more inert, droplet core. This breakup is similar to that observed in previous studies of aerodynamic drop shattering. Figure 6.3e, taken at 90 μs, indicates that the drops have essentially disintegrated into a fine mist.

At lower gas speed and higher drops' surface tension, formation of bag structures and breakup into smaller droplets have been observed following the initial drop deformation into a disk [5]. In the current experiment, these smaller drops are not observed, primarily due to the large Weber number. A mist with scales smaller than the camera resolution, which was 1.2 μm/pixel, was visible both

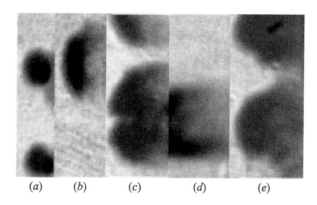

(a) (b) (c) (d) (e)

Figure 6.4 (a) Prior to the shock arrival; (b) at 16 μs after the shock impact, the drops are deformed and initial material is removed at the periphery of drop (M = 2.27; We = 4854); (c) at 30 μs, a substantial deformation of the drop with significant material removal (M = 2.15; We = 3897); and (d) and (e) at 52 (M = 2.25; We = 4686) and 150 μs (M = 2.31; We = 5198), respectively, indicate complete disintegration of the drops within the field of view.

in the pure JP-10 and the JP-10 + additives mixtures. Based on acceleration estimates, the axial displacement of the drops at these experimental conditions was 0.85 mm for the first 50 μs after the shock impact, which represents a 1.3 drop diameter in these experiments [7].

Droplets of heterogeneous composition appear to break faster than single compounds. Figure 6.4 shows a sequence of JP-10 with 5% ethyl-hexyl nitrate. The Mach number ranged from 2.15 to 2.31, and the Weber numbers ranged from 3900 to 5200 based on mixture measured surface tension.

Figure 6.4a shows the drop prior to the shock arrival. Due to the experiment timing, these drops have acquired a spherical configuration. Figure 6.4b shows drops at 16 μs after the shock impact. The drops are deformed, and initial material removal at the periphery of the drop where the shear is stronger has begun. Figure 6.4c, taken at 30 μs from the impact, shows a substantial deformation of the drop with significant material removal. This situation is more evident in the half portion of the top drop, whereas the lower drops have coalesced following deformation. Figures 6.4d and 6.4e, taken at 52 and 150 μs, respectively, indicate complete disintegration of the drops within the field of view. Within the 70 μs after the shock reached the drop, the motion of the drop, based on acceleration estimates, was 1.2 mm, which is a 1.5 drop diameter in these experiments.

A similar situation is noted in Fig. 6.5 for mixtures of 10% ethyl-hexyl nitrate in JP-10. At 22 μs, the deformation and initial breakup are evident, while at later times, 60 and 135 μs, respectively, the drops have experienced a significant breakup, becoming, essentially, clouds of fine mist.

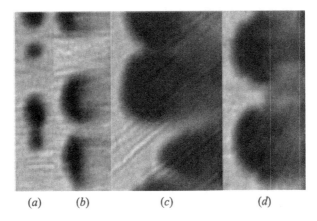

(a)   (b)   (c)   (d)

Figure 6.5 Mixtures of 10% ethyl-hexyl nitrate in JP-10: (a) before shock arrival; (b) at 22 μs, deformation and initial breakup has begun (M = 2.11, We = 3512); and (c) and (d) at 60 (M = 2.24, We = 4489) and 135 μs (M = 2.3, We = 4982), respectively, the drops experience a significant breakup, becoming clouds of fine mist.

6.4   CONCLUDING REMARKS

A study of effects of additives to JP-10 droplet breakup under Mach 2 shock impact indicated the following. (1) The droplet breakup at these conditions follows the deformation – material stripping – disintegration phases due to aerodynamic shear. (2) The addition of 5% to 10% ethyl-hexyl nitrate to JP-10 accelerates the droplet breakup in comparison with pure JP-10 at these experimental conditions to a certain extent; however, the effect is not significant. (3) Higher Mach/smaller drop diameter analyses are needed, including shock–spray interactions at conditions more realistic to PDE operation. (4) Although drop breakup has not dramatically increased, the effects of HED can be materialized through increased energy deposition that would contribute to detonation sustainability. In that regard, specifically designed HED compounds may be required.

ACKNOWLEDGMENTS

This work was performed under ONR contract N00014-99-1-0745.

REFERENCES

1. Buzukov, A. A. 1963. Droplets and jets destruction by air shock wave. *PMTF* 2: 154–58.

2. Ranger, A. A., and J. A. Nicholls. 1969. The aerodynamic shattering of liquid drops. *AIAA J.* 7(2):285–90.

3. Fuller, R., P.-K. Wu, K. Kirkendall, and A. Nejad. 1997. Effects of injection angle on the breakup processes of liquid jets in subsonic crossflows. AIAA Paper No. 97-2966.

4. Volynsky, M. S. 1948. On droplets breakup in an air stream. *Doklady AN USSR* 67(2):301–4; and: 1949. Study of droplets breakup in a gas stream. *Doklady AN USSR* 68(2):237–40. (Cited in: Livingston, T., C. Segal, M. Schindler, and V. A. Vinogradov. 2000. Penetration and spreading of liquid jets in an external-internal compression inlet. *AIAA J.* 38(6):989–94.)

5. Chou, W.-H., L.-P. Hsiang, and G. M. Faeth. 1997. Dynamics of drop deformation and formation during secondary breakup in the bag breakup regime. AIAA Paper No. 97-0797.

6. http://www.aem.umn.edu/research/Aerodynamic_Breakup.

7. Mikolaitis, D. W., C. Segal, and A. Chandy. 2003. Ignition delay for jet propellant 10/high-energy density fuel/air mixtures. *J. Propulsion Power* 19(4):601–6.

8. Varatharajan, B., and F. A. Williams. 2001. Chemistry of hydrocarbon detonations. *ONR 14th Propulsion Meeting Proceedings.* Minneapolis, MN. 105–10.

Comments

Anonymous: Does your mixture stay mixed throughout?

Segal: Yes, it does.

Lindstedt: The use of additives should be considered from safety and cost efficiency points of view.

Segal: I agree. However, there can exist applications where additives are allowed.

Schmidt: With a choice of additives, you can change the volatility characteristics.

Segal: We have not looked for optimal additives. The additive we study is being used in diesel fuel.

Waesche: It appears that effects only exist under a combination of certain conditions, is that right*?

Segal: I think so.

---

*Experimental findings of Segal *et al.* are important for a better understanding of the effect of additives on two-phase detonations. It is well-known that characteristic times of drop breakup behind strong shock waves are small fractions of the ignition-delay times in two-phase detonation fronts. The ignition delay in such conditions includes physical time (parent drop breakup, partial evaporation of drop fragments, mixing of fuel vapor with oxidizer, etc.) and chemical time (development of a chain-thermal ignition process in the gas phase). Segal *et al.* show that the additive does not really affect the parent drop breakup time and the effect of additive should be explained by other stages of the shock-induced ignition process. (*Editor's remark.*)

Chapter 7

IMPULSE PRODUCTION BY INJECTING FUEL-RICH COMBUSTION PRODUCTS IN AIR

A. A. Borisov

The idea of the use of a high-pressure jet produced by self-ignition or burning of a monopropellant in a small closed volume and injected in the main combustion chamber filled with air as a source of a high-intensity reactive shock wave in the chamber carrying a large impulse is tested both experimentally and by numerical modeling. Fuel-rich energetic materials, such as nitromethane and isopropyl nitrate, are used as model monopropellants. Experiments show that the injected products react with air, supporting reactive shock waves that spread at a velocity of up to 1400 m/s. Numerical modeling suggests that efficient burning of the jet material in air can be achieved with the introduction of obstacles in the chamber, providing much better mixing than in an unobstructed chamber. It also demonstrates that unsteady reactive shock waves can be more efficient impulse generators than detonation of a homogeneous monopropellant mixture with air. Measured impulses also provide evidence of incomplete burning of the injected material. Possible modifications of the chamber geometry and injection procedure are discussed.

7.1 INTRODUCTION

Chapman–Jouguet (CJ) detonation is believed to be an ideal regime used in pulsed detonation engines (PDEs). However, this regime brings about serious problems. Among these problems are: (i) filling of the chamber with air and fuel within a very short time period; (ii) provision of nearly perfect mixing between the components (which is needed because detonation can be initiated and propagated within quite a narrow equivalence ratio range); (iii) preevaporation of liquid fuels (experiments show that fuels such as kerosene cannot be detonated

329

in a duct if the vapor phase pressure in the mixture is insufficient); (*iv*) initiation of detonation within short distances available in engines (apart from the necessity of very large energy inputs for direct initiation of detonation or special, very sophisticated measures speeding up the deflagration-to-detonation transition (DDT) process, steady CJ detonations in real engines would hardly be attained because CJ detonation in fuel–air mixtures forms within distances of no less than 1.5 m); and finally, (*v*) critical diameter of detonation which rapidly increases as the ambient pressure drops (e.g., at 0.25 bar, propane–air mixtures cannot detonate in tubes less than 150 mm in diameter). Thus, even if all the above-listed problems are successfully solved, the burning regime in the PDE chamber is inevitably unsteady.

Are any other approaches that allow one to avoid the aforesaid difficulties and, at the same time, are as efficient as detonation of the mixture in the chamber? The present work explores the possibility of combining mixing and reactive-shock generation in a single process. This can be done by injection of a preconditioned fuel in the main chamber filled with air. Preconditioning means preheating of the fuel to a temperature that would provide its fast spontaneous reaction with the ambient air. If the pressure in the jet is high enough to drive a strong shock wave at the initial stage of discharge that would be supported at later stages by the fuel reaction within the mixing layer, the burning process would be similar, at least to some extent, to detonation.

This approach eliminates most of the difficulties inherent in detonation engines, namely, the jet initiates the reaction, so that the initiation problem is no longer critical. The same applies to fuel preevaporation and detonation limits, because the injected material is preheated and reacts with air with no limitations. The combustion chamber needs refilling with only air, which is much easier to arrange than filling it with a fuel–air mixture. As to the essentially unsteady nature of the flow, it is, as mentioned above, inevitable in any short combustion chambers. The only problem left is mixing, but its solution requires other approaches than those in the case of detonation of premixed components.

The questions to be answered are:

– How efficient is the unsteady process produced by a jet?

– What is the flow structure, reaction intensity, and efficiency?

– What should be done to enhance the mixing process?

High-pressure jets can be generated by self-igniting (in a preheated volume) or igniting with a spark. This is achieved either by injecting a partially reacted material in air into a fuel-rich liquid monopropellant or by injecting the products of decomposition of a small amount of monopropellant together with a conventional hydrocarbon fuel. The intention of this study is to answer the above-formulated questions.

7.2   EXPERIMENTAL STUDY

Experiments were conducted in tubes schematically shown in Figs. 7.1a and 7.1b. In the first version of the setup, the tube is 3 m long and 120 mm in inner diameter. The tube was equipped with five pressure gauges spaced 50 cm apart to monitor the wave velocity and pressure profiles. The injector is a thick-walled steel cylinder screwed in the end flange. The opposite tube end is open. Liquid nitromethane (NM) in an amount of 4 to 9 g with small additives of Al powder (0.3 to 0.5 g) is poured in the injector closed with a diaphragm. The mixture is ignited with a pyrotechnic primer. The injector diameter-to-length ratio is varied from 1/5 to 1/12 to find an optimal value at which the wave velocity and pressure amplitude are the greatest. Inasmuch as NM contains too much oxygen

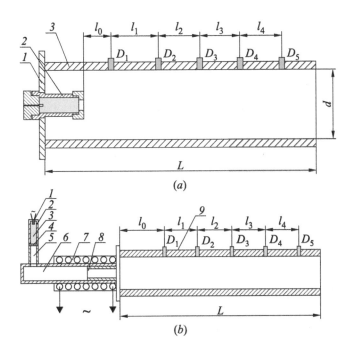

Figure 7.1 (a) Schematic of the experimental setup for studying propagation of reactive shock waves produced by injection of hot fuel into a combustion chamber: 1 — butt flange, 2 — injector, and 3 — steel tube. Distances between the gauges are $l_0 = 25$ cm, $l_1 = l_2 = 50.5$ cm, $l_3 = l_4 = 50$ cm. Tube diameter $d = 120$ mm, and tube length $L = 3$ m. (b) Experimental apparatus in which NM is self-ignited in the injector: 1 — primer; 2 — container; 3 — propellant; 4 — separating diaphragm; 5 — duct; 6 — injector; 7 — furnace; 8 — orifice; and 9 — tube. $L = 135$ cm; $l_0 = 30$ cm; $l_1 = 22.5$ cm; $l_2 = 23$ cm; $l_3 = 22.5$ cm; and $l_4 = 22.5$ cm.

in its molecule and the heat of its combustion in air is low, in the next set of tests, isopropyl nitrate (IPN) was used as a fuel. Isopropyl nitrate cannot be ignited with a primer. Therefore, it is ignited either by gradually heating the injector until self-ignition occurs or by rapidly admitting the liquid in the preheated injector. Experiments with self-ignition are conducted in a tube 1.35 m long and 95 mm in diameter equipped with five pressure gauges.

To assess the efficiency of the process suggested to produce thrust, the shorter tube is suspended and the impulse is measured by the pendulum technique. Since the measured impulse depends on the discharge conditions, the efficiency of heterogeneous jets in producing impulse is assessed in comparative tests in which experiments with heterogeneous jets are compared with detonation of homogeneous mixtures. Detonation in homogeneous mixtures is initiated by detonating a small volume of a stoichiometric propylene–oxygen mixture in the tube. The impulse produced by the initiator is measured in a run where the tube is filled with air and subtracted from the impulse measured in runs with a fuel–air mixture present in the tube.

The saturated vapor pressure of IPN is too low to allow preparing a stoichiometric IPN–air mixture in the tube. Moreover, IPN is easily adsorbed by the tube walls. Therefore, detonated was a lean mixture, and the actual IPN concentration was estimated by the average measured detonation velocity using a calculated dependence of the CJ detonation velocity on fuel concentration.

7.3 EXPERIMENTAL RESULTS

In experiments with NM, the highest wave velocities (up to 1400 m/s) and pressure of about 30 atm are observed at injector diameter-to-length ratios (L/d) ranging between 1/12 and 1/8. The wave parameters measured near the injector are higher because the amount of NM is insufficient to make a stoichiometric mixture with air in the tube. The representative pressure records in the tube are shown in Fig. 7.2. At smaller L/d ratios, only low-velocity regimes (with velocities of about 600 m/s) were observed. These results and the long pressure pulses provide evidence of an intense reaction between the injected material and air. To increase the amount of fuel to its stoichiometric content, 1.8 cm^3 of kerosene (JP type) was poured on the diaphragm closing the injector. As seen in Fig. 7.3, burning a small amount of NM in the injector resulted in spraying and preheating of kerosene and caused spontaneous reaction of kerosene with air.

Experiments with IPN are performed with the hope that the reaction of its decomposition products with air would produce stronger reactive shock waves. Figure 7.4 shows the recorded pressure profiles. The amount of IPN in this case is stoichiometric, and the charge is initiated by gradual heating of the injector.

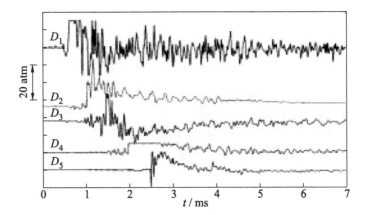

Figure 7.2 Representative records of the reactive shock generated in the tube upon injection of a reacting jet. Fuel: 8 cm^3 NM + 0.5 g Al. Initiator — primer cap. Measured velocity between gauges D_1 and D_2, $U_1 = 1260$ m/s; D_2–D_3 — $U_2 = 1215$ m/s; D_3–D_4 — $U_3 = 1040$ m/s; and D_4–D_5 — $U_4 = 1040$ m/s.

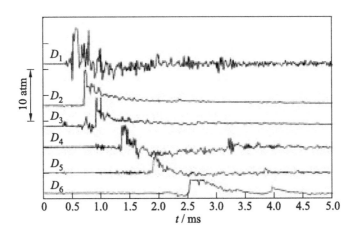

Figure 7.3 Representative records of the reactive shock generated in the tube upon injection of a reacting jet. Fuel: Al + nitromethane (4 cm^3 NM + 0.2 g Al) + 1.8 cm^3 kerosene. Initiator, primer cap. Measured velocity between gauges D_1 and D_2, $U_1 = 1390$ m/s; D_2–D_3 — $U_2 = 1170$ m/s; D_3–D_4 — $U_3 = 1165$ m/s; D_4–D_5 — $U_4 = 950$ m/s; and D_5–D_6 — $U_5 = 820$ m/s.

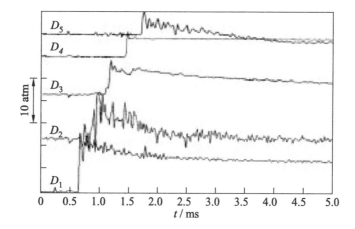

Figure 7.4 Pressure records for a reactive shock caused by IPN self-ignition (3.5 g IPN): $U_1 = 1125$ m/s; $P_1 = 13.3$ atm; $U_2 = 920$ m/s; $P_2 = 8.9$ atm; $U_3 = 680$ m/s; $P_3 = 4.3$ atm; $U_4 = 805$ m/s; and $P_4 = 6.8$ atm.

The pressure records indicate that only a small fraction of the injected fuel reacts with air and mostly near the injector.

The detonation velocity measured in a propylene mixture is indicated in Fig. 7.5. As seen, the process is unsteady even with a very high initiation energy used. The second important point is that the steady CJ detonation does not set in (the final velocity is below the calculated level). The measured impulse for a stoichiometric propylene–air mixture ranges between 1600–1730 s. The detonation wave initiated in a homogeneous IPN–air mixture is also unsteady (Fig. 7.6). The measured specific impulse of IPN ranges between 600–700 s for the homogeneous mixtures and between 250–300 s for heterogeneous IPN jets. The specific impulse of NM measured in jet experiments is close to 150 s. To understand why the impulse produced by homogeneous mixtures is about twice as high as the impulse generated by jets and to find ways how to make the injection process efficient, numerical modeling of the unsteady flow produced by reactive jets was performed.

7.4   NUMERICAL MODELING

A special two-dimensional (2D) gasdynamic code is developed to model multiphase flows. It allows for an arbitrary number of condensed and gaseous components, homogeneous and heterogeneous chemical reactions, interactions between condensed particles, and an arbitrary number of mesh refinement levels.

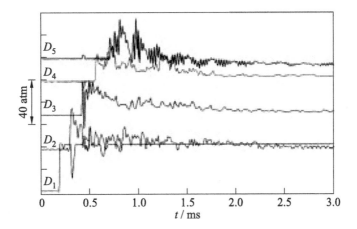

Figure 7.5 Measurements of the specific impulse produced by detonation of a stoichiometric gaseous mixture (C_3H_6 + air, initiator — $C_3H_6 + O_2$ mixture): $U_1 = 2000$ m/s; $P_1 = 42$ atm; $U_2 = 1840$ m/s; $P_2 = 35$ atm; $U_3 = U_4 = 1635$ m/s; $P_3 = P_4 = 28$ atm; and $I = 1730$ s.

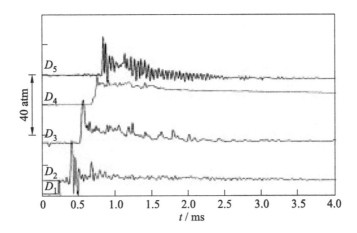

Figure 7.6 Detonation of gaseous IPN–air mixture (1.15 g IPN): $U_1 = U_3 = 1330$ m/s; $P_1 = P_3 = 18$ atm; $U_2 = 1700$ m/s; $P_2 = 3$ atm; $U_4 = 1660$ m/s; and $P_4 = 29$ atm.

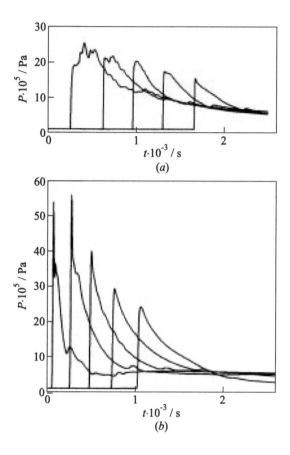

Figure 7.7 Pressure–time histories at various tube sections: (*a*) NM + Al injection and (*b*) instantaneous mixture reaction within the injector.

The traditional burning-time equation is used for decomposition and burning of condensed particles, and an Arrhenius-type global equation is used for homogeneous reactions. To analyze the flow pattern and compare computations (wave velocities and impulses) with the experiment, the following problems are solved:

(1) Injection of a NM + Al (95/5) mixture through the injector orifice in a tube of the same size as in the experiment, charge weight is 9.0 g;

(2) Detonation of the same mixture homogeneously distributed in the tube;

(3) Injection of 3.5 g of IPN in a 1.2-meter-long tube; and

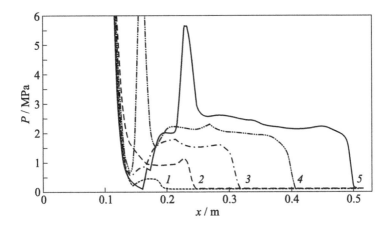

Figure 7.8 Axial pressure profiles at the initial stage of injection at $t = 0.075$ ms (1), 0.15 ms (2), 0.225 ms (3), 0.3 ms (4), and 0.375 ms (5).

(4) Detonation of a homogeneous IPN spray in the same tube.

Modeled also are detonations of stoichiometric hydrogen–air and propane–air mixtures in a 0.8-meter-long tube.

A simple analysis suggests that to generate an intense initial shock wave the pressure in the injector must be high. However, in contrast to low-pressure jets, in the high-pressure case the products expand at almost equal velocities in the axial and radial directions. Thus, the products are expected to expel a significant portion of air from the tube.

Numerical computations are performed to assess the fuel fraction not participating in the reaction with air and to find ways of increasing the burning efficiency. Unfortunately, the process within the injector is not amenable to modeling. Therefore, in order to fit the pressure at which the injector diaphragm bursts, it was assumed that initially 10% of the propellant reacts at a constant volume. Further on, it was also assumed that the initial size of liquid drops in the jet was 25 μm. This allows the wave velocities to be approximately fitted to the experiment.

Figure 7.7 shows the pressure profiles in the tube for a NM + Al jet and detonation of a homogeneous mixture. The reaction between the products and air takes place only at the jet head and is illustrated in Fig. 7.8, where the pressure peaks locally at each time instant. The low efficiency of fuel oxidation by air oxygen is seen in Fig. 7.9, where time histories of the mass fractions (referred to as the amount of the component to be formed in the propellant decomposition reaction) of fuel components are plotted. Thus, only about 15%

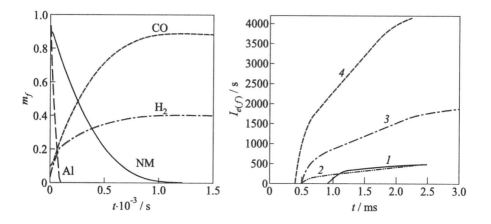

Figure 7.9 Mass fractions of the mixture components integrated over the tube volume.

Figure 7.10 Time histories of the impulse at the open tube end: *1, 2* — injection of NM/Al open and closed ends; *3* — propane–air; and *4* — hydrogen–air.

of CO formed is oxidized. The unburnt fraction of H_2 and CO is greater in the IPN jet.

The basic characteristic of engine efficiency is the impulse produced by the fuel combustion. The impulse depends on the chamber and nozzle geometry. As in the experiments, in calculations the impulses produced by detonation of a homogeneous propellant–air mixture and by injecting the propellant in air under identical conditions were compared. The impulse is calculated by the thrust at the closed and open tube ends. Time histories of the specific impulse are illustrated in Fig. 7.10. The results of calculations are listed in Table 7.1.

Table 7.1 Predicted specific impulse produced by heterogeneous jets and gaseous detonations.

	I_{sp}, s	
	Open end	Closed end
NM + Al (injection)	442	—
NM + air (homogeneous)	455	495
IPN (injection)	241	273
IPN (homogeneous)	671	686
Propane + air	1867	1911
Hydrogen + air	4145	

7.5   DISCUSSION

As computations show, the major reason of incomplete burning of the injected material is that the jet expands too fast to produce a plug flow at the beginning of the tube. Thus, the mixing layer area is reduced to the jet head only, and therefore, the major fuel fraction is not oxidized in air. The experiment suggests even poorer mixing efficiency and a lesser fraction of decomposed propellant. This indicates that, on the one hand, the jet expansion should be restricted, and, on the other, the jet must be split into several smaller jets to drastically enhance the mixing process, keeping the jet velocity at a high level. This can be done either by confining the jet in a tube of a smaller diameter with perforations to eject the propellant and products of its decomposition into the main chamber as the jet spreads through the smaller tube or by injecting the decomposition products through several orifices distributed over the chamber. Turbulizing obstacles can increase the mixing rate. Their design and arrangement in the chamber could be ascertained by numerical modeling. However, the most important finding that follows from computations is that the impulse produced by a heterogeneous NM jet is nearly equal to that of the gaseous detonation, in spite of the fact that only a small fraction of the jet material is oxidized by air. This is attributed to the longer pressure pulses resulting from the lesser energy left in the reaction products and the higher density and velocity of the fluid discharged from the tube as compared to the detonation wave issuing from the tube.

7.6   CONCLUDING REMARKS

Thus, the study has demonstrated the possibility of generating a strong reactive shock wave by injecting partially burnt monopropellant in air. Being discharged from the open tube end, the reactive shock wave produces an impulse to produce thrust. A comparison of experiments with calculations and analysis of computation results demonstrates that high efficiency of direct injection of hot heterogeneous fuel-rich products in air in producing impulse can be achieved only if special measures are undertaken to enhance mixing between the jet material and air. It is important that the unsteady process induced by injection of a reacting mixture can even be more efficient than detonation of the same amount of propellant homogeneously distributed over the combustion chamber. In the case of successful optimization of the mixing process, the approach offers the following benefits: (i) compactness of the design (with no valves), (ii) no necessity of strong ignition devices (specifically when the injector operates in the self-ignition regime), and (iii) no restrictions associated with detonation limits. The experiment demonstrates also that self-igniting monopropellants (or liquid oxidizers) can be used in small amounts to disperse, evaporate, and preheat the basic fuel.

ACKNOWLEDGMENTS

The work was performed under ONR contract N68171-01-M-5997.

Chapter 8

THERMODYNAMIC EVALUATION OF THE DUAL-FUEL PDE CONCEPT

S. M. Frolov and N. M. Kuznetsov

Three approximate approaches to determine the total pressure and gasphase composition in water – hydrogen peroxide (HP) two-phase systems depending on solution composition and temperature are presented. Chemical composition is assumed to vary (e.g., due to HP decomposition) slowly as compared with the rate of physical relaxation processes. Although the approaches are based on different prerequisites, all of them are in good agreement with each other in terms of predictions of total pressure, activity coefficients, and equilibrium gas-phase composition. The approaches have been generalized on three- and four-component systems containing low-volatility (nonsolvable jet propulsion fuel (JPF)) and high-volatility (air) components. The results are discussed in view of the dual-fuel concept of the pulse detonation engine (PDE) for advanced propulsion.

8.1 INTRODUCTION

The operational ability of an air-breathing, liquid-fueled PDE is dependent on the fuel used. In the authors' previous publications, a concept of a dual-fuel PDE has been suggested and substantiated [1–3]. The concept particularly implies the use of two liquid energetic materials: conventional JPF and HP*. In terms of the critical initiation energy, the gas-phase JPF–air mixtures with 5% and 20% of HP vapor were shown to be equivalent to stoichiometric ethylene–air and hydrogen–air mixtures, respectively [3]. As the dual-fuel, air-breathing PDE [1–3] implies the use of liquid sprays of JPF and HP, it is important to know the

---

*Contrary to other PDE concepts utilizing one fuel and two oxidizers (air and oxygen), this concept applies two liquid fuels (JPF and HP) and one oxidizer (air). The use of HP increases the energy density of the burning material and the detonation sensitivity of the JPF–HP blend. (*Editor's remark.*)

thermodynamic properties (activity coefficients, gas-phase composition, etc.) of the multiphase multicomponent mixture containing JPF, aqueous solution of HP, and air at high temperature and pressure. There is no general approximation that would be readily applicable to such two-phase systems [4–6].

This chapter suggests three approximate analytical approaches for calculating total pressure $P(x, T)$ and activity coefficients of the $H_2O_2–H_2O$ two-phase systems. The effect of JPF and air on the total pressure is also considered.

8.2 LIQUID–VAPOR PHASE EQUILIBRIUM CURVES FOR INDIVIDUAL COMPONENTS

For individual components, a two-parameter approximation [7–9] of pressure vs. temperature dependence along the phase equilibrium curve is used:

$$P(T) = \left[\left(\frac{T}{\alpha} \right)^{1/8} - A \right]^8 \tag{8.1}$$

where T is temperature in K, P is pressure in atm, and α and A are parameters. Equation (8.1) provides very accurate results at relatively high pressure, including the critical point. Parameters α and A are determined by using experimental data for saturated vapor pressure vs. temperature. Equation (8.1) differs from other available approximations of the $P(T)$-curve by its simplicity and a possibility to resolve it in terms of a simple temperature vs. pressure dependence:

$$T(P) = \alpha[Z + A]^8, \quad Z \equiv \left(\frac{P}{P_0} \right)^{1/8}, \quad P_0 = 1 \text{ atm} \tag{8.2}$$

For water, HP, and n-tetradecane (to simulate JPF), the liquid–vapor phase equilibrium curves have the following explicit formulae:

$$P_w(T) = \left[(2.8836 \cdot 10^6 T)^{1/8} - 12.4575 \right]^8 \tag{8.3}$$

$$P_{HP}(T) = \left[(2.6566 \cdot 10^6 T)^{1/8} - 12.5302 \right]^8 \tag{8.4}$$

$$P_{JPF}(T) = \left[(7.5324 \cdot 10^5 T)^{1/8} - 10.8801 \right]^8 \tag{8.5}$$

Parameters α and A for water and n-tetradecane were determined by using tables of thermodynamic data [11, 10]; for HP they were calculated by using nonlinear regression and eight points on a four-parameter $P(T)$-curve approximation [11]:

$$\log P_{HP}(\text{mm Hg}) = C_0 + \frac{C_1}{T} + C_2 \log T + C_3 T$$

Table 8.1 Comparison between liquid–vapor phase equilibrium curves (Eqs. (8.3) to (8.5)) and experimental data [10, 12] for water and n-teradecane, with the approximation of experimental data [13] by Eq. (8.2) for HP. $\Delta P/P$ is the error of approximation. The last lines of each block correspond to the critical points.

	Water		
T (K)	P (kgf/cm^2) [14]	P (kgf/cm^2), Eq. (8.3)	$|\Delta P/P|$, %
73.15	1.0332	1.0364	0.31
423.15	4.854	4.8616	0.16
473.15	15.857	15.76 0	0.62
523.15	40.56	40.326	0.58
573.15	87.61	87.641	0.03
623.15	168.63	169.15	0.31
633.15	190.42	190.72	0.16
638.15	202.21	202.25	0.02
643.15	214.68	214.31	0.17
647.30	225.65	224.72	0.41
	Hydrogen peroxide		
T (K)	P (atm), Eq. (8.2)	P (atm), Eq. (8.4)	$|\Delta P/P|$, %
423.15	0.994	1.012	1.8
473.15	4.004	4.069	1.6
523.15	11.99	12.06	0.6
573.15	29.19	29.19	0
623.15	61.23	61.17	0.1
673.15	115.2	115.2	0
723.15	199.9	199.8	0.03
730.15	214	214.66	0.3
	n-tetradecane		
T (K)	P (mm Hg) [15]	P (mm Hg), Eq. (8.5)	$|\Delta P/P|$, %
423.15	34.57	34.386	0.5
433.15	50.73	50.714	0.03
453.15	102.28	102.55	0.2
473.15	190.99	191.32	0.7
493.15	334.34	334.33	0
513.15	553.9	553.49	0.07
533.15	875.52	875.69	0.02
695.15	$\sim$ 16 atm	16.6 atm	—

$$C_0 = 44.5760; \quad C_1 = -4025.3; \quad C_2 = -12.996; \quad C_3 = 0.0046055 \qquad (8.6)$$

Table 8.1 shows the accuracy of Eqs. (8.3) to (8.5) as compared to the experimental data (for water and n-tetradecane) and to Eq. (8.2) (for HP). The error of the approximations at pressure 0.003 atm $\leq P \leq P_c$ (P_c is the critical pressure) is typically a fraction of a percent*.

---

*It would be interesting to apply Eq. (8.1) for determining the phase equilibrium curve for JP-10. (*Editor's remark.*)

Considering HP as an individual substance and a component of solution, the authors digress from considering the kinetic issues of its stability in terms of decomposition and other chemical transformation. This is acceptable if the local characteristic time of attaining the conditional thermodynamic equilibrium (at "frozen" chemical composition of the solution) is less than the characteristic times of chemical relaxation. The probability of violation of this condition increases with temperature. Here, quantitative criteria for the existence of this type of conditional equilibrium are not formulated.

8.3   CALCULATION OF THE TOTAL PRESSURE OF TWO-PHASE SYSTEM AT ISOTHERMS

Correlation Based on Redlich–Kister Method

To provide the correlation for the total pressure $p(x, T)$, the authors [13] use the formula:

$$P(x, T) = \gamma_{HP} P_{HP}(T)x + \gamma_w P_w(T)X \tag{8.7}$$

where x and $X = 1 - x$ are the molar fractions of HP and water, respectively, and γ_{HP} and γ_w are the corresponding activity coefficients. To calculate γ_{HP} and γ_w, the following relationships [14] are used:

$$\gamma_w = \exp\left\{\left(\frac{x^2}{RT}\right)[B_0 + B_1(1 - 4X) + B_2(1 - 2X)(1 - 6X)]\right\} \tag{8.8}$$

$$\gamma_{HP} = \exp\left\{\left(\frac{X^2}{RT}\right)[B_0 + B_1(3 - 4X) + B_2(1 - 2X)(5 - 6X)]\right\} \tag{8.9}$$

Parameters B_0, B_1, and B_2 in Eqs. (8.8) and (8.9) were determined by fitting the results provided by Eq. (8.7) with the experimental data. For this purpose, total pressure of aqueous solutions of HP at several values of x at 5 isotherms ($T = 317.65$ K, 333.15 K, 348.15 K, 363.15 K, and 378.15 K) was measured [13]. The data for phase equilibrium curves $P_{HP}(T)$ and $P_w(T)$ have been taken from the literature. As a result, the following expressions for B_0, B_1, and B_2 have been obtained (dimension in cal/mol):

$$B_0 = -1017 + 0.97T, \quad B_1 = 85, \quad B_2 = 13 \tag{8.10}$$

When substituting Eqs. (8.8) and (8.9) for γ_w and γ_{HP} and Eqs. (8.3) and (8.4) for $P_w(T)$ and $P_{HP}(T)$ into Eq. (8.7) and using Eqs. (8.10), one obtains the explicit analytical dependence $P(x, T)$. Application of Eqs. (8.8) and (8.9), that are based on low-temperature data ($T \leq 378$ K), to high temperatures can be considered as extrapolation. Its accuracy can be estimated by comparing the results obtained by different methods (see below).

Correlation Based on Similarity of Component Properties

For systems containing components with similar thermodynamic properties, the values of α and A in Eq. (8.1) appear close (e.g., for water and HP they differ by 8.5% and 0.6%, respectively). It is reasonable to assume that for such systems, Eq. (8.1) can be used for estimating $P(x, T)$. In this case, parameters α and A in Eq. (8.1) can be approximated as linear combinations:

$$\alpha(x) = \alpha_1(1 - x) + \alpha_2 x , \quad A(x) = A_1(1 - x) + A_2 x \qquad (8.11)$$

with indices 1 and 2 corresponding to components of a binary solution, and $1 - x$ and x representing their molar fractions. Within this approximation, Eqs. (8.1) and (8.2) provide isothermal dependence of pressure and isobaric dependence of temperature on solution composition, respectively. Application of Eqs. (8.1) and (8.11) to aqueous solutions of HP results in the following approximation for $P(x, T)$:

$$P(x, T) = \left[\left(\frac{T}{\alpha(x)} \right)^{1/8} - A(x) \right]^8$$
$$\alpha(x) = 3.4679 \cdot 10^{-7}(1 - x) + 3.7642 \cdot 10^{-7}x \qquad (8.12)$$
$$A(x) = 12.4575(1 - x) + 12.5302x$$

Correlation Based on the Boiling Temperature of Binary Solution

The other approach for estimating $P(x, T)$ is based on Eq. (8.2). If one knows the dependence $T_b(x, P)$ of the solution boiling temperature T_b on solution composition at $P = const$, then the procedure of determining $P(x, T)$ is straightforward.

Figures 8.1a and 8.1b show the measured dependences of T_b on HP concentration in aqueous solutions [16] at total pressure $P = 0.04$ atm (Fig. 8.1a) and $P = 1$ atm (Fig. 8.1b). It follows from Fig. 8.1 that function $T_b(x, P)$ at $P = const$ is almost linear, at least within the range $P = 0.04$–1 atm. This finding allows one to assume that the isobaric function $T_b(x, P)$ remains approximately linear for higher pressures. Although this assumption is insufficiently substantiated due to the lack of available experimental data, a general trend of balancing the thermodynamic properties of liquids and dense gases with temperature can serve as the indirect indication of its validity. With this in mind, function $T_b(x, P)$ is represented as a linear combination:

$$T_b(x, P) = T_w(P)(1 - x) + T_{HP}(P)x \qquad (8.13)$$

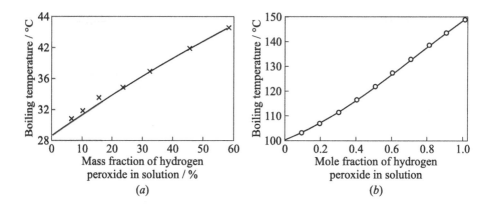

Figure 8.1 Experimental dependencies [15] of the boiling temperature on the mass and mole fraction of HP in aqueous solution at $P = 0.04$ atm (a) and $P = 1$ atm (b).

where $T_w(P)$ and $T_{HP}(P)$ are given by (see Eq. (8.2)):

$$T_w(P) = 3.4679 \cdot 10^{-7}[Z + 12.4575]^8$$

$$T_{HP} = 3.7642 \cdot 10^{-7}[Z + 12.5302]^8 \tag{8.14}$$

$$Z \equiv \left(\frac{P}{P_0}\right)^{1/8}, \quad P_0 = 1 \text{ atm}$$

Substituting Eqs. (8.14) into Eq. (8.13) results in an equation relating T_b, P, and x. Thus, for a given $T = T_b$, one obtains the equation determining $P(x,T)$ implicitly:

$$T = 3.4679 \cdot 10^{-7}[Z + 12.4575]^8(1 - x) + 3.7642 \cdot 10^{-7}[Z + 12.5302]^8 x \tag{8.15}$$

8.4   RESULTS OF TOTAL PRESSURE CALCULATIONS

Figure 8.2 shows the example of total pressure calculations at isotherm $T = 573$ K for aqueous solutions of HP. The curves are marked with abbreviations corresponding to various approximations: RK stands for Redlich–Kister approach (Eqs. (8.7) to (8.10)), CS stands for "Component Similarity" approach (Eqs. (8.12), and BT stands for "Boiling Temperature" approach (Eq. (8.15)). To distinguish between curves CS and BT, the CS curve is plotted as a dashed curve. In addition, the "Ideal Solution" (IS) curve is plotted in Fig. 8.2. Within the ideal solution approximation, the total pressure is determined by Eq. (8.7) with $\gamma_w = \gamma_{HP} = 1$, i.e.,

$$P(x, T) = P_{HP}(T)x + P_w(T)X \qquad (8.16)$$

The difference between the calculated results for nonideal solutions attains a maximum value at $x \approx 0.3$–0.4 for all isotherms within the temperature range $373 \leq T \leq 623$ K. The maximum difference in the total pressure predicted by the approaches of pages 344–346 is 13% at $T = 373$ K, 8% at $T = 423$ K, 5% at $T = 573$ K, 3.5% at $T = 523$ K, 3% at $T = 573$ K, and 2.6% at $T = 623$ K. The remarkable feature of the comparison is that a good agreement between the predicted values of total pressure exists even at 573 and 623 K, i.e., in the domain where the gas-phase density is high and interaction between molecules in the gas phase becomes significant. Equations (8.12) and (8.15) provide almost identical dependencies $P(x, T)$. The maximum difference of less than 2% is attained in the vicinity of $x = 0.3$.

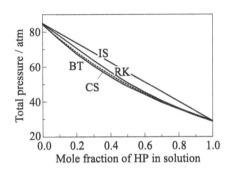

Figure 8.2 Predicted dependencies of total pressure on the mole fraction of HP in aqueous solution at $T = 573$ K.

The value of total pressure predicted by the ideal-solution relationship (8.16) differs considerably from the values provided by other approaches.

8.5   CALCULATION OF ACTIVITY COEFFICIENTS AND GAS-PHASE COMPOSITION

Parameters B_0, B_1, and B_2 entering Eqs. (8.8) and (8.9) for the activity coefficients were obtained [13] based on the limited set of low-temperature ($T \leq 378$ K) experimental data on the total pressure of binary water–HP system for several values (x, T). As pages 345 and 346 provide the approximate expressions for $P(x, T)$ that are applicable at temperatures ranging from 330–370 K to $T \leq T_c$, it becomes possible to determine B_0, B_1, and B_2 more precisely for the extended temperature range. For this purpose, Eq. (8.7) combined with Eqs. (8.3) and (8.4) was applied to determine B_0, B_1, and B_2 by attaining the best least square fit with the $P(x, T)$-dependence of Eq. (8.12). Two approximations were used: in the first (referred to as II-parameter approximation), parameter B_2 in Eqs. (8.8) and (8.9) was taken as zero; in the second (III-parameter approximation), all three parameters B_0, B_1, and B_2 were determined. It was found that the dependencies of B_0, B_1, and B_2 on T are very regular. The following exponential interpolations were obtained for the II-parameter approximation:

$$B_0^{(II)} = -431.31 - 225 \exp \frac{423.15 - T}{125.54}$$

$$B_1^{(II)} = 201.0 + 247.1 \exp \frac{423.15 - T}{121.3} \tag{8.17}$$

$$B_2^{(II)} = 0$$

and for the III-parameter approximation:

$$B_0^{(III)} = -376.69 - 197.41 \exp \frac{438.39 - T}{112.81}$$

$$B_1^{(III)} = 99.21 + 110.77 \exp \frac{445.66 - T}{140.69} \tag{8.18}$$

$$B_2^{(III)} = -106.62 - 189.07 \exp \frac{438.58 - T}{111.79}$$

Maximum interpolation errors are 0.2% for $B_0^{(II)}$ and 0.7% for $B_1^{(II)}$ (Eqs. (8.17)), and 0.2% for $B_o^{(III)}$, 0.2% for $B_1^{(III)}$, and 0.8% for $B_2^{(III)}$ (Eqs. (8.18)).

Equations (8.8) and (8.9) in combination with Eqs. (8.17) or (8.18) can be used for calculating γ_{HP} and γ_w. Then, one can determine the equilibrium composition of the gas phase in the water–HP system by applying the following formulae:

$$Y = \frac{\gamma_w P_w(T) X}{P(x, T)} \tag{8.19}$$

$$y = \frac{\gamma_{HP} P_{HP}(T) x}{P(x, T)} \tag{8.20}$$

where Y and y are the molar fractions of water and HP vapor, respectively.

Alternatively, for the gas phase obeying Dalton's law, y can be obtained directly (i.e., without calculating γ_{HP} and γ_w) using the Duhem's equation [6]:

$$\left(\frac{\partial \ln P}{\partial y} \right)_T = \frac{y - x}{y(1 - y)} \tag{8.21}$$

If the total pressure P is known as a function of y (along the isotherm), then Eq. (8.21) provides algebraic or transcendental dependence $y(x)$ at a given T. Usually, total pressure is known as a function of solution composition, x. By changing variables from y to x in Eq. (8.21), one can arrive at the differential equation for water concentrations in the two-phase water–HP system:

$$\frac{dY}{dX} = \frac{Y(1-Y)}{Y-X} Z(X,T), \quad Z(X,T) = \left(\frac{\partial \ln P}{\partial X}\right)_T \qquad (8.22)$$

To find a unique integral curve of Eq. (8.22), it is necessary to specify a point on the curve. For this purpose, one can choose a singular point at the edge of interval $0 \leq X \leq 1$, where the molar fraction of one component in solution and in gas phase is zero. It can be shown that the singular point $X = 0$, $Y = 0$ is of the saddle type, and Eq. (8.22) should be integrated from this point.

Thus, the gas-phase composition in the water–HP system can be determined by using Eqs. (8.19) and (8.20) or by direct integration of Eq. (8.22). It is instructive to compare the results provided by both approaches. Although the latter approach does not require evaluation of γ_w and γ_{HP}, the calculated dependencies $Y(X)$ or $y(x)$ can be used to calculate γ_w and γ_{HP} by using Eqs. (8.19) and (8.20). The activity coefficients thus obtained are compared in Fig. 8.3 for $T = 523$ K. Both II- and III-parameter approximations of Eqs. (8.17) and (8.18) for B_0, B_1, and B_2 were applied. The discrepancy in γ_w values obtained by direct integration of Eq. (8.22) and using Eqs. (8.8) and (8.9) with II-

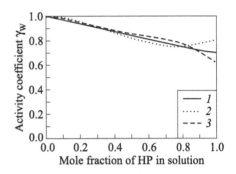

Figure 8.3 Predicted dependencies of the water activity coefficient γ_w on the mole fraction of HP in aqueous solution at $T = 523$ K: 1 — integration of Duhem's equation; 2 — II-parameter approximation; and 3 — III-parameter approximation.

and III-parameter approximations for B_0, B_1, and B_2 does not exceed 4% within interval $0 \leq x \leq 0.9$. However, at $0.9 \leq x \leq 1$, i.e., at the interval where water concentration in solution is small ($X \ll 1$), the discrepancy is getting higher and is both quantitative and qualitative (see dotted line in Fig. 8.3 depicting the result relevant to the II-parameter approximation). The use of the III-parameter approximation does not diminish the discrepancy (see dashed line in Fig. 8.3). However, discrepancies relevant to II- and III-parameter approximations exhibit different signs within a wide range of x. In view of it, one can naturally suggest a better approximation for the activity coefficients that is based on the arithmetic means:

$$\gamma_w = \frac{\gamma_w^{(II)} + \gamma_w^{(III)}}{2}, \quad \gamma_{HP} = \frac{\gamma_{HP}^{(II)} + \gamma_{HP}^{(III)}}{2} \qquad (8.23)$$

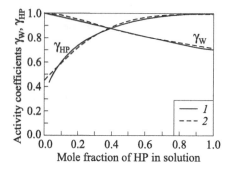

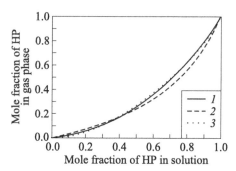

Figure 8.4 Predicted dependencies of activity coefficients γ_{HP} and γ_w on the mole fraction of HP in aqueous solution at $T = 523$ K; 1 — integration of Duhem's equation and 2 — approximation given by Eq. (8.23).

Figure 8.5 Predicted dependencies of HP mole fractions in the gas phase on the HP mole fraction aqueous solution at $T = 523$ K: 1 — integration of Duhem's equation; 2 — ideal solution approximation; and 3 — approximation given by Eq. (8.23).

where indices II and III correspond to II- and III-parameter approximations, respectively.

The accuracy of the approximation given by Eq. (8.23) in relation to the numerical solution of Eq. (8.22) is illustrated in Figs. 8.4 and 8.5. Solid curves in Fig. 8.4 show the results of calculations of γ_w and γ_{HP} based on function $Y(X)$ obtained by direct integration of Eq. (8.22) along isotherm $T = 523$ K. Dashed curves in Fig. 8.4 correspond to the approximation given by Eq. (8.23). The solid curve in Fig. 8.5 shows the results of direct integration of Eq. (8.22). The dotted curve in Fig. 8.5 corresponds to the approximation given by Eq. (8.23); the dashed curve corresponds to ideal solution approximation.

It follows from Fig. 8.4 that at $x > 0.02$, mean and maximum discrepancies $|\Delta\gamma_i|/\gamma_i$ are, respectively, fractions of percent and 4%–6%. However, gas-phase HP concentrations determined by Eqs. (8.20) and (8.23) differ from those obtained by direct integration of Eq. (8.22) no more than by 0.011 at $T = 423$ K and 0.006 at $T = 523$ K at all $0 \leq x \leq 1$ (see Fig. 8.5). Such accuracy allows one to apply Eqs. (8.20) and (8.23) for practical calculations of gas-phase composition at given x and T rather than directly integrate Duhem's equation (8.22).

8.6   IDEAL SOLUTION APPROXIMATION

The fact that coefficients γ_w and γ_{HP} can be considerably less than unity (see Figs. 8.3 and 8.4) indicates that aqueous solutions of HP do not obey the ideal solution laws, in particular, at small x or X. Nonideality of the solutions has a relatively insignificant effect on the total pressure. Nevertheless, it results in a considerable variation of the deficient gas-phase component, i.e., the component that can be considered as a small additive in the gas phase. With increasing temperature, the activity coefficients tend to unity. The accuracy of formulae used for evaluation of γ_w and γ_{HP} decreases with the gas-phase density and, correspondingly, with the degree of gas phase departing from the ideal gas law. However, as the gas-phase density approaches the liquid density along the phase equilibrium curve, one can expect that realistic activity coefficients are closer to unity than the estimated values.

8.7   TERNARY SYSTEM WATER – HYDROGEN PEROXIDE – AIR

Consider the effect of air on the phase equilibrium of aqueous solutions of HP. When the volume V_G occupied by a three-component gas phase is large as compared to the volume V_L occupied by the liquid solution, and air pressure P_A is not too high (e.g., $V_G > V_L$ and $P_A < 10$ atm), the fraction of air solved in the solution is negligibly small as compared to both the mole fraction of air in the gas phase and to mole fractions of other components. Due to the low compressibility of liquid, variation of gaseous pressure caused by the presence of air does not affect significantly the thermodynamic state of solution and, in particular, the chemical potentials of main solution components. Chemical potentials of main components in the gas phase remain also unaffected if the molar fraction of air is small or if the total pressure does not exceed several dozens of atmospheres. If the chemical potentials of main components in the solution and in the gas phase are independent of the air partial pressure, all above formulae and equations determining $P(x, T)$ and binary system composition remain valid (with the original normalization: $x + X = 1$, $y + Y = 1$). The total pressure of the three-component system (denoted as $\Pi(x, T, \rho_A)$) within the ideal gas approximation will be given as a sum:

$$\Pi(x, T, \rho_A) = P(x, T) + P_A , \quad P_A = \frac{\rho_A RT}{\mu_A}$$

where ρ is the density, μ is the molecular mass, and subscript A denotes properties of air.

8.8 TERNARY SYSTEM WATER – HYDROGEN PEROXIDE – JET PROPULSION FUEL

Herein, JPF is modeled by n-tetradecane (NTD). NTD differs from water and HP by low volatility. For example, at temperature $T = 423$ K the saturated vapor pressure of NTD is 34.6 mm Hg, that is 103 times less than the corresponding water vapor pressure and 22 times less than the HP vapor pressure. NTD is not soluble in aqueous solutions of HP and may form emulsions similar to water-in-oil emulsions. In such emulsions, NTD affects thermodynamics of the water–HP binary subsystem through the gas-phase pressure (with the accuracy of the order of inter-phase surface effects in emulsion), as the partial pressure of NTD vapor contributes to the total pressure. However, taking into account the above examples for the vapor pressure of individual components, this contribution is insignificant, and the effect of NTD on the equilibrium composition of the water–HP subsystem is negligible. As for the effect of water and HP vapors on the phase equilibrium of NTD itself, it is also insignificant, at least at $T \leq 523$ K. At $T = 523$ K, the density of the gas phase in the water–HP system is approximately two orders of magnitude less than the liquid density, and Dalton's law relating the partial pressures of components is still valid. However, at higher temperatures and, correspondingly, higher densities of the gas phase, the chemical potential of NTD vapor changes considerably (increases) due to intermolecular interaction (with domination of repulsive forces). As a result, phase equilibrium of NTD is shifted towards lower vapor pressure. A quantitative description of these effects is beyond the scope of this chapter.

8.9 CONCLUDING REMARKS

Approximate approaches to estimate the total pressure in the water–HP system are presented. Chemical composition of the system is assumed to be either "frozen" or varying slowly as compared with the rate of physical relaxation processes. Although the approaches are based on different prerequisites, they are in good agreement with each other in terms of predictions of total pressure, activity coefficients, and equilibrium gas-phase composition. In the approach of the subsection on page 344, total pressure was calculated as a sum of partial pressures of water and HP vapors. On pages 345 and 346, the concept of partial pressure was not used at all. The expression for the total pressure derived on page 345 does not incorporate the ideal gas approximation and is applicable within the entire temperature range where each of the solution components exists in both liquid and gas phases. It is hardly possible that a good agreement (within a wide temperature range including the critical temperature of water) between the predictions is occasional. More preferably, all the approaches give realistic approximations for $P(x, T)$. The extended validity of the approach on page 344 (based on the ideal gas law) for calculating $P(x, T)$ is most probably caused by compensation of errors in the approximations for the binary system under consideration.

The activity coefficients and the equilibrium gas-phase composition have been calculated by using relationships based on Dalton's law. This also applies to Duhem's equation in the form of Eq. (8.22). For the water–HP system this approach is approximately valid at $T \leq 520$ K. Comparison of activity coefficients and gas-phase composition obtained within the Redlich–Kister approximations and by numerical integration of Eq. (8.22) indicates that the modified approach of Eq. (8.23) agrees well with the solution of Eq. (8.22). With this modification, Eqs. (8.8), (8.9), (8.17), (8.18), and (8.23) provide the explicit and fairly accurate dependencies of activity coefficients on solution composition and temperature. Based on the activity coefficients, gas-phase composition can be readily obtained by using the approximations for the total pressure and Eqs. (8.3) and (8.4). Application of these formulae for estimating activity coefficients and gas-phase concentrations at higher temperatures ($T > 520$ K) should be considered as extrapolation. The accuracy of this extrapolation can be worse as compared to total pressure calculations.

The approaches for calculating equilibrium gas-phase composition in a two-phase system containing aqueous solution of HP, air, and JPF are also suggested. The further step in evaluating the performance of the dual-fuel, air-breathing PDE [1–3] is to incorporate chemical kinetics of HP decomposition and JPF oxidation. Preliminary results on simulation of JPF (or HP) liquid drop ignition and combustion in air with HP (or JPF) vapor have been reported [16].

ACKNOWLEDGMENTS

The work was performed under ONR contract.

REFERENCES

1. Frolov, S. M., and V. Ya. Basevich. 1999. Application of fuel blends for active detonation control in a pulsed detonation engine. AIAA Paper No. 99-34130.

2. Frolov, S. M., V. Ya. Basevich, A. A. Belyaev, and M. G. Neuhaus. 1999. Application of fuel blends for controlling detonability in pulse detonation engines. In: *Gaseous and heterogeneous detonations: Science to applications.* Eds. G. Roy, S. Frolov, K. Kailasanath, and N. Smirnov. Moscow, Russia: ENAS Publ. 313.

3. Frolov, S. M., V. Ya. Basevich, and A. A. Vasil'ev. 2001. Dual-fuel concept for advanced propulsion. In: *High-speed deflagration and detonation: Fundamentals and control*. Eds. G. Roy, S. Frolov, D. Netzer, and A. Borisov. Moscow, Russia: ELEX-KM Publ. 315.

4. Dodge, B. F. 1944. *Chemical engineering thermodynamics*. 1st ed. New York–London: Mc-Graw Hill.

5. Kogan, V. B., and V. M. Fridman. 1957. *Handbook on liquid–vapor equilibrium in binary and multicomponent systems*. Leningrad: St. Sci.-Tech. Publ.

6. Reid, R. C., J. M. Prausnitz, and T. K. Sherwood. 1977. *The properties of gases and liquids*. New York: Mc-Graw Hill.

7. Kuznetsov, N. M. 1981. Two-phase water–steam mixture: Equation of state, sound velocity, isentropes. *Doklady USSR Acad. Sci.* 257:858.

8. Kuznetsov, N. M. 1982. Equation of state and phase equilibrium curve for the liquid-vapor system. *Doklady USSR Acad. Sci.* 266:613.

9. Kuznetsov, N. M., E. N. Alexandrov, and O. N. Davydova. 2002. Analytical presentation of liquid–vapor phase equilibrium curves for saturated hydrocarbons. *J. High Temperature* 40(3):111.

10. Vukalovich, M. P. 1950. *Thermodynamic properties of water and water vapor*. Moscow: Mashgiz.

11. Schumb, W. C., C. N. Satterfield, and R. L. Wentworth. 1955. *Hydrogen peroxide*. New York: Reinhold Publ. Corp.

12. Vargaftik, N. B. 1963. *Handbook on thermophysical properties of gases and liquids*. Moscow: Fizmatgiz.

13. Scatchard, G., G. M. Kavanagh, and L. B. Ticknor. 1952. Hydrogen peroxide + water. *J. American Chemical Society* 74:3715–20.

14. Redlich, O., and A. K. Kister. 1948. Thermodynamics of non-electrolyte solutions, x–y–t relations in a binary system. *Ind. Engineering Chemistry* 40:341–45.

15. Seryshev, G. A., ed. 1984. *Chemistry and technology of hydrogen peroxide*. Leningrad: Khimiya.

16. Frolov, S. M., V. Ya. Basevich, A. A. Belyaev, and V. S. Posvyanskii. 2000. Combustion and detonation control by in-situ blending of liquid fuels. *13th ONR Propulsion Meeting Proceedings*. Eds. G. D. Roy, and P. J. Strykowski. Univ. Minnesota, MN. 207.

Chapter 9

THERMAL DECOMPOSITION OF JP-10 STUDIED BY MICROFLOW TUBE PYROLYSIS–MASS SPECTROMETRY

R. J. Green, S. Nakra, and S. L. Anderson

A microflow tube, guided-ion beam, quadrupole mass spectrometer instrument was used to study the pyrolysis chemistry of exotetrahydrodicyclopentadiene (exo-THDCP or JP-10) and related compounds, adamantane, cyclopentadiene (CPD), dicyclopentadiene, and benzene from room temperature up to > 1700 K. Products of pyrolysis were studied by chemical ionization (CI) and electron impact ionization (EI) mass spectrometry. On a millisecond time scale, JP-10 begins to decompose above 900 K and has decomposed by 1300 K. In the initial decomposition, the principal products are CPD, benzene, C_4H_x, and C_3H_4. At high temperatures, the CPD decomposes, and the principal species observed are benzene, C_3H_4, C_4H_x, ethylene, and acetylene.

9.1 INTRODUCTION

JP-10 is a synthetic, high volumetric-energy-density fuel used in missile applications and in many combustion research studies. Good data are available on thermochemical, rheological, and other physical properties [1–10]. On the other hand, little is known regarding the combustion mechanism. Rate constants have been reported for reaction of JP-10 with OH, O_2, and O_3 [3, 5, 7], but only at room temperature. Shock-tube measurements have been performed on JP-10 ignition, providing information on the kinetics of the combustion but not detailed information about the chemical species involved [11, 12]. Williams et. al. [11] have suggested a mechanism in which the first step of JP-10 breakdown involves breaking a C–C bond common to two five-membered rings, leading eventually to acetylene and ethylene. Davidson *et al.* [12] conclude from shock-tube measurements and kinetic modeling that C_2 species probably play an important role in the decomposition of JP-10. Information regarding product speciation as a function of temperature is important, both in guiding development of combustion models and in development and interpretation of optical diagnostics for JP-10 ignition/combustion.

9.2 EXPERIMENTAL SETUP

The micro-Flow Tube Reactor/Tandem Guided-Ion Mass Spectrometer (micro-FTRMS) is described in detail in previous publications [13, 14]. The only modification here is to use an alumina tube (2.39 mm in inner diameter) heated by external tantalum windings, capable of reaching at least 1800 K. The temperature calibration method is also improved. During reaction runs, the temperature is measured continuously by a thermocouple embedded against the outside wall of the alumina tube. To calibrate this external thermocouple, separate calibration runs are carried out periodically, in which a second thermocouple is inserted inside the flow-tube bore.

Argon or helium at a pressure of $\sim$ 100 Torr is bubbled through JP-10, generating a reactant flow that is $\sim$ 5% JP-10. The reactant mix is metered into the flow-tube through a leak valve where the pressure drops to a few toricelli. The flow properties of the reactor have been discussed in detail previously. Briefly, the total pressure in the heated zone of the flow-tube varies from $\sim$ 3 to $\sim$ 1.2 Torr over the temperature range from 300 to 1700 K, while the residence time drops from $\sim$ 4.5 to $\sim$ 1.5 ms. The flow-tube exhaust is dumped into a temperature-stabilized ($\sim$ 200 °C) ionization source where the pressure is $\sim$ 1 Torr. The analyte molecules can be ionized either by EI or by methane CI. The ions are then mass analyzed by the tandem guided ion-beam quadrupole mass spectrometer described in the above-referenced publications.

9.3 RESULTS AND DISCUSSION

Pyrolysis studies were conducted on JP-10 and also on related compounds that are potential products of JP-10 pyrolysis. Figure 9.1 shows a series of methane CI mass spectra of the material leaving the flow-tube reactor, after JP-10 has passed through at different temperatures. This is a new set of spectra, taken with better control over the ion source conditions. The spectra were actually scanned over a mass range up to more than twice the JP-10 mass (136) to check for adducts or species that might result from polymerization on the flow-tube walls. Only the range up to mass 140 is plotted, as no significant peaks are

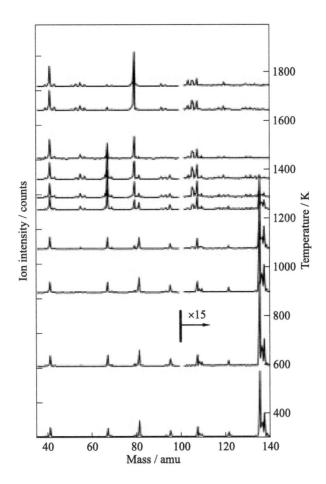

Figure 9.1 Chemical ionization fingerprint spectra of JP-10 passed through the flow-tube reaction at different temperatures.

observed at higher masses. The spectra at the lowest temperatures (right-hand scale) show the CI fingerprint of JP-10, itself. For JP-10, little signal is observed for $(M + H)^+$ (137 daltons), and the CI mass spectrum is, instead, dominated by a characteristic set of fragment ion peaks. The dominant CI fragment peaks, and the corresponding net atom losses from $(M + H)^+$, are 135 ($-H_2$), 95 ($-C_3H_6$), 81 ($-C_4H_8$), 67 ($-C_5H_{10}$), and 41 ($-C_7H_{12}$). In some cases, the net atom losses may correspond to the loss of more than one neutral fragment (e.g., mass 41 might correspond to the loss of $C_7H_{10} + H_2$ rather than C_7H_{12}).

In Fig. 9.1, note that above 1000 K, some peaks characteristic of JP-10 (e.g., mass 81, the group of peaks at 135–137) begin to decrease in intensity, disappearing completely by 1360 K. The disappearance of the JP-10 "fingerprint" CI mass spectrum indicates that JP-10 is decomposing on the millisecond time scale over this temperature range. At the same time, new peaks grow in, indicating the presence of pyrolysis products. There is a group around mass 55, corresponding to C_4 species; a large peak at mass 79, corresponding to $C_6H_7^+$; and peaks at 91 and 93, corresponding to C_7 products. There are also groups of peaks corresponding to C_8, C_9, and C_{10} products; however, these contribute a negligible fraction of the total product signal (note $15\times$ scale change at mass 100). Some mass peaks clearly appear in the spectra of both JP-10 and its pyrolysis products. For example, mass 67, corresponding to $C_5H_7^+$, is a major peak in the JP-10 fingerprint spectra at low temperatures, but increases in intensity over the 1000–1300 K temperature range where JP-10 decomposes. The implication is that $C_5H_7^+$ is produced both in proton-induced fragmentation of JP-10 and in CI of the pyrolysis product(s).

From fitting the temperature dependence of the disappearance of the JP-10 fingerprint peaks, it is straightforward to determine the breakdown vs. temperature behavior of JP-10. In addition, it is desirable to determine the product distribution as a function of temperature. For this purpose, pyrolysis mass spectra of various species that might be expected to appear as products, as well as species whose presence is suggested by the mass spectral patterns, were also studied. As discussed in previous reports, the main products have been identified as CPD, benzene, C_3H_4, and C_4H_x, along with minor amounts of heavier fragments. Previous work has also tracked pyrolysis of CPD and benzene at high temperatures.

The authors' efforts in the past year have focused on two issues. One is the product distribution at the highest temperatures, where primary JP-10 pyrolysis products like benzene and CPD are undergoing further decomposition. This is a particularly difficult problem as these secondary decomposition products have masses that are interfered with by fragmentation of heavier species and by background peaks from the methane CI source. The other issue is to verify that the major products are really CPD and benzene, without significant contribution from C_5H_x or C_6H_x, $x > 6$.

To address these problems, JP-10 pyrolysis was studied under identical flow-tube conditions, detecting products with low-energy EI, rather than CI. The problem with EI is that to get high detection efficiency, electron energies well in excess of molecular dissociation energies are used (typical EI uses 50–70 eV electrons). For typical hydrocarbons, EI results in extensive production of fragment ions, dependent on both the molecule and the electron energy. Indeed, for JP-10 under standard conditions, essentially no signal is observed for ion masses greater than 95, i.e., $> 99\%$ of the JP-10 fragments upon ionization. This high degree of fragmentation is consistent with the photoionization experiments of Fedorova et al. [1, 15], where it was found that a variety of C_5, C_6, C_7, C_8, and C_9 fragment ions have appearance energies within 1 eV of the JP-10 ionization

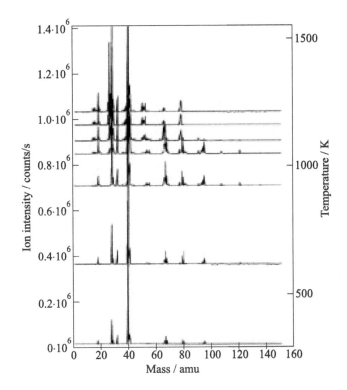

Figure 9.2 Electron impact ionization mass spectra of JP-10 in argon as a function of flow-tube temperature.

energy. Given that EI tends to generate ions with significant internal energy, it is not surprising that there is extensive fragmentation.

While EI fragmentation patterns are very useful as fingerprints for different isomers, etc., extensive fragmentation complicates the analysis of a mixture of hydrocarbons, such as the material exiting the flow-tube at high temperatures. The authors collected their EI data at an electron energy of 25 eV, reducing the degree of fragmentation substantially, at the cost of reduced total ionization efficiency. The variable-temperature EI spectra for JP-10 diluted in argon are shown in Fig. 9.2. Figure 9.3 gives the low-mass region of the EI spectra for JP-10 diluted in helium. The motivation for the helium data set was to avoid mass 40 background that might mask the presence of a $C_3H_4^+$ product.

Consider the room temperature spectrum, where one knows that JP-10 is stable. Note that even for low electron energies, the JP-10 molecular mass peak (136) is very weak, i.e., there is extensive fragmentation even at the lower electron energy. JP-10 is quite unusual in this regard. For example, of the 40 $C_{10}H_{16}$

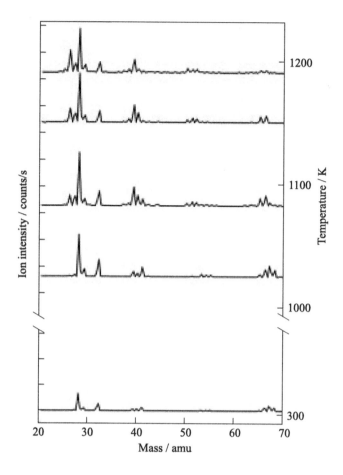

Figure 9.3 Electron impact ionization mass spectra of JP-10 in helium.

isomers in the National Institute of Standards and Technology (NIST) on-line
mass spectral database [16], all but a few, and all the cyclic and polycyclic
isomers, give mass spectra with substantial molecular mass peaks, even for 70 eV
electron energy. Despite the fragmentation, the EI spectra make several points
clear. First, the JP-10 breakdown temperature dependence inferred from the CI
spectra is in good agreement with that inferred from the temperature dependence
of the EI data. The EI data also corroborate the assignments of CPD and
benzene as the only significant C_5 and C_6 pyrolysis products. In particular,
the EI data definitively rule out any significant contributions to the pyrolysis
product distribution from C_5H_8 or C_6H_8 isomers. Consider C_5H_8. All 17 of the
C_5H_8 isomers listed in the NIST mass spectral database give strong peaks at

mass 67 and 68 under standard 70 eV EI conditions. Under the low-energy EI conditions, these near-molecular mass peaks would be even stronger, owing to reduced fragmentation. In the EI spectra of JP-10 at high flow temperatures, where JP-10 is completely decomposed, the only peaks in the C_5 mass range are at 65 and 66, eliminating the possibility that any of the C_5H_8 isomers are present in significant concentration. Based on the analogous data for C_5H_{10} and for C_6H_x ($x = 8$, 10), one can conclude that the only significant C_5 and C_6 products are CPD and benzene, respectively.

The EI spectra also provide some insight into the low-molecular-weight region of the product distribution. Low-mass fragment ions are observed even at low temperatures, resulting from EI-induced fragmentation of JP-10. Note, however, that there is no signal at mass 26 in the JP-10 low-temperature EI spectrum. Thus, the appearance of this mass in the high-temperature spectrum indicates that it originates from a pyrolysis product. The mass 26 signal could conceivably result from EI-induced fragmentation of a heavier pyrolysis product, such as CPD, benzene, C_4H_6, or C_3H_4, all identified in the CI spectra at high temperatures. Note, however, that in the high-temperature EI spectra, mass 26 is considerably more intense than 27. For CPD, benzene, and all common isomers of C_4H_6 and C_3H_4, the EI fragmentation pattern has considerably more intensity for 27 than 26, eliminating these species as the major carrier of the 26 signal.

Instead, one concludes that the mass 26 peak indicates the production of acetylene as a significant high-temperature JP-10 pyrolysis product. The major source of acetylene is probably the secondary decomposition of CPD at high temperatures. As shown by separate experiments on CPD pyrolysis, CPD decomposes to C_3H_4 (mass 41 in CI), implying acetylene as the other product. The appearance of low-intensity 26 peaks in the JP-10 EI spectrum near 1100 K, prior to the onset of strong CPD decomposition, may indicate a second route to acetylene.

The other major low-mass peak that grows in the temperature range where JP-10 decomposes is mass 28, $C_2H_4^+$. In this case, the absence of strong mass 27 and 29 peaks suggests that EI-induced fragmentation of higher mass pyrolysis products is not the dominant contribution to the mass 28 signal. Standard mass spectra of likely C_3, C_4, C_5, and C_6 species indicated that many such molecules do give significant mass 28 EI fragment peaks; however, they also tend to give strong peaks at mass 27 and/or mass 29 [16]. The large 28:27 and 28:29 ratios in the high-temperature JP-10 EI spectra indicate that EI of a C_2H_4 pyrolysis product must be the dominant contributor to the mass 28 signal.

To extract quantitative JP-10 breakdown vs. temperature results, the set of JP-10 pyrolysis CI mass spectra was fit, using a contracting grid, least-squares program developed for this purpose. Recall that the CI spectra do not cover the mass range corresponding to C_1 or C_2 products, because of high background in this mass range. The fitting analysis, therefore, only provides information regarding C_n, $n \geq 3$, products, although the EI data discussed above clearly shows that C_2H_2 and C_2H_4 are major high-temperature products, as well. Currently, the authors are working on a procedure to integrate the information from the EI and CI spectra, although this is a rather complex task. The experimental spectra are first fit to a series of voight peak profiles, whose areas are used to generate stick spectra. These stick spectra are fit as linear combinations of basis spectra, where the basis spectra are similarly generated stick spectra for known or suspected product species. Recall that the ion source is held at constant temperature and pressure, independent of flow-tube temperature, and that there are sufficient collisions with buffer gas to equilibrate the gas entering the source prior to ionization. As a consequence, the basis spectrum for any given compound should be flow-tube-temperature independent, over the range where that compound is stable. The fitting routine allows penalties to be assigned for nonphysical fits (e.g., for negative residual peaks) and, thus, allows one to see the contribution of each basis spectrum to the JP-10 spectrum at a given temperature and also the residual spectrum not captured by the basis spectra.

The raw fits give the fraction of the total ion signal at each temperature, contributed by each compound. To get the desired fraction of the neutral composition, one needs to correct the ion ratios for differences in chemical ionization efficiency for different compounds. Chemical ionization efficiency depends on two

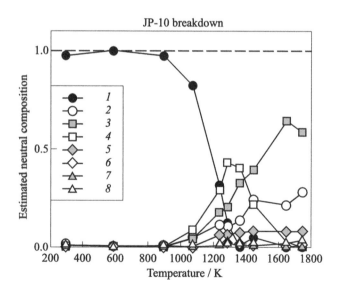

Figure 9.4 Corrected JP-10 breakdown curves: 1 — JP-10; 2 — C_3H_4; 3 — benzene; 4 — CPD; 5 — C_4H_x; 6 — C_7H_x ($x = 6.8$); 7 — C_7H_{10}; and 8 — unfit residual.

factors: the collision cross-section and the proton transfer probability per collision. Both factors depend in predictable ways on molecular properties. The ion–molecule collision cross-section depends on the polarizability and dipole moment of the neutral species. Both parameters either are known or can be calculated *ab initio*. The proton transfer probability depends on the relative proton affinities for the analyte molecules and methane, which is chosen precisely because it has a very low proton affinity. For all the molecules of interest here, the equilibrium constant for proton transfer is large, i.e., one expects proton transfer on every collision.

The fitting results are given in Fig. 9.4, corrected for collision cross-section effects. Note that JP-10 begins to decompose between 900 and 1000 K and is completely decomposed by $\sim$ 1350 K. Initially, the major product is CPD; however, CPD is unstable at high temperatures and is replaced by other products. The major high-temperature products are benzene, propyne, and C_4H_x ($x = 4$, 6, 8), with smaller amounts of C_7H_x ($x = 6, 8, 10$) and a variety of C_8 and C_{10} species, grouped under "unfit residuals" in the figure. Bear in mind that there is substantial contribution from C_2H_2 and C_2H_4 at temperatures above 1000 K, not shown in the figure.

9.4 CONCLUDING REMARKS

The thermal decomposition of JP-10 over a time and temperature range relevant to combustion has been probed. As expected for a complex hydrocarbon like JP-10, there are numerous decomposition pathways. The initial decomposition produces mostly C_5 products, but at higher temperatures, the C_5 intensity decreases, and a range of C_2, C_3, C_4, and C_6 products are observed. At the highest temperatures, the product distribution shifts progressively to the lighter hydrocarbon products.

ACKNOWLEDGMENTS

This work was performed under ONR contract N00014-01-1-0541.

REFERENCES

1. Fedorova, M. S., Y. V. Denisov, and V. K. Potapov. 1973. Mass-spectrometric study of the photoionization processes of tricyclo[5.2.1.02,6]decane and its alkyl derivatives. *Russ. J. Physical Chemistry* 47:1498.

2. Smith, N. K., and W. D. Good. 1979. Enthalpies of combustion of ramjet fuels. *AIAA J.* 17:905.

3. Popov, A. A., N. N. Blinov, N. S. Vorob'eva, G. E. Zaikov, and S. G. Karpova. 1981. Kinetics of the reaction of ozone with bi- and tricycloparaffins. *Rus. J. Kinet. Katal.* 22:139.

4. Inman, R. C., and M. P. Serve. 1982. The fragmentation behavior of the endo- and exo-octahydro-4,7-methano-1H-indene systems. *Organic Mass Spectrometry* 17:220.

5. Atkinson, R., S. M. Aschmann, and W. P. L. Carter. 1983. Rate constants for the gas-phase reactions of hydroxyl radicals with a series of bi- and tricycloalkanes at 299 ± 2 K: Effects of ring strain. *Int. J. Chem. Kinet.* 15:37.

6. Herzschuh, R., H. Kuehn, and M. Muehlstaedt. 1983. Mass spectrometric analysis of hydrocarbons. XV. Effect of strain energies on the ionization and fragmentation of tricyclo[5.2.1.02,6]decanes, -decenes and -decadienes. *Russ. J. Prakt. Chem.* 325:256.

7. Nesterov, M. V., V. A. Ivanov, V. M. Potekhin, and A. I. Grigor'ev. 1984. Initiated oxidation of exo-tricyclodecane by molecular oxygen. *Russ. J. Prikl. Khim.* 57:1102.

8. Lin, Y. T., C. Lin, K. Liou, S. S. Cheng, and M. J. Chang. 1986. High energy fuels. II. Gas chromatographic separation of energetic compounds derived from dicyclopentadiene. *J. Chinese Chemical Society* 33:341.

9. Brossi, M., and C. Ganter. 1988. The adamantane rearrangement of syn- and anti-tricyclo[4.2.1.12,5]decane. Part II. Rearrangements initiated by regioselective formation of carbocations at C(3) and C(9). *Helv. Chim. Acta* 71:848.

10. Steele, W. V., R. D. Chirico, S. E. Knipmeyer, and N. K. Smith. 1989. High-temperature heat-capacity measurements and critical property determinations using a Differential Scanning Calorimeter: Results of measurements on toluene, tetralin, and JP-10. Natl. Inst. Pet. Energy Res. Bartlesville, OK, USA. Report.

11. Williams, F. A., R. K. Hanson, and C. Segal. 1999. Fundamental investigations of pulse-detonation phenomena. *CPIA Publication* 692:151.

12. Davidson, D. F., D. C. Horning, J. T. Herbon, and R. K. Hanson. 2000. Shock tube measurements of JP-10 ignition. *Proc. Combustion Institute* 28:1687.

13. Li, Z., J. Eckwert, A. Lapicki, and S. L. Anderson. 1997. Low energy high resolution scattering mass spectrometry of strained molecules and their isomers. *Int. J. Mass Spectrometry Ion Processes* 167/168:269.

14. Li, Z., and S. L. Anderson. 1998. Pyrolysis and isomerization of quadricyclane, norbornadiene, and toluene. *J. Physical Chemistry A* 102:9202.

15. Fedorova, M. S., Y. V. Denisov, and V. K. Potapov. 1973. Mass-spectrometric study of the photoionization of tricyclo[5.2.1.02,6]decane and its alkyl derivatives. *Sov. J. Fiz. Khim.* 47:2667.

16. Stein, S. E. 2000. IR and mass spectra. In: *NIST Chemistry WebBook, NIST Standard Reference Database Number 69.* Eds. W. G. Mallard and P. J. Linstrom. (http://webbook.nist.gov). Gaithersburg, MD: National Institute of Standards and Technology, NIST Mass Spec Data Center.

Chapter 10

LASER DIAGNOSTICS AND COMBUSTION CHEMISTRY FOR PULSE DETONATION ENGINES

R. K. Hanson, D. W. Mattison, L. Ma, D. F. Davidson, and S. T. Sanders

Recent progress in laser-based diagnostic development and shock-tube studies of fuel decomposition related to pulse detonation engines (PDEs) is presented. Newly developed diode-laser-based sensors for *in situ* measurements of flow properties used for simulation validation, fuel-charge monitoring, active PDE control, and spray characterization have been applied to facilities at both the Naval Postgraduate School (NPS) and Stanford University. The diode-laser techniques presented include a wavelength-agile temperature and pressure sensor, a propane concentration sensor, an ethylene-based active control scheme, and a two-phase flow diagnostic. Shock-tube measurements of JP-10 ignition times and decomposition products have been performed to aid in the development of reduced mechanisms.

10.1 INTRODUCTION

Typical methods of studying pulse detonation engines (PDEs) use pressure, thrust, and qualitative imaging techniques to examine PDE performance. While many of these techniques and results are important to PDE development, they do not provide the full range of data necessary to validate computational simulations or characterize the performance of specific PDE components such as the fuel-loading schemes. Newly developed diode-laser schemes provide a practical method of extracting important system information such as temperature, pressure, burned-gas velocity, fuel-charge distribution, and fuel-spray properties in PDE experiments [1–3]. These diagnostics can thus play an important role in PDE development and advancement.

The development of ignition time correlations and kinetic models for use in PDE simulations requires accurate, reproducible ignition time data and individual species concentration time history profiles. Ignition times and species profiles have been measured in a shock tube for a wide range of fuel components of interest in PDE research, employing *in situ* absorption techniques to accurately measure fuel concentrations. Using these data, ignition time correlations have been developed and current chemical kinetic models have been tested with a view to developing reduced mechanisms to describe the fuels individually and in groups.

Progress is reported on the development of a wavelength-agile temperature and pressure sensor, a propane concentration sensor, an ethylene-based active control scheme, and a two-phase flow diagnostic. In the area of shock-tube kinetics studies, progress is reported on JP-10 ignition times and decomposition products to aid in the development of reduced chemical mechanisms.

10.2 WAVELENGTH-AGILE TEMPERATURE AND PRESSURE SENSOR

Previously, a multiple, fixed-wavelength sensor for measuring the gas temperature of detonation products was reported [1]. While this sensor provided important results, the rapid, broad-wavelength-scanning capabilities of advanced diode-laser technologies allow for a simpler and more robust technique to measure both temperature and pressure simultaneously [4, 5]. This technique, described below, provides gas temperature and pressure histories, spanning 2000–4000 K and 0.5–30 atm, respectively, with microsecond time response.

Figure 10.1 depicts a single-wavelength sensor used to measure PDE burned-gas temperature and pressure. For this technique, a vertical cavity surface emit-

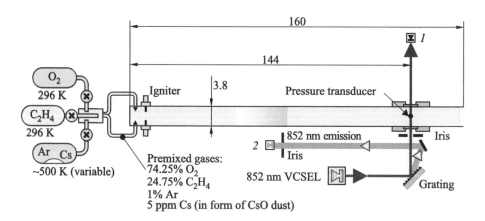

Figure 10.1 Schematic of the Stanford PDE facility, with VCSEL-absorption sensor applied to measure gas temperature and pressure near the exit. Detector *1* monitors Cs absorption lineshapes and detector *2* monitors thermal emission from Cs. Dimensions are in cm.

ting laser (VCSEL) probes the absorption lineshape of the $\sim$ 852-nanometer D_2 transition of atomic cesium which is seeded (5 ppm seeding fraction) into the feedstock gases of the C_2H_4/O_2-based Stanford PDE facility. Using aggressive injection current modulation, the VCSEL is scanned across a 10 cm^{-1} spectral window at a 1-megahertz rate, providing absorption lineshapes with microsecond time resolution. As the detonation wave passes the measurement station, detector 1 monitors Cs-absorption lineshapes, and detector 2 simultaneously monitors Cs emission from the volume probed by the VCSEL beam. By performing a two-line Voigt fit on the Cs-absorbance lineshape, the spectroscopic parameters including integrated area and collisional width can be extracted.

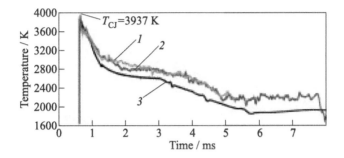

Figure 10.2 Measured (1 — $I_{Cs,kin}$ (collisional linewidth) and 2 — $T_{Cs,el}$ (population ratio)) and computed (3 — NRL simulation) gas temperatures for detonation of stoichiometric C_2H_4/O_2 mixture.

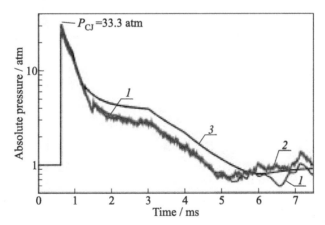

Figure 10.3 Measured (1 — pressure transducer and 2 — spectroscopic) and computed (3 — NRL simulation) pressures for detonation of stoichiometric C_2H_4/O_2 mixture.

Temperature can be measured by two methods using this technique; the results are shown in Fig. 10.2 compared to a simulation performed by the Naval Research Laboratory (NRL) [6]. The electronic temperature, $T_{Cs,el}$, is obtained from the ratio of the Cs population in the excited state (emission signal) to the Cs population in the ground state (integrated absorbance area). The kinetic temperature, $T_{Cs,kin}$, is calculated using a pressure measurement obtained from a sidewall transducer and the temperature-dependent collisional width of the absorption lineshape. Spectroscopically measured pressure, shown in Fig. 10.3 compared to both NRL simulations and transducer measurements, can be calculated using the electronic temperature and the measured collisional width of the Cs-absorption lineshape. For both the temperature and pressure results, trends in measurements agree with the NRL simulations, although discrepancies in the magnitude are apparent. These results are an important component of an ongoing effort to advance PDE development by validating numerical simulations.

10.3   PROPANE SENSOR

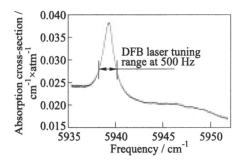

Figure 10.4 High-resolution absorption spectrum of propane near 1.68 μm measured by temperature-tuning a distributed feedback (DFB) diode laser: $P_{total} = 1$ atm and $T = 291$ K.

Previously, a 3.39-micrometer He–Ne absorption scheme was used to measure propane concentration [7]. Sensors based on this scheme have the advantage of offering strong absorption because many fundamental bands of propane are located near the 3.39-micrometer wavelength region, but the fixed-wavelength lasers available in this region limit this sensor's applications in practical systems where interference absorption and noise are common problems.

Recently, development of a more accurate and robust propane sensor by replacing the fixed-wavelength He–Ne laser with a tunable diode laser was initiated. The strongest absorption of propane in the near-infrared region occurs near 1.68 μm. Figure 10.4 shows the first known high-resolution spectral data of propane near 1.68 μm, from which one can identify the optimum detection strategy. In this work, the diode laser was scanned across the absorption peak at 1.6837 μm at 500 Hz to measure propane concentration. This sensor was applied to the head-end of the PDE at Stanford University, as shown in Fig. 10.5. The initial results of propane measurements are shown in Fig. 10.6. These results contain a wealth of information

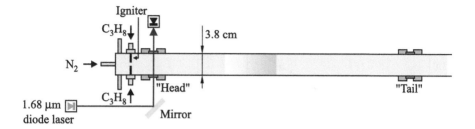

Figure 10.5 Schematic of propane sensor applied to the Stanford PDE.

for improving engine operation, optimizing valve timing, and actively controlling the engine.

10.4 ETHYLENE-BASED ACTIVE CONTROL

Figure 10.7 depicts the scanned-wavelength ethylene sensor that is used to optimize fuel consumption and maximize impulse through an active PDE control scheme. This technique has been detailed previously [3]. In brief, the diode laser is modulated

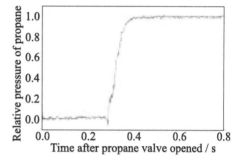

Figure 10.6 Record of propane concentration at head-end measurement station.

across the C_2H_4 combination band Q-branch near 1.62 μm. The Stanford PDE is filled using finite-volume supply tanks of ethylene and oxygen. At the beginning of the experiment, the supply gas valves are opened. When fuel is detected

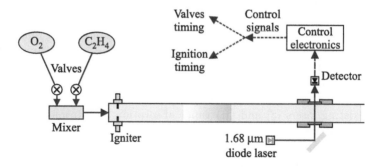

Figure 10.7 Schematic of ethylene-based active control scheme applied to the Stanford PDE.

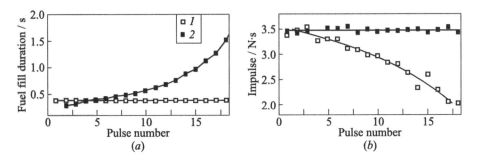

Figure 10.8 Active control experiments to realize full tube fills in the Stanford PDE: (a) indicates fuel filling duration, and (b) indicates cycle impulse for the cases of fixed valve opening duration and variable opening for active control: 1 — fixed duration and 2 — active control.

at the tail-end using this sensor, a control signal is sent to close the mixture filling valves and fire the ignitor. After a fixed-duration cooling cycle, the control scheme is repeated until the gas supply tanks have emptied. Figure 10.8a shows the results of the gas filling duration for this set of experiments. As the supply tanks are depleted, the control scheme adjusts the filling duration to ensure full tube fills. As seen in Fig. 10.8b, this active control maintains constant impulse compared to fixed valve timing.

Similarly, the ethylene-based control system was used to actively control the spark timing of the NPS predetonator tube. When ethylene is detected at the tail-end, a signal is sent to actuate the ignitor ensuring full tube fills and minimizing wasted fuel. As shown in Fig. 10.9, the missing peaks in the equivalence ratio histories are due to detonation failure due to pulse-to-pulse interference. The actively controlled spark is able to reduce this performance-degrading behavior.

10.5 TWO-PHASE MIXTURE DIAGNOSTIC

Research is underway to develop a rapid-response diagnostic for simultaneous measurements of line-of-sight-averaged fuel droplet size, droplet volume fraction, and fuel vapor concentration. Laser diagnostics for monitoring both vapor and liquid droplets are complex, in that (i) the vapor of most liquid hydrocarbon fuels has diffuse spectral features which cannot be covered by the rapid-tuning range of existing lasers [8] and (ii) vapor absorption and droplet extinction of the probe beam are usually coupled. The initial approach is a sensor based on wavelength-multiplexing of five laser beams: three are used to measure droplet size (Sauter mean diameter, D_{32}) based on transmission measurements; and the other two are used to monitor vapor concentration based on differential absorption [3]. Successful initial results in a laboratory-scale ethanol spray, as

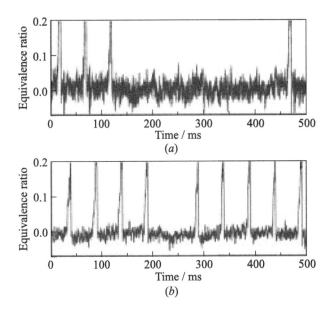

Figure 10.9 Results of active control experiments at the NPS PDE facility using a 3.8-centimeter diameter, 20-centimeter-long tube running on ethylene/oxygen with a continuous air purge. The equivalence ratio at the tail-end is shown for (*a*) fixed spark timing (control off) and (*b*) actively controlled spark timing using the fuel diagnostic (control on).

shown in Fig. 10.10, motivated further work and extended applications of this technique in propulsion systems.

10.6 SHOCK-TUBE STUDIES

The primary goals of the shock-tube studies are to characterize ignition times, to measure selected individual species concentration time histories, and to develop and validate reduced mechanisms for fuels used in PDE models and experiments. Previously, the ignition times and OH concentration time histories for propane and other *n*-alkanes were investigated, and the behavior of current oxidation mechanisms for these fuels was characterized. Currently, a database of ignition times and species concentration time histories for JP-10 is under development by using a variety of diagnostics to assist in characterizing JP-10 decomposition pathways and in developing a reduced mechanism for this fuel.

Existing shock-tube ignition time measurements are known to have relatively large scatter and poor facility-to-facility reproducibility. Several experimental methods have been developed to improve the quality of ignition time data

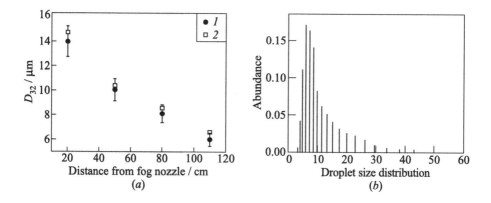

Figure 10.10 (*a*) Comparison of droplet size, D_{32}, measured with diode-laser sensor (*1*) and with Malvern Particle Sizer (MPS) (*2*). (*b*) Inset shows droplet size distribution measured by MPS 50 cm from the air-blast injector [3].

and alleviate some of the deficiencies in earlier measurements. Ignition time is determined from time histories of CH-emission traces, as well as pressure. Concentration of condensable fuels is now determined by *in situ* laser absorption at 3.39 μm and gas chromatography, in addition to manometric measurement.

The measured ignition times and activation energies of propane and other *n*-alkanes are in good agreement with earlier work, but now show substantially smaller scatter. The activation energies of propane and the *n*-alkanes (*n*-butane, *n*-heptane, *n*-decane) and JP-10 were found to be similar and substantially different from that for ethylene [9].

JP-10 Decomposition

To develop a kinetic mechanism for JP-10 oxidation, information about the initial decomposition pathway is needed. The initial JP-10 decomposition steps have been studied in a shock tube using a recently developed ultraviolet kinetic spectrograph [10]. Figure 10.11 shows the absorption cross-sections of JP-10 products at early times. At 5 and 15 μs, the JP-10 product spectra are nearly identical to 0.5× and 0.9× the cyclopentene products spectra, respectively. This allows one to suggest that JP-10 decomposes into cyclopentene at a very early stage and that the yield for this path approaches unity. At 50 μs, the JP-10 decomposition product absorption cross-section is nearly identical to that of twice the decomposition product spectra of cyclopentadiene. Both these observations are consistent with an initial decomposition pathway of JP-10 to cyclopentene and a C_5-compound. This C_5-compound has secondary decomposition products similar to cyclopentadiene (i.e., C_5H_5 and others). Using this technique, several

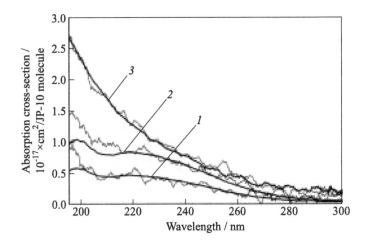

Figure 10.11 Absorption cross-sections of JP-10 products (thin curves). Reflected shock conditions: 1462 K, 1.24 atm, 250 ppm JP-10/Ar. Thick curves: $1 — 0.5 \times C_5 H_8$ products; $2 — 0.9 \times C_5 H_8$ products; and $3 — 2.0 \times C_5 H_6$ products.

other suggested major decomposition product pathways for JP-10 can be ruled out based on their absorption spectra. Thus, there is evidence that only small quantities of 1,3-butadiene and benzene are expected to form at early times.

The Stanford JP-10 mechanism is constructed from two parts: the initial JP-10 decomposition pathway derived from the results of kinetic spectrograph experiments and a C_4–C_6 mechanism taken from Laskin *et al.* [11].

A comparison of the modeled and measured ignition times is shown in Fig. 10.12. Using the proposed JP-10 product distribution and the selected C_4–C_6 mechanism, there is good agreement with both the activation energy and the absolute magnitude of the ignition times over a wide range of temperatures (1350–1650 K) and pressures (1–6 atm). The ease in which the overall mechanism was developed once the decomposition pathway was known confidently is a strong validation for this approach to the study of fuel chemistry.

Other researchers have recently seen evidence that is supportive of the JP-10 decomposition pathway to cyclopentene. Burcat and Dvinyaninov [12] reported that they found cyclopentene in their single-pulse shock-tube experiments of JP-10 oxidation. Green and Anderson [13] reported that they found cyclopentadiene in their recent flow-reactor measurements of JP-10 pyrolysis.

10.7 CONCLUDING REMARKS

Diode-laser sensors have proven useful for measurements in harsh measurement environments such as the PDE. They have been used to extract a number of

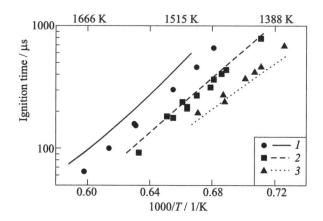

Figure 10.12 Comparison of modeled (curves) and measured (symbols) ignition times (for the stoichiometric JP-10/O_2/Ar mixture with 0.2% JP-10): *1* — 1 atm; *2* — 3 atm; and *3* — 6 atm.

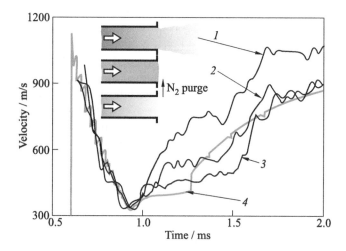

Figure 10.13 Measured and simulated burned-gas velocity near the tube exit of the Stanford PDE facility for various exit boundary conditions for C_2H_4/O_2 detonations: *1* — overfill; *2* — uniform fill; *3* — underfill; and *4* — NRL simulation.

performance-important parameters that both help designers optimize system performance and help simulators validate numerical codes.

Future diode-laser measurements will begin to investigate the effects of changing exit boundary conditions such as changes caused by incorporating nozzles into system geometry. Other work will also include expanding fuel-sensing capabilities to probe additional species and multiphase conditions.

For the exit boundary condition studies, it will be important to study system parameters such as burned-gas velocity in the nozzle and the possibility of nonequilibrium product composition. The ability to measure burned-gas velocity near the exit of a straight tube has been demonstrated [2]. The results of exit boundary condition effects on burned-gas velocity is shown in Fig. 10.13. The same technique used to perform these measurements can be extended to investigate virtually any nozzle geometry.

A new PDE tube that will allow for expanded optical and mechanical access to the entire length of the tube will also be developed at Stanford University.

ACKNOWLEDGMENTS

This work was performed under ONR contract N00014-99-0475. The authors gratefully acknowledge Prof. C. M. Brophy and Prof. D. W. Netzer at NPS, Monterey, CA, for hosting their visits to demonstrate diode-laser sensors in the Rocket Propulsion and Combustion Laboratory. The authors also thank Dr. K. Kailasanath of NRL in Washington, D.C., for providing detonation simulations and for helpful technical discussions. The authors also acknowledge the PDE team members including Kevin Hinckley and Eric Romo for assistance in performing experiments.

REFERENCES

1. Sanders, S. T., J. A. Baldwin, T. P. Jenkins, D. S. Baer, and R. K. Hanson. 2000. Diode-laser sensor for monitoring multiple combustion parameters in pulse detonation engines. *Combustion Symposium Proceedings* 28:587–94.

2. Sanders, S. T., D. W. Mattison, J. B. Jeffries, and R. K. Hanson. 2003. Time-of-flight diode-laser velocimetry using a locally-seeded atomic absorber: Application in a pulse detonation engine. *Int. J. Shock Waves* 12:435–41.

3. Ma, L., S. T. Sanders, J. B. Jeffries, and R. K. Hanson. 2002. Monitoring and control of a pulse detonation engine using a diode-laser fuel concentration and temperature sensor. *Combustion Institute Proceedings* 29:161–66.

4. Sanders, S. T., D. W. Mattison, L. Ma, J. B. Jeffries, and R. K. Hanson. 2002. Wavelength-agile diode-laser sensing strategies for monitoring gas properties in optically harsh flows: Application in cesium-seeded pulse detonation. *Optics Express* 10:505–14. (http://www.opticsexpress.org/abstract.cfm?URI=OPEX-10-12-505.)

5. Sanders, S. T., D. W. Mattison, J. B. Jeffries, and R. K. Hanson. 2002. Sensors for high-pressure, harsh combustion environments using wavelength-agile diode lasers. *Combustion Institute Proceedings* 29:2661–67.

6. Kailasanath, K., and G. Patnaik. 2000. Performance estimates of pulsed detonation engines. *Combustion Institute Proceedings* 28:595–601.

7. Horning, D. C. 2001. A study of the high-temperature autoignition and thermal decomposition of hydrocarbons. Ph.D. Thesis, Stanford, CA: Stanford University.

8. Ma, L., J. B. Jeffries, E. A. Romo, and R. K. Hanson. 2003. Two-phase fuel measurements using a diode-laser sensor. AIAA Paper No. 2003-0401.

9. Davidson, D. F., D. C. Horning, J. T. Herbon, and R. K. Hanson. 2000. Shock tube measurements of JP-10 ignition. *Combustion Institute Proceedings* 28:1687–92.

10. Davidson, D. F., D. C. Horning, M. A. Oehlschlaeger, and R. K. Hanson. 2001. The decomposition products of JP-10. AIAA Paper No. 2001-3707.

11. Laskin, A., H. Wang, and C. K. Law. 2000. Detailed kinetic modeling of 1,3-butadiene oxidation at high temperatures. *Int. J. Chemical Kinetics* 32:589–614.

12. Burcat, A., and H. Dvinyaninov. 1997. Detailed kinetics of cyclopentadiene decomposition studied in a shock tube. *Int. J., Chemical Kinetics* 29:505–15.

13. Green, R. J., and S. L. Anderson. 2000. Pyrolysis chemistry of JP-10. *13th ONR Propulsion Meeting Proceedings*. Minneapolis, MN. 271–76.

Comments

Frolov: At one of your viewgraphs, I have seen the comparative Arrhenius plot of measured ignition-delay times for JP-10, propane, *n*-heptane, and *iso*-octane. At this plot, *iso*-octane has considerably lower ignition delays than *n*-heptane that is in contradiction with a bulk of existing data. Could you comment on this?

Hanson: Our measurements show that the ignition times for *iso*-octane (at a fixed temperature, stoichiometry, and pressure) vary strongly with oxygen concentration. At low oxygen concentration $\sim 0.5\%$, the ignition times for *iso*-octane are longer than those for *n*-heptane, while at high oxygen concentration $\sim 21\%$, the ignition times for *iso*-octane are shorter than those of *n*-heptane. In the viewgraph in question, the ignition times measured over a wide range of concentrations were correlated to 21% oxygen. Ignition time measurement at high *iso*-octane concentration by Burcat *et al.* (1991. *Conference (International) Shock Waves Proceedings* 18:771–80) and Niemitz *et al.* (In: Westbrook *et al.*, 1988. *Proc. Combustion Institute* 22:893–901) support this trend. This rapid falloff in ignition time with oxygen concentration is also found in several recent *iso*-octane mechanisms (Curran *et al.*, 1998. *Proc. Combustion Institute* 27:379–87; and Ranzi *et al.*, 1997. *Combustion Flame* 108:24–42) and should be the subject of further experimental investigations.

Chapter 11

COMPUTATIONAL STUDIES
OF PULSE DETONATION ENGINES

K. Kailasanath, C. Li, and S. Cheatham

Computational studies of pulse detonation engines (PDEs) include assessing the impact of chemical recombination and detonation initiation conditions on the computed performance and estimating the theoretical performance of an ideal PDE. A wide range of cases involving partial fuel-fill effects have been investigated to quantify the effects, elucidate the underlying mechanism, and derive a limiting value for the performance enhancement. The general expression for a quick estimation of the performance of an idealized PDE has been extended to a wider range of fuel–oxygen and fuel–air mixtures and modified to account for partial fuel-fill effects. Progress to date on liquid-fueled detonations and easing the transmission of an ethylene–air detonation from a small initiator to the main detonation tube is also reported.

11.1 INTRODUCTION

There are several reasons for investigating PDEs, such as simplicity of the device, ability to operate over a range of speeds from zero to hypersonic, and overall performance. As part of the U.S. Office of Naval Research (ONR) program, computational tools are being developed and applied to gain a better understanding of the operation and performance of PDEs. The basic technical issues to be addressed and the various computational tools needed have already been discussed in previous years' reports [1]. Some of the key issues that remain unresolved are the ideal performance estimate of the PDE, rapid initiation of detonations with minimum oxygen, and operation with multiphase fuels. Here, the last accomplishments are presented, and their implications are discussed.

In [2], it was shown that numerical predictions of the pressure, temperature, and fluid velocity behind an ethylene–oxygen detonation in a PDE thrust tube were in excellent agreement with experimental data from Stanford University. The performance predictions for an ethylene–oxygen mixture were compared to various available data and good agreement was shown. When there were differences, additional simulations performed were able to explain them. A general expression that can be used to quickly estimate the performance of an idealized PDE consisting of a tube closed at one end and open at the other was also presented. The investigations have been continued to a range of fuel–oxygen and fuel–air mixtures, and a slightly revised general expression is presented. This expression is also compared to similar expressions based on theoretical considerations and experimental data that have become available.

In spite of the general agreement noted above for an idealized system, performance estimates continue to be a controversial issue. The impact of chemical recombination on the computed performance and the maximum possible theoretical performance are two of the controversial issues that have been addressed recently [3] and will be discussed in more detail here.

The advantages of partially filling the PDE thrust tube with a fuel–air mixture and the rest with air were discussed in [4]. The results were compared to experimental data [5], and they were found to be in good qualitative and quantitative agreement. Additional simulations of PDE tubes of various lengths and various extents of fuel fill have been conducted, and a general expression for the performance enhancement from partial fuel filling was derived [6].

Experimentally, it is difficult to directly initiate most fuel–air mixtures of interest in PDE tubes of realistic sizes. Therefore, one of the approaches adopted is to use a more detonable fuel–oxygen mixture to initiate a detonation wave and then transit the detonation into the desired fuel–air mixture. However, for this approach to be practically feasible, the amount of oxygen needed must be reduced. The results on the detonation transition from ethylene–oxygen to ethylene–air mixtures have been presented and compared to experimental observations from the Naval Postgraduate School, Monterey, CA [7, 8]. Currently, the investigations are being continued with a view to reducing the amount of oxygen needed. Progress to date on this topic is presented in this chapter.

Pulse detonation engines operating on liquid fuels are the overall emphasis of the ONR program. A computational tool capable of accounting for radial and axial variations in liquid-fuel distribution has been developed. Preliminary results using this simulation tool are presented.

The problems highlighted above, as well as their overall implications on the performance and operation of PDEs, are discussed in more detail in the rest of this chapter.

11.2   PERFORMANCE ESTIMATES OF AN IDEALIZED PULSE DETONATION ENGINE

The idealized PDE consists of a tube closed at one end and open at the other end. This is a simple geometry both for computational simulations as well as for obtaining detailed experimental diagnostics. This does not represent an actual flying engine, but can be considered as an idealization to the thrust producing component of a more complex engine consisting of multiple tubes with inlets and nozzles. The details of calculating the various performance measures (thrust, impulse, fuel, and mixture-based specific impulses) of this PDE have been discussed earlier [1, 2] and are not repeated here.

Effects of Chemical Recombination

One of the controversies mentioned earlier was the effect chemical dissociation and recombination will have on the performance estimates [9]. Earlier simulations [10, 11] of a hydrogen–air PDE with detailed chemistry that included the appropriate dissociation and recombination effects implicitly did not show a significant effect. However, those simulations were for short tubes (10–20 cm in length). It is possible that for longer tubes, where more time is available for the detonation products to undergo recombination reactions before being evacuated from the tube, the result could be more significant. To explore this issue, detonation in tubes of various lengths (10–60 cm) filled with a stoichiometric hydrogen–air mixture, initially at 298 K and 1 atm, were simulated using both a detailed chemistry model as well as a two-step global chemistry model. The two-step model includes dissociation effects in the shocked flow, but assumes that the mixture is chemically frozen at the Chapman–Jouguet (CJ) state. Hence, all further energy addition due to recombination is neglected. The detailed chemistry simulations include these effects. The impulse from simulations of a 60-centimeter long PDE tube using the two models is shown in Fig. 11.1. The peak impulse is different by 5.3% and translates into an equivalent gain in specific impulse due to chemical recombination effects. These observations are similar to those made recently in another computational study [12]. A similar study has been conducted with ethylene, but the results are less reliable because of the uncertainties still present in the reaction mechanism and the various rates.

Detonation Initiation

The effects of parameters used for initiating detonations in simulations have been discussed previously [10, 11] and estimated to be as much as 27%. Clearly, this

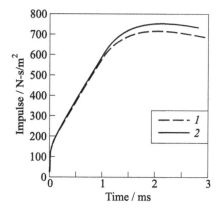

Figure 11.1 Chemistry effects on the time histories of the impulse from simulations of a 60-centimeter-long PDE operating on a stoichiometric hydrogen–air mixture: *1* — two-step induction parameter model and *2* — detailed chemistry.

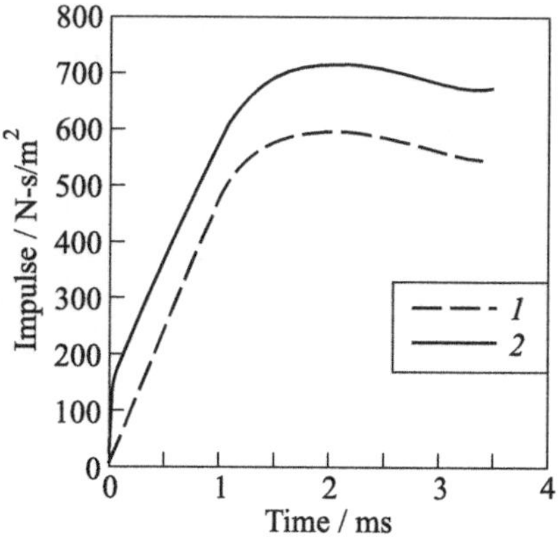

Figure 11.2 Effect of initiation energies on the time histories of the impulse (per unit area) from simulations of a 60-centimeter-long PDE operating on a stoichiometric hydrogen–air mixture: *1* — low-energy initiator and *2* — high-energy initiator.

is an unacceptable uncertainty and needs to be reduced. Use of lower energy initiators and longer tubes has been suggested [11]. Typically, a high-pressure and high-temperature zone of a certain width near the closed-end of the tube is used for initiating detonations in simulations. A series of global-chemistry simulations was carried out with various temperatures, pressures, and zone widths. The impulse from two cases (both for 60-centimeter long tubes) using a high- and a low-energy initiator is shown in Fig. 11.2, and the peak values differ by about 17%. However, the peak specific impulse for the two cases is 4439 and 4161 s, differing by only 6.3%. This is because the mass of fuel used for initiation is also larger for the case with the higher impulse. A 2-centimeter zone near the closed-end of the tube at 50 atm and 3000 K was used as the high-energy initiator, while a 0.2-centimeter zone at 20 atm and 2000 K was used for the low-energy initiator. These simulations again emphasize the importance of considering the contribution from initiators in making performance estimates. These effects could overshadow the differences due to neglecting chemical recombination effects.

Generalization of the Performance Results

In [2], it was reported that good estimates for the propulsive performance measures of a stoichiometric ethylene–oxygen mixture initially at 1 atm are about

2100 N·s/m^3 for the impulse (independent of length), about 163–168 s for the mixture-based I_{sp}, and 725–745 s for the fuel-based I_{sp}. These are in good agreement with experimental data [13]. With the confidence gained from the excellent agreement on the overall performance, simulations have been carried out of PDEs operating on other fuel–oxygen and fuel–air mixtures. Rather than go through the results for each of the fuels, the impulse from the various cases has been normalized using the predicted overpressure, $P_3 - P_0$ (P_3 is the detonation plateau pressure, and P_0 is the ambient pressure), and the residence time of the detonation, t_{CJ} (length of the thrust chamber divided by the CJ-detonation velocity of the mixture). This generalized result is shown in Fig. 11.3. More than one data point for a mixture indicates data from simulations with different tube lengths.

What this generalization implies is that one can estimate the impulse from an idealized PDE knowing the plateau pressure, P_3, and the detonation velocity. That is, the impulse per unit area is given by

$$\frac{I}{A} = 4.65(P_3 - P_0)t_{CJ}$$

This expression derived from the authors' numerical simulations is similar to the expressions for I_{sp} obtained from theoretical analysis [14] and experiments [15]. The constants of proportionality are slightly different in the various studies, suggesting some dependence on the details of the particular configuration such as initiators or tube lengths used for deriving the correlation. Estimates at the Naval Research Laboratory (NRL) have changed by 4% or less as data from new simulations are added to the correlation. This provides an estimate for the level of uncertainty inherent in such correlations.

Partial Fuel-Fill Effects

The effect of filling the thrust tube partially with a fuel–air mixture and filling the rest of the tube with air was investigated numerically using two-dimensional simulations and has been reported earlier [4]. Subsequently, an experimental study [5] showed similar trends. In [2], direct comparisons with experimental data were presented, and the mechanism responsible for the observed performance enhancement was discussed. The emphasis of the work on this topic has been on further clarification of the mechanism involved and derivation of a limiting value for the performance enhancement.

Two sets of multidimensional numerical simulations have been conducted: (i) a fixed-length PDE tube with fuel sections of varying length (L_f), and (ii) PDE tubes of various lengths (L_t) with a fuel section of fixed length. As noted before, the interface and exit expansion waves control development of the

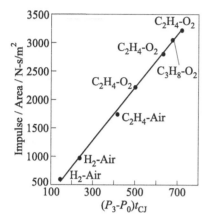

Figure 11.3 The generalized impulse from a series of simulations of PDEs operating on different fuels.

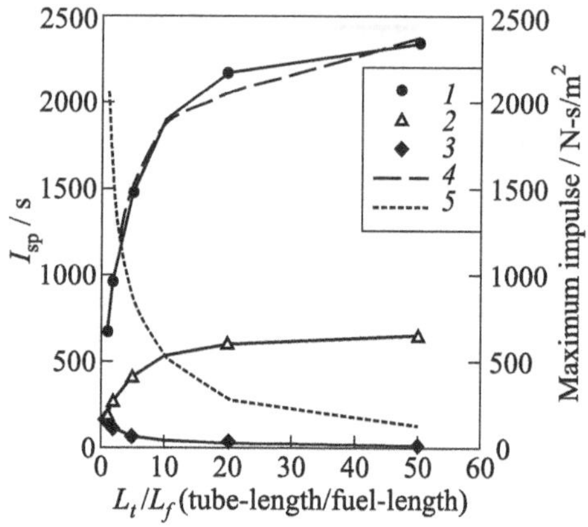

Figure 11.4 Effects of partial fuel fill on the computed performance measures: 1 — I_{sp}^{f} (fixed amount of fuel); 2 — maximum impulse (fixed amount of fuel); 3 — I_{sp}^{m} (both classes); 4 — I_{sp}^{f} (fixed tube length of 1 m); and 5 — maximum impulse (1-meter tube).

flow field [6]. The long tube simulations also clearly show the presence of a "quasi-steady" shock suggested previously.

The maximum impulse, the fuel-based specific impulse (I_{sp}^{f}), and the mixture-based specific impulse (I_{sp}^{m}) are all shown in Fig. 11.4 as a function of the ratio, L_t/L_f. The maximum impulse for a tube of fixed length (1 m) decreases as L_t/L_f increases because the amount of fuel available for detonation decreases. For a fixed amount of fuel, the maximum impulse increases as L_t/L_f increases because of the additional shock compression of the air in the tube.

The specific impulses from both sets of simulations collapse onto one curve when viewed as a function of the ratio L_t/L_f. I_{sp}^{f} increases with L_t/L_f, implying that for air-breathing applications, the fuel efficiency of a PDE improves with partial fuel filling. On the other hand, I_{sp}^{m} decreases with increasing L_t/L_f, indicating that for rocket applications, the fuel efficiency declines with partial fuel filling. Based on these results, a general expression for I_{sp} can be derived to be [6]

$$\frac{I_{sp}^{f}}{I_{sp\,\text{fully-filled}}^{f}} = a - (a - 1) \left/ \exp\left(\frac{L_t/L_f - 1}{8}\right)\right.$$

The constant a is the asymptotic limit of the benefit multiplier, representing the maximum benefit that can be obtained by partial fuel filling. From the cases simulated, a has a value between 3.2 and 3.5. From Fig. 11.4, one also can see that most of the benefits of partial fuel filling occur for L_t/L_f less than 10. Over this range, the enhancement due to partial fuel filling can be estimated from the simple expression

$$\frac{I_{sp}^f}{I_{sp\ \text{fully-filled}}^f} = \left(\frac{L_t}{L_f}\right)^{0.45}$$

11.3 THERMODYNAMIC CYCLE ANALYSIS

The performance measures discussed so far have been derived from numerical simulations or experimental data. The correlations given before are useful for a quick estimate, but are specific for the simple configuration considered. Many attempts have been made to derive performance measures from classical cycle analysis. So far, no such analysis has credibly shown that it is a good representation of the inherent dynamics evident in simulations of PDEs. However, classical thermodynamic cycle analysis may be useful in deriving an upper bound for the performance that can be obtained from engines based on detonative combustion. The results from these analyses must be viewed as representing an ideal detonation wave engine, not necessarily a PDE.

A recent comprehensive study [16] shows that the ideal detonative cycle is indeed more efficient than the ideal constant-volume Humphrey and the ideal constant-pressure Brayton cycles under all conditions. The relative advantage does decrease with increasing inlet compression temperature ratios and, hence, will decrease with increasing flight Mach numbers. The thermodynamic cycle efficiencies have been related to overall performance measures using conventional steady-state analysis. The appropriateness of this approach to an inherently unsteady device is debatable but is worth considering as an additional performance estimate.

The maximum I_{sp} shown in Fig. 7 in [16] for static conditions (and inlet temperature ratio of unity) is only about 3500 s. Since the results are said to be valid for hydrogen and hydrocarbon fuels and are thought to be an upper estimate, it is curious that the computed performance measures reported in this chapter and other experimental and computational studies for a stoichiometric hydrogen–air mixture are all higher than this reported maximum value. Therefore, an independent cycle analysis similar to that reported by Heiser and Pratt [16] was carried out for the mixtures of interest.

The current thermodynamic cycle analysis gives an efficiency of 0.46 for a stoichiometric hydrogen–air mixture initially at 298 K and 1 atm. Using the approach given in [16], this translates into a force for unit mass flow rate (Specific Thrust) of 1802 N·s/kg and a fuel-based I_{sp} of 6257 s for the ideal cycle. The differences between two estimates using identical analysis are probably due to differences in the fuel–air ratios and heat input values used. Consistent with the definition of the ideal cycle, heat input corresponding to the lower heating value of hydrogen and a fuel–air ratio corresponding to the stoichiometric value has been used in the current analysis.

Effects of Forward Flight

Detailed simulations of the effects of forward flight require the specification of an engine geometry, including inlets and exhaust nozzles. With the current level of understanding, the authors believe this is premature. The thermodynamic cycle analysis discussed above can be used to estimate the effects of forward flight. The conditions corresponding to a Mach 2.1 flight at an altitude of 9.3 km have been specified in [17]. These include a flight velocity of 640 m/s, an ambient temperature of 228 K, and a combustor inflow static temperature of 428 K. Using these values, an ideal thermodynamic cycle efficiency of 0.69, a specific thrust of 1660 N·s/kg, and an I_{sp} of 5764 s have been calculated for the stoichiometric hydrogen–air mixture. While the thrust and I_{sp} are lower, they do not drop off rapidly with increasing flight speeds due to the increase in the thermodynamic cycle efficiency. Similar calculations carried out for a stoichiometric ethylene–air mixture show a reduction in specific thrust from 1688 to 1544 N·s/kg and an I_{sp} reduction from 2531 to 2316 s for a change in operating conditions from static to Mach 2.1.

Comparison of Theoretical Performance Estimates

The value for a fuel-based specific impulse obtained from numerical simulations of the idealized PDE operating on a stoichiometric hydrogen–air mixture is 4161 s, while the value from the ideal cycle analysis is 6257 s. There may be several reasons for this difference. While the geometry has been idealized as a straight tube abruptly open to the outside, the thermodynamic processes have not been explicitly idealized. Indeed, it is difficult to correlate the various station conditions in the classical cycle analysis with those observed in the numerical simulations. The simulations capture the inherent unsteadiness of the PDE and are only for a single cycle. It is possible that a more realistic PDE consisting of multiple tubes and a common inlet and nozzle can better approximate a "steady engine." The thermodynamic cycle analysis and performance measures derived from it are really for a "steady" engine that happens to have heat addition using a detonation wave.

The flow expansion process at the exit of the tube may be more gradual with the addition of a nozzle than that observed in the current simulations where the tube opens abruptly into the ambient atmosphere. This observation is consistent with the authors' earlier reports of a better performance with boundary conditions that allow for a slower relaxation rate at the open-end of the tube [11]. The addition of a nozzle will also enhance the performance due to the partial fuel-fill effect discussed above. Separating the gain in performance from the better expansion of the detonation products from the gain due to the partial fuel-fill effect could be a challenging task. The effective heat added to the flow in the numerical simulations will be less than the theoretical value used in the cycle analysis due to the dissociation effects included in the simulations. This could also account for the observed differences in the performance measures.

11.4 DETONATION TRANSITION

In [2], results were presented on the effect of filling a straight tube with various extents of an ethylene–oxygen mixture and using the detonation in this mixture to initiate a detonation in the ethylene–air mixture filling the rest of the tube. Although this approach is adequate for laboratory studies, it requires too much oxygen for flight applications. The amount of oxygen needed can be decreased by reducing the diameter of the section of the tube filled with the fuel–oxygen mixture (Fig. 11.5a). Ideally, this initiator section of the tube should also contain only the fuel–air mixture. Previous studies [7, 8] have shown that if the initiator tube diameter is "too small" or does not have enough oxygen, the expansion waves generated at the interface between the tubes weaken the detonation front and prevent successful transition into the fuel–air mixture. Currently, approaches to aid the transition without significantly increasing the amount of oxygen required are being explored.

The use of jets of oxygen-rich, nonreactive, or reactive mixtures located at the corner of the tube interface to assist the detonation transition (Fig. 11.5b) is investigated. The potential benefits of the corner jets are (1) the oxygen content in the jets (nonreactive or reactive) may enhance the combustion process in the weakened part of the detonation near the interface corner and/or facilitate the formation of new transverse waves near the corner; and (2) the momentum carried by the corner jets may weaken the expansion waves and, therefore, aid the detonation transition.

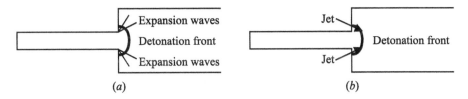

Figure 11.5 Concept of using corner jets to enhance detonation transition.

11.5   MULTIPHASE DETONATIONS

The discussion above has focused on gaseous fuels. However, for most volume-limited applications, liquid fuels will have to be used. The primary fuel of interest to the Navy is JP-10. Detailed simulations of PDEs operating on JP-10 have been hampered by the lack of information on the chemistry as well as the multiphase properties of this fuel.

A numerical simulation tool has been developed that is capable of accounting for radial and axial variations in liquid-fuel distribution. In the absence of definite information on the chemistry of JP-10, a two-step global model analogous to that used in the gas-phase detonation simulations has been constructed. In this model, the residence time of the fuel vapor is tracked, and energy release is allowed to occur only after the elapse of a specified induction time. The amount of energy release is based on the lower heating value of the fuel and is calibrated to result in the CJ-detonation velocity for a fully vaporized fuel.

For effective use of the developed model, information on the induction time and droplet evaporation rates, as a function of the local conditions in the shock-heated mixture, is needed. Currently, in the absence of such information, parametric studies with various constant induction times and droplet evaporation rates have been carried out. The predicted detonation velocity as a function of the initial droplet size is shown in Fig. 11.6 for a nominal JP-10–oxygen mixture with an equivalence ratio of 0.12. A d^2-law evaporation with a rate of 0.1 cm^2/s and an induction time for the fuel–vapor of 1 μs was used for this series of simulations. The velocity deficit observed previously in many experimental studies of multiphase detonations is predicted by the numerical model. In the absence

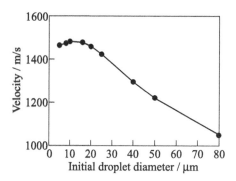
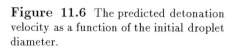

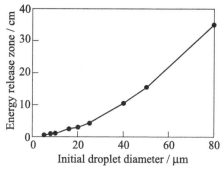

Figure 11.6 The predicted detonation velocity as a function of the initial droplet diameter.

Figure 11.7 Average thickness observed for the energy release zone shown as a function of the initial droplet diameter.

of direct experimental confirmation, these results must be considered qualitative and further refinements in the model as well as more information on the input parameters may be needed to obtain quantitative predictions. With this limitation in mind, further analysis has shown that as the droplet size increases, the zone where energy release occurs becomes wider, as shown in Fig. 11.7. It is possible that for larger droplet sizes, some of the energy release may not be able to interact with the leading shock/detonation front in the time available for propagation in tubes of length of the order of 1 m. Further parametric studies on longer tubes are currently underway.

11.6 CONCLUDING REMARKS

The computational studies of PDEs show that the impact of chemical recombination is small ($\sim 5\%$) and introduces comparable or less uncertainty in performance predictions than the conditions used to initiate detonations. A wide range of cases involving partial fuel-fill effects have been investigated to quantify the effects, elucidate the underlying mechanism, and derive a limiting value for the performance enhancement. The general expression for a quick estimation of the performance of an idealized PDE has been extended to a wider range of fuel–oxygen and fuel–air mixtures and modified to account for partial fuel-fill effects. Ideal thermodynamic cycle analysis has also been carried out for the mixtures of interest. Comparison of these analytical results to the results of simulations suggests that these estimates are likely to provide an upper bound for the performance available from a detonation engine. Progress to date on liquid-fueled detonations and easing the transition of an ethylene–air detonation from a small initiator to the main detonation tube are also reported.

ACKNOWLEDGMENTS

The authors wish to thank Dr. Gopal Patnaik of the NRL for previous help with the development and improvement of the codes used in the computations presented here. The work was performed under ONR contract N0001402WX20595 and the NRL Computational Physics Task Area.

REFERENCES

1. Kailasanath, K., C. Li, E. Chang, and G. Patnaik. 2000. Computational studies of pulse detonation engines. *13th ONR Propulsion Meeting Proceedings*. Minneapolis, MN: University of Minnesota. (See also 1999. *12th ONR Propulsion Meeting*

Proceedings. Salt Lake City, UT: University of Utah; 1998. *11th ONR Propulsion Meeting Proceedings*. West Palm Beach, FL.)

2. Kailasanath, K., C. Li, and G. Patnaik. 2001. Computational studies of pulse detonation engines. *14th ONR Propulsion Meeting Proceedings*. Chicago, IL: University of Illinois.

3. Kailasanath, K. 2002. On the performance of pulsed detonation engines. In: *Advances in Confined Detonations*. Eds. G. Roy, S. Frolov, R. Santoro, and S. Tsyganov. Moscow, Russia: TORUS PRESS. 207–12.

4. Li, C., K. Kailasanath, and G. Patnaik. 2000. A numerical study of flow field evolution in a pulse detonation engine. AIAA Paper No. 2000-0314.

5. Sanders, S. T., T. P. Jenkins, and R. K. Hanson. 2000. Diode-laser sensor system for multi-parameter measurements in pulse detonation engine flows. AIAA Paper No. 2000-3592.

6. Li, C., and K. Kailasanath. 2002. Performance analysis of pulse detonation engines with partial fuel filling. AIAA Paper No. 2002-0610.

7. Li, C., and K. Kailasanath. 2000. Detonation transmission and transition in channels of different sizes. *Combustion Institute Proceedings* 28:603–9.

8. Sinibaldi, J. O., C. M. Brophy, C. Li, and K. Kailasanath. 2001. Initiator detonation diffraction studies in pulsed detonation engines. AIAA Paper No. 2001-3466.

9. Povinelli, L. A. 2001. Impact of dissociation and sensible heat release on pulse detonation and gas turbine engine performance. *ISABE-2001-1212 Proceedings*. Bangalore, India.

10. Kailasanath, K., C. Li, and G. Patnaik. 1999. Computational studies of pulse detonation engines: A status report. AIAA Paper No. 99-2634.

11. Kailasanath, K., and G. Patnaik. 2000. Performance estimates of pulsed detonation engines. *Combustion Institute Proceedings* 28:595–601.

12. Povinelli, L. A., and S. Yungster. 2002. Airbreathing pulse detonation engine performance. NASA/TM-2002-211575.

13. Kailasanath, K. 2002. Recent developments in the research on pulse detonation engines. AIAA Paper No. 2002-0470.

14. Wintenberger, E., J. Austin, M. Cooper, S. Jackson, and J. E. Shepherd. 2001. An analytical model for the impulse of a single-cycle pulse detonation engine. AIAA Paper No. 2001-3811.

15. Falempin, F., D. Bouchaud, B. Forrat, D. Desbordes, and E. Daniau. 2001. Pulsed detonation engine: Possible application to low cost tactical missile and to space launcher. AIAA Paper No. 2001-3815.

16. Heiser, W. H., and D. T. Pratt. 2002. Thermodynamic cycle analysis of pulse detonation engines. *J. Propulsion Power* 18(1):68–76.

17. Yang, V., Y. H. Wu, and F. H. Ma. 2001. Pulse detonation engine performance and thermodynamic cycle analysis. *14th ONR Propulsion Meeting Proceedings*. Chicago, IL.

Chapter 12

SIMULATION OF DIRECT INITIATION OF DETONATION USING REALISTIC FINITE-RATE MODELS

K.-S. Im and S.-T. J. Yu

This chapter reports the progress of high-fidelity simulation of the direct initiation process of cylindrical detonation waves by concentrated energy deposition. The purpose is to understand various mechanisms in the initiation process and to estimate the critical energy required for successful detonation initiation. Model equations are solved by the space–time conservation element and solution element (CESE) method, including realistic finite-rate models of multiple species and multiple reactions steps, correctly derived Jacobian matrices of convection terms and source terms, and comprehensive thermodynamics relations. In the setting of the CESE method, stiff source terms in species equations due to chemical reactions are treated based on a unique space–time volumetric integration with subtime-step integrations for high resolution. Detailed results of subcritical, critical, and supercritical initiation process are reported. This work is a steppingstone for further parametric studies to understand the competing mechanisms of heat release, finite-rate reactions, wave curvature, and unsteadiness in the direct detonation initiation process.

12.1 INTRODUCTION

The present work focuses on high-fidelity simulation of the detonation initiation process using realistic finite-rate chemistry models. In general, there are three experimental methods to initiate detonation: (i) flame initiation, (ii) shock-wave initiation, and (iii) direct initiation. In all three cases, shock waves occur prior to detonation initiation. This chapter focuses on the third initiation mode, which is relevant to the detonation initiation process in a Pulse Detonation Engine (PDE). For background information, a brief account of the above three initiation modes is provided in the following.

In the flame initiation mode [1], a weak spark ignites an explosive gas mixture, which is usually confined in an enclosure. The generated flame propagates towards the unburned gas mixture. As hotter burned gas has higher specific volume than that of the unburned gas, the fluid motion acts like a hot-gas piston and generates a compression wave, which imparts a downstream velocity to the unburned gases ahead of the flame. Under suitable conditions, these traveling compression waves will produce a shock wave ahead of the flame. With enough transition distance, the accelerating flame will catch up to the shock. As a result, a detonation is initiated. Two mechanisms are responsible for flame acceleration: (i) increasing flame area that increases the volumetric heat-release rate, and (ii) induced turbulence in the moving unburned mixture ahead of the flame which allows the flame to leap ahead. This process is referred to as deflagration-to-detonation transition (DDT) or self-initiation because the detonation is initiated solely by the energy release from the combustion of the mixture itself. The most important parameter in this process is the run-up distance, which depends on the geometry of the tube, location of the igniter, thermodynamic conditions, etc.

In the shock initiation mode [2], either an incident or reflected shock wave is the primary means to produce the detonation. Basically, the shock rapidly heats the gas by compression. Under suitable conditions, an explosion occurs behind the shock wave. This explosion generates accelerating pressure waves, which quickly become a detonation wave before or after catching up with the initially applied shock wave. In the present U.S. Office of Naval Research (ONR) project, the simulation of shock initiation mode has been used to access the numerical accuracy and to validate the code. In the last mid-year review, detailed simulation of detonation initiation by a reflected shock was reported. The results were also summarized and reported in the AIAA Sciences Meeting held in 2002 [3].

In the direct initiation mode, a large amount of energy is instantaneously deposited to a small region of unconfined combustible mixture. Immediately, a strong blast wave is generated. This shock wave expands and decays while it continues heating the gas mixture. Due to shock heating, chemical reactions occur and chemical energy is released. Under suitable conditions, detonation is initiated. The blast wave generated by the igniter plays an important role because it produces the critical states for the onset of the detonation. Therefore, it is often referred to as the blast initiation mode.

Zel'dovich et al. [4] were the first to study the direct spark initiation process. They pointed out that the amount of the deposited energy, or the critical energy, is the key parameter controlling the initiation process. In general, three regimes could be discerned: (i) the supercritical regime for successful detonation initiation, (ii) the subcritical regime for failed initiation, and (iii) the critical regime for marginally sustainable detonation initiation.

In addition to the classical models, many attempts have been made to predict the critical energy for initiating detonation under various circumstances. He and

Clavin [5] performed quasi-steady analysis of the direct initiation process. They developed the critical curvature model, which states that the failure mechanism of the detonation is mainly caused by the nonlinear curvature effect of the wave front. Eckett *et al.* [6] proposed the critical decay-rate model and pointed out that the critical mechanism of a failed detonation initiation process is due to the unsteadiness of the reacting flow. Their theory for spherical detonation initiation has been supported by numerical simulation and experimental data.

Most of the numerical calculations have been based on the use of (i) single-step irreversible reaction models and (ii) the assumption of a polytropic gas mixture. In order to catch the essential features of real detonations and detonation initiation phenomena, Lee and Higgins [7] suggested that one should use (i) chemical mechanisms with at least two or three reaction steps and (ii) thermodynamics relations for gas mixtures at high temperatures.

This chapter focuses on direct initiation of cylindrical detonation in an $H_2/O_2/Ar$ mixture. A finite-rate model of twenty reaction steps and nine species is adopted in the present work. Three values of initiation energy are used to simulate the supercritical, the subcritical, and the critical processes.

12.2   THEORETICAL MODEL

Flow Equations

The equations of the mass, momentum, energy, and species concentrations can be written in the following vector form:

$$\frac{\partial \mathbf{U}}{\partial t} + \frac{\partial \mathbf{F}(\mathbf{U})}{\partial r} = \mathbf{G}(\mathbf{U}) + \mathbf{S}(\mathbf{U}) \tag{12.1}$$

where

$$\mathbf{U} = (\rho, \rho u, \rho E, C_1, C_2, \ldots, C_{N-1})^T$$

$$\mathbf{F} = (\rho u, \rho u^2 + p, (\rho E + p)u, C_1 u, C_2 u, \ldots, C_{N-1} u)^T$$

$$\mathbf{G} = -\frac{j}{r} (\rho u, \rho u^2, (\rho E + p)u, C_1 u, C_2 u, \ldots, C_{N-1} u)^T$$

$$\mathbf{S} = (0, 0, 0, \dot{\omega}_1, \dot{\omega}_2, \ldots, \dot{\omega}_{N-1})^T$$

where ρ, u, p, E, and C_i are density, velocity, pressure, specific total energy, and densities of species i, respectively; $j = 0$, 1, and 2 for planar, cylindrical, and spherical flows, respectively; $\dot{\omega}_i$ is the production rate of species i; and ρ is the summation of all species densities,

$$\rho = \sum_{i=1}^{N_s} C_i \tag{12.2}$$

The total energy E is defined as

$$E = e + \frac{u^2}{2} \tag{12.3}$$

where e is the internal energy of the gas mixture per unit mass and is calculated based on a mass-weighted average of the specific internal energy of each species e_i, i.e.,

$$e = \sum_{i=1}^{N_s} Y_i e_i \tag{12.4}$$

$Y_i = C_i/\rho$ is the mass fraction of species i. Since the internal energy e and the total energy E include the heat of formation of each species in their definitions, no source term exists in the energy equation.

Initial and Boundary Conditions

The initial conditions are taken from [8]. A specific amount of energy, E_s, in the form of high temperature and high pressure (with a subscript s) is deposited instantaneously into the driver section of a reactive gas mixture.

$$\begin{aligned} 0 \leq r < r_s, \quad p = p_s, \quad T = T_s, \quad Y_i = Y_{i,0}, \quad u = 0, \\ r \geq r_s, \quad p = p_0, \quad T = T_0, \quad Y_i = Y_{i,0}, \quad u_0 = 0 \end{aligned} \tag{12.5}$$

Refer to Fig. 12.1. The radius of the driver section, r_s, is about 15 times smaller than the critical radius R_c [8]. Inside the driver section, pressure is set at about 15–20 times higher than the peak values of the corresponding Chapman–Jouguet (CJ) detonation. Essentially, the initial condition provides a strong cylindrical expanding blast wave. The species compositions at both sides are $H_2 + O_2 + 7Ar$. The pressure and temperature of the driven section are 0.2 atm and 293 K, respectively. The deposited energy, E_s, is calculated based on the internal energy equation for a perfect gas:

$$E_s = \frac{\sigma_j r_s^{j+1} p_s}{(\gamma - 1)\rho_s}$$

$$\sigma_j = \frac{2j\pi + (j-1)(j-2)}{j+1} \tag{12.6}$$

At $j = 1$, three values of E_s are selected in the present calculations: $E_s = 33.9$, 53.0, and 76.3 J/cm, corresponding to the initiation radius $r_s = 0.4$, 0.5, and 0.6 cm, respectively; $p_s = 200$ atm for all calculations.

At $r = 0$, the boundary condition is derived based on a limiting form of Eq. (12.1) for r approaching zero. At $r = \infty$, the standard nonreflecting boundary conditions are employed.

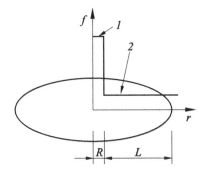

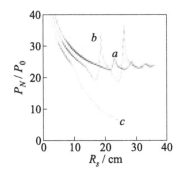

Figure 12.1 A schematic of the initial condition of the direct detonation initiation process: *1* — driver section $(T = T_s, \ p = p_s, \ Y_i = Y_s,$ and $u = 0)$ and *2* — driven section $(T = T_0, \ p = p_0, \ Y_i = Y_0,$ and $u = 0).$

Figure 12.2 Time histories of pressure peaks of the direct initiation processes of cylindrical detonations in a mixture of $H_2 + O_2 + 7Ar$ with three different propagation regimes: (*a*) supercritical $(E_s = 76.3$ J/cm), (*b*) critical $(E_s = 53.0$ J/cm), and (*c*) subcritical $(E_s = 33.9$ J/cm), depending on the deposited initiation energy.

12.3 RESULTS AND DISCUSSIONS

Figure 12.2 shows the evolving peak pressures of the leading shock as a function of the radius of the advancing wave front. The detonation initiation fails when $E_s = 33.9$ J/cm. This propagation mode is in the subcritical regime. For $E_s = 76.3$ J/cm, a strong detonation is formed and continues expanding at about the corresponding CJ detonation values. This propagation mode is in the supercritical regime. $E_s = 53.0$ J/cm is in the neighborhood of the critical state. As a result, the flow is highly unstable. As shown in Fig. 12.2, a galloping wave pattern is observed. After a while, the flow is stabilized, and a self-sustained detonation is formed. These results are compared well with the experimental observation reported by Bach *et al.* [9]*.

Figure 12.3 shows the evolving pressure profiles of the detonation initiation process with the above three different values of the deposited initial energy. With enough deposited energy, the initial blast wave develops to be a stationary CJ detonation as shown in Figs. 12.3*a* and 12.3*b*. In the subcritical regime the detonation initiation fails as shown in Fig. 12.3*c*.

---

*One-dimensional calculations of critical detonation initiation energy are known to significantly overestimate compared to the experimental values. One of the reasons is the essentially nonunidimensional nature of mixture ignition behind a lead shock front. Therefore, it is suggested that the reported results can be treated only as qualitative. (*Editor's remark.*)

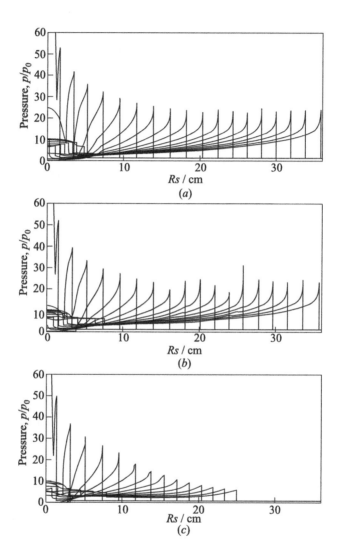

Figure 12.3 Spatial pressure profiles of three regimes of the direct detonation initiation processes: (a) supercritical ($E_s = 76.3$ J/cm), (b) critical ($E_s = 53.0$ J/cm), and (c) subcritical ($E_s = 33.9$ J/cm).

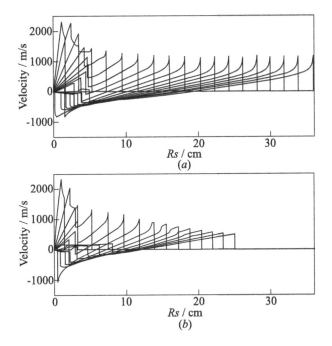

Figure 12.4 Velocity profiles of (a) supercritical ($E_s = 76.3$ J/cm) and (b) subcritical ($E_s = 33.9$ J/cm) propagation regimes at direct detonation initiation.

Figures 12.4a and 12.4b show the evolving velocity profiles corresponding to the successful and failed initiation processes, respectively. In both cases, the fluid velocity suddenly jumps to a high value at the beginning of the process. While the leading shock wave travels outwardly, part of the fluid behind the shock moves inwardly towards the center. The inwardly moving fluid quickly forms a strong shock wave at a very high speed. This wave converges to the point of $r = 0$ and generates a tremendous pressure spike. In the calculation, special treatment is implemented to suppress numerical instability caused by the converging shock.

12.4 CONCLUDING REMARKS

High-fidelity numerical results of the direct detonation initiation processes have been reported. For a cylindrical detonation process in an $H_2/O_2/Ar$ mixture, propagating waves in three different regimes, i.e., supercritical, critical, and subcritical, are calculated according to the value of the deposited energy in the initial conditions. Calculation has been done based on the use of realistic finite-rate chemistry models and comprehensive thermodynamics models. The numerical method employed is the space–time CESE method. The results here are a steppingstone for further investigation of the competing mechanisms of heat release, finite-rate chemistry, curvature effects, and the flow unsteadiness in the detonation processes.

ACKNOWLEDGMENTS

The research is supported by ONR contract N00014-01-1-0643. The authors are indebted to Prof. J. E. Shepherd of the California Institute of Technology for fruitful discussions and the general direction of this research work. Useful interactions with other members in the ONR PDE program are also acknowledged.

REFERENCES

1. Lee, J. H. S. 1977. Initiation of gaseous detonation. *Annual Review Physical Chemistry* 28:75–104.

2. Strehlow, R. A., and R. Cohen. 1962. Initiation of detonation. *Physics Fluids* 5(1):97–101.

3. Im, K.-S., C.-K. Kim, and S.-T. J. Yu. 2002. Numerical simulation of detonation with detailed chemistry by the space–time conservation element and solution element method. AIAA Paper No. 2002-1020.

4. Zel'dovich, Ya. B., S. M. Kogarko, and N. N. Simonov. 1956. An experimental investigation of spherical detonation of gases. *Sov. J. Technical Physics* 1(8):1689–713.

5. He, L., and P. Clavin. 1994. On the direct initiation of gaseous detonations by an energy source. *J. Fluid Mechanics* 277:227–48.

6. Eckett, C. A., J. J. Quirk, and J. E. Shepherd. 2000. The role of unsteadiness in direct initiation of gaseous detonations. *J. Fluid Mechanics* 421:147–83.

7. Lee, J. H. S., and A. J. Higgins. 1999. Comments on criteria for direct initiation of detonation. *Philos. Trans. Royal Society London* A 357:3503–21.

8. He, L. 1996. Theoretical determination of the critical conditions for the direct initiation of detonation in hydrogen–oxygen mixtures. *Combustion Flame* 104: 401–18.

9. Bach, G. G., R. Knystautas, and J. H. Lee. 1969. Direct initiation of spherical detonations in gaseous explosives. *12th Symposium (International) Combustion Proceedings*. Pittsburg, PA: The Combustion Institute. 853–64.

Chapter 13

SYSTEM PERFORMANCE AND THRUST CHAMBER OPTIMIZATION OF AIR-BREATHING PULSE DETONATION ENGINES

V. Yang, F. H. Ma, and J. Y. Choi

This research deals with the system performance and thrust chamber dynamics of air-breathing pulse detonation engines (PDEs). The work accommodates all the essential elements of an engine, including inlet, manifold/valve, injector, combustor, and nozzle. Emphasis is placed on multitube configurations with repetitive flow-distribution. The primary outcome is a general framework, in a form suitable for routine practical applications, for assessing the effects of all known processes on engine dynamics. It also helps designers and researchers to optimize the overall system performance and to identify the major technological barriers at minimal computational expense. Major variables and phenomena of consideration include fueling strategy, injection scheme, chamber conditions, and configuration geometry.

13.1 INTRODUCTION

The present work represents an extension of establishing a system performance and thermodynamic cycle analysis for air-breathing PDEs by means of a modular approach, as shown schematically in Fig. 13.1. Each module represents a specific component of the engine, and its dynamic behavior was formulated using complete conservation equations. The governing equations and their associated boundary conditions were numerically solved using a recently developed space–time method which circumvents the deficiencies of existing computational methods for treating detonation waves and shock discontinuities. Both one-

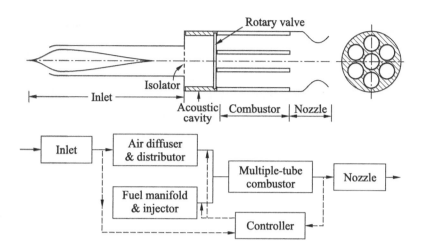

Figure 13.1 Supersonic air-breathing pulse detonation engine.

and two-dimensional simulations were conducted to study the various fundamental scientific and practical engineering issues involved in the development of
PDEs.

The research was conducted at several levels of complexity in a hierarchical
manner to establish its accuracy and reliability: (1) supersonic inlet dynamics;
(2) detonation in single-tube, multicycle environments; and (3) system performance and thermodynamic cycle analysis. The progress made in each of the
above three areas was reported at the 14th U.S. Office of Naval Research (ONR)
Propulsion Meeting [1]. The work performed focused on the following three areas: (1) effect of nozzle configuration on PDE performance, (2) single-tube thrust
chamber dynamics, and (3) multitube thrust chamber dynamics.

A brief summary of major results obtained to date is given below.

13.2 EFFECT OF NOZZLE CONFIGURATION
ON PDE PERFORMANCE

Nozzle plays a decisive role in determining the propulsive performance of a
PDE [1–3]. In the ONR PDE program, a unified numerical analysis accommodating both the combustion chamber and the nozzle was developed. Three
nozzle configurations were tested, i.e., a converging, a diverging, and a plug nozzle. Figure 13.2 shows a snapshot of the density-gradient field for a straight
tube. The stagnation pressure and temperature at the combustor entrance are

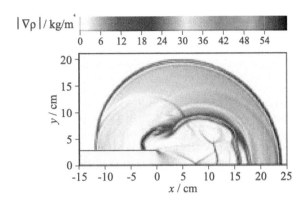

Figure 13.2 Snapshot of density-gradient field. (Refer color plate, p. XXIV.)

2.12 atm and 428 K, respectively, corresponding to the flight altitude of 9.3 km and the freestream Mach number of 2.1 [1]. The diverging nozzle offers the best performance. The plug nozzle functions equally well to the diverging nozzle, although the expansion pattern at the exit is much more complex. In all the cases studied here, the performance gain appears limited, mainly due to the lack of a physical mechanism for preserving the chamber pressure during the blow-down and refilling processes.

In view of this deficiency, a choked convergent–divergent nozzle is employed. The nozzle has a length of 20 cm, with a 45-degree convergent angle and a 15-degree divergent angle. A comprehensive parametric study was conducted based on a quasi-one-dimensional Zel'dovich–Neumann–Döring (ZND) model. Figure 13.3 shows the effect of the valve close-up time (τ_{close}) on the fuel-based specific impulse for three different cycle periods (τ_{cycle}): 3, 4, and 5 ms. It is clear that an optimum τ_{close} exists for each operation frequency considered. When τ_{close} increases from its optimum

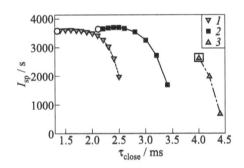

Figure 13.3 Effect of τ_{close} on specific impulse: 1 — $\tau_{\text{cycle}} = 3$ ms; 2 — 4 ms; and 3 — 5 ms. The circles indicate inlet over pressurized and the square indicates combustor overfilled.

value, the specific impulse decreases exponentially. This is attributed to the facts that (i) the head-end pressure decreases as τ_{close} increases, and (ii) the amount of reactant refilled per cycle decreases, thus resulting in a lower mean chamber pressure. Tests also found that if τ_{close} becomes even larger, the specific impulse may become negative. As τ_{close} decreases from its optimum value, two lower limits may soon be reached. The first lower limit is inlet overpressurization, i.e., the head-end pressure is higher than the total pressure of the incoming air when the valve is open. A reverse flow then occurs and may cause engine unstart. The second lower limit is overfilling, i.e., the fresh reactant flows out of the engine before being burned. Therefore, it is essential to keep τ_{close} as small as possible to obtain higher performance. Figure 13.3 also demonstrates that an optimum frequency exists for a given configuration.

13.3 SINGLE-TUBE THRUST CHAMBER DYNAMICS

In order to estimate the accuracy of the quasi-one-dimensional results, a two-dimensional model was employed to study the single-tube PDE dynamics. The work also provides a basis for multitube PDE analysis.

Figure 13.4 shows a snapshot of the pressure-gradient field at $t = 0.80$ ms for the case with a cycle period of 3 ms and a valve close-up period of 2.1 ms. At this time instance, the primary shock wave resulting from the detonation wave has moved out of the nozzle and transformed to a weakened bow shock. Other common features, including the formation of an oblique shock train in the chamber due to shock reflection, the presence of vortices at the nozzle exit due to shock diffraction, and the attachment of secondary shocks onto the vortices, are all clearly resolved. Figure 13.5 shows the effect of valve

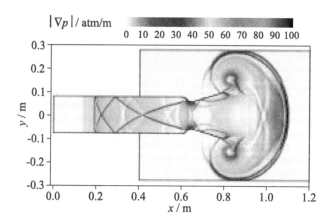

Figure 13.4 Snapshot of pressure-gradient field at $t = 0.80$ ms; $\tau_{\text{cycle}} = 3$ ms, $\tau_{\text{close}} = 2.1$ ms, and $\tau_{\text{purge}} = 0.1$ ms. (Refer color plate, p. XXIV.)

close-up time on specific impulse. The cycle period and purging time remain fixed. The results agree well with those predicted by the quasi-one-dimensional model.

13.4   MULTITUBE THRUST CHAMBER DYNAMICS

In addition to the study of single-tube PDE system dynamics, much effort was expended to investigate the intricate combustion and gasdynamic processes in multitube thrust chambers. As a specific example, a thrust chamber consisting of three detonation tubes connected downstream with a common convergent–divergent nozzle is considered herein. This configuration helps preserve the chamber pressure during the blow-down and refilling stages and, consequently, improves the propulsive performance of the engine. Figure 13.6 presents the time evolution of the density-gradient field within one cycle of operation. The frequency is 333 Hz

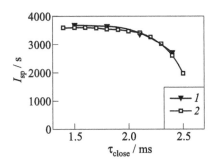

Figure 13.5 Effect of valve close-up time on specific impulse for a single-tube PDE; $\tau_{\text{cycle}} = 3$ ms, and $\tau_{\text{purge}} = 0.1$ ms: 1 — two-dimensional and 2 — quasi-one-dimensional.

for each tube. Initially, the bottom tube is partially filled with a stoichiometric hydrogen/air mixture. Detonation is initiated, propagates downstream, and eventually degenerates to a nonreacting shock wave. The resultant shock wave then proceeds further downstream, diffracts at the exit of the tube, reflects on the inner walls, and causes complex waves propagating upstream into all three detonation tubes and downstream into the nozzle (see Fig. 13.6b). During this period, the middle tube undergoes the purging and refilling processes. After a one-third cycle period, detonation is initiated and propagates in the middle tube, while the top tube begins to purge burnt gases and refill fresh mixtures (see Fig. 13.6c). The detonation wave then degenerates to a shock wave after passing through the interface between the reactant and purged gases. Further interactions between the shock wave and the local flowfield result in an extremely complex flow structure as shown in Fig. 13.6d. After another one-third cycle period, detonation is initiated and propagates in the top tube (see Fig. 13.6e).

Stable cyclic operation is reached at the fifth cycle. The cycle-averaged specific impulse and specific thrust obtained are 3543 s and 896 m/s, respectively. They are about 5% higher than those achieved by the single-tube PDE, demon-

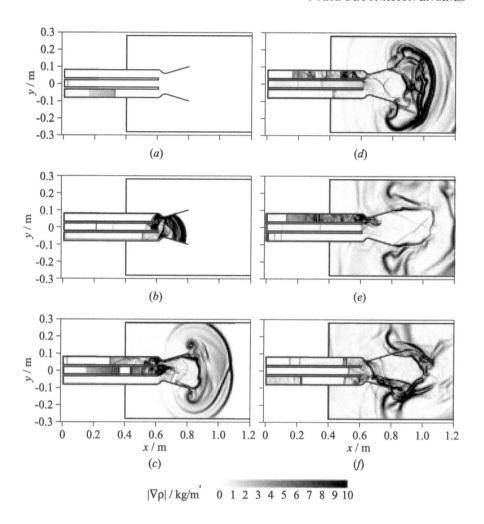

$|\nabla\rho| / kg/m^4$ 0 1 2 3 4 5 6 7 8 9 10

Figure   13.6   Time evolution of density-gradient field during the first cycle of operation; $\tau_{cycle} = 3$ ms, $\tau_{close} = 2.1$ ms, $\tau_{purge} = 0.1$ ms, without free volume: (a) $t = 0.15$ ms; (b) 0.60 ms; (c) 1.15 ms; (d) 1.60 ms; (e) 2.15 ms; and (f) $t = 2.60$ ms.

strating the improvement by implementing a multitube design. Figure 13.7 shows the time evolution of the density-gradient field during the fifth cycle of operation.

In another configuration, the length of the detonation tube decreases to 45 cm, leaving a free volume of 15 cm long between the detonation tubes and the nozzle. The flowfield exhibits a structure similar to the case without free volume. The cycle-averaged specific impulse and specific thrust are 3411 s and

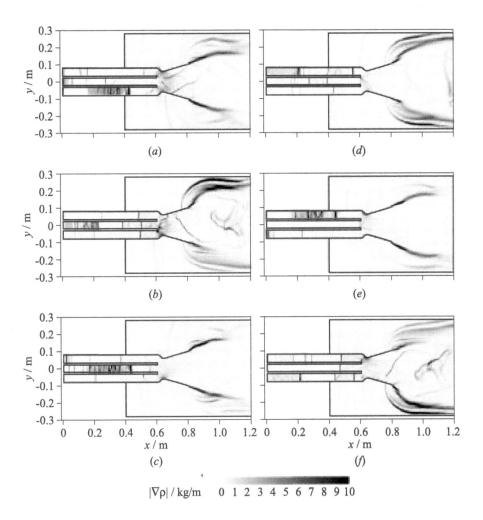

Figure 13.7 Time evolution of density-gradient field during the fifth cycle of operation; $\tau_{cycle} = 3$ ms, $\tau_{close} = 2.1$ ms, $\tau_{purge} = 0.1$ ms, without free volume: (a) $t = 12.15$ ms; (b) 12.60 ms; (c) 13.15 ms; (d) 13.60 ms; (e) 14.15 ms; and (f) $t = 14.60$ ms.

864 m/s, respectively, which are slightly (3%) lower than those of the previous case. For comparison, Fig. 13.8 shows the time history of the instantaneous specific thrust during the fifth cycle for the single- and triple-tube and the triple-tube with free volume cases. A very high peak exists for the single-tube case. The deviation between the peak and the cycle-averaged values represents the number of peaks increases, but the peak values are significantly reduced.

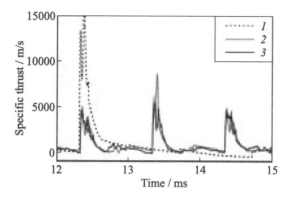

Figure 13.8 Time history of specific thrust during the fifth cycle of operation; $\tau_{\text{cycle}} = 3$ ms, $\tau_{\text{close}} = 2.1$ ms, $\tau_{\text{purge}} = 0.1$ ms: *1* — single tube; *2* — triple tube without free volume; and *3* — triple tube with free volume.

Inclusion of a free volume further reduces the degree of unsteadiness, although the system performance may not necessarily increase. It should be noted that there may exist lateral thrust in the vertical direction for multitube PDEs due to their unsymmetric operations. The present triple-tube PDE may produce a degree of unsteadiness of the engine operation. In the triple-tube cases, maximum lateral thrust of 1000 N per 1 kg/s air mass flow rate, thereby causing unnecessary vibration of the vehicle. This undesired effect can be harnessed by introducing the concept of tube pair.

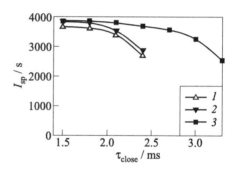

Figure 13.9 Effects of valve close-up time on specific impulse; no free volume: *1* — $\tau_{\text{cycle}} = 3$ ms, single tube; *2* — $\tau_{\text{cycle}} = 3$ ms; and *3* — $\tau_{\text{cycle}} = 4$ ms.

Each tube pair includes two detonation tubes, which are located at symmetric positions and operate synchronously in time, to diminish the lateral thrust*.

To address the timing effect on the system performance, a parametric study was conducted by varying τ_{cycle} and τ_{close}, with τ_{purge} fixed at 0.1 ms. Results for the configuration without free volume are presented here. Figure 13.9 shows the effects of τ_{close} on the specific impulse (cycle-averaged) for two different cycle pe-

---

*In addition to the problem of lateral thrust, an issue of nozzle durability in the presence of asymmetric loads is worth mentioning. (*Editor's remark.*)

riods: 3 and 4 ms. The single-tube results at $\tau_{\text{cycle}} = 3$ ms are also displayed for comparison. Clearly, the multitube design helps improve the system performance. The lower frequency operation offers a better performance margin. This can be explained as follows. At a lower frequency, the refilling time becomes longer and more reactant is delivered into the tube, thus maintaining a higher mean chamber pressure and increasing the performance. However, an exceedingly long refilling time may result in overfilling and decrease the performance. Therefore, there exists an optimum frequency for a given configuration.

13.5   CONCLUDING REMARKS

The present work provides detailed information about the thrust chamber dynamics of single-tube and multitube PDEs. The effects of various operating parameters and chamber configurations on engine performance have been investigated systematically. Results can be effectively utilized to optimize the engine design and to identify the various loss mechanisms limiting the PDE performance.

ACKNOWLEDGMENTS

This work was performed under ONR contract N00014-99-1-0744.

REFERENCES

1. Yang, V., Y. H. Wu, and F. H. Ma. 2001. Pulse detonation engine performance and thermodynamic cycle analysis. *14th ONR Propulsion Meeting Proceedings*. Chicago, Illinois.

2. Wu, Y. H, F. H. Ma, and V. Yang. 2002. System performance and thermodynamics cycle analysis of air-breathing pulse detonation engines. AIAA Paper No. 2002-0317.

3. Ma, F. H., J. Y. Choi, Y. H. Wu, and V. Yang. 2002. Modeling of multitube pulse detonation engine dynamics. In: *Advances in confined detonations*. Eds. G. Roy, S. Frolov, R. Santoro, and S. Tsyganov. Moscow, Russia: TORUS PRESS. 231–34.

Comments

Cambier: In your simulations, you vary cycle frequency; however, cycle frequency is not a free parameter in practice.

Yang: That is correct. There are many practical constraints and cycle frequency is one of them.

Shepherd: You use a common nozzle in your PDE configuration. How about if you use individual nozzles?

Yang: With individual nozzles gain in efficiency is about 3%.

Shepherd: But there are also performance considerations.

Gore: How did you validate the results of calculations?

Yang: We performed model validation at the component level, i.e., Chapman–Jouguet state, supersonic inlet, DDT, etc.

Chapter 14

SOFTWARE DEVELOPMENT FOR AUTOMATED PARAMETRIC STUDY AND PERFORMANCE OPTIMIZATION OF PULSE DETONATION ENGINES

J.-L. Cambier and M. R. Amin

The continuing development and application of a new software for Pulse Detonation Engine (PDE) configuration design and performance evaluation is described in this chapter. This multiplatform software provides the core of a performance tool for cross-comparisons of engine configurations and numerical methods.

14.1 INTRODUCTION

The design and optimization of a PDE propulsion system is difficult, due to the unsteady nature of the propulsion cycle and the strong coupling of the propulsive flow with vehicle configuration and ambient environment (e.g., transient forces and inlet flow perturbations). Numerical modeling [1] and the use of design and analysis software tools can play a major role towards the successful development of this technology. Specialized software tools based on Computational Fluid Dynamics (CFD) methods would facilitate the setup and analysis of the PDE-based propulsion system. The problem's complexity may also require the efficient interaction of multiple R&D teams. This imposes compatibility and portability constraints in the design software tools. For example, if the performance evaluations obtained by two teams are slightly different, it would not be *a priori* clear whether this is a consequence of different methods, physical models, or different accounting procedures for the balance of forces, unless a common software platform could be used. The latter should therefore be (*a*) portable, (*b*) flexible, and (*c*) scalable. Portability would guarantee that all software modules, including Graphical User Interfaces (GUI) and visualization procedures, are operational on many computing platforms. Code flexibility implies easy maintenance and extension, i.e., implementation of new modules, and interface with existing software (legacy codes). Detailed modeling of the propulsion system would require the use of large computational resources, i.e., parallel computers. Scalability would provide efficient code parallelization and platform independency.

The goal of this research effort is to provide the core of a software tool with the required capability for rapid design and analysis of PDE propulsion systems.

14.2   OBJECT-ORIENTED DESIGN

To satisfy the three requirements of portability, flexibility, and scalability, the code is designed through the extensive use of Object-Oriented (OO) programming. Another aspect of the research project, besides the development and application of the PDE performance analysis tool in itself, is the use of Java as the programming language. Java is a strong OO language designed for complete portability and is therefore very attractive for multiteam development efforts. The source code does not need to be exchanged or recompiled for a specific platform. Furthermore, Java includes its own set of GUI, visualization tools, and database management that is also completely portable.

Other advantages of Java include a simpler syntax than C++, automatic checking of out-of-bound errors, the absence of pointer-arithmetic (which can be very error-prone), and automatic garbage collection. Java has not yet been fully embraced for computationally intensive applications, although there is noticeable progress in that direction [2, 3]. Computing performance has usually been the most problematic feature of Java codes, but the situation is rapidly improving. New releases of the Java Virtual Machine (JVM) ("HotSpot," by Sun) appear to yield performance almost comparable to C/C++. A simple benchmark yielded the same performance for a Java code with the HotSpot VM, as for a level-2 optimized C code [4]. Multithreading, an asynchronous parallelization approach, can also greatly benefit performance. Easily implemented in Java, it does not require a set of specialized instructions and is completely portable (process creation and scheduling are handled by the OS). Remote Method Invocation (RMI) and *JavaSpaces* are implementations of distributed processing (simpler alternatives to CORBA), which can make efficient use of existing resources. These techniques are also fully scalable.

14.3   VIRTUAL DESIGN ENVIRONMENT

The Java code developed so far has been modeled after the MOZART code [5]. The PDE system being modeled is composed of several engine components, such as igniter, fuel and oxidizer injectors, thrust chamber (PDE tube), nozzle, inlet, and possible afterbody structures for engine–vehicle interactions. Each engine component can be described by one or more grids, assembled together with the appropriate choice of boundary conditions. There are also external components, e.g., ambient environment, and afterbody exhaust regions (for a more accurate description of the boundary conditions). Module dimensionality is also variable: for example, the PDE chamber could be modeled in one-dimensional, the inlet as two-dimensional, and the injector with a zero-dimensional approximation; this approach can maximize computing efficiency. The user would assemble the engine components using a combination of GUI and numerical data entry. Java applets could also eventually be used for remote setup. Virtual "sensors," measuring pressure, heat transfer, or any other quantities of interest, are also implemented. These are used to control the engine operation (e.g., valve timing). Time-traces of the measured variables can be used for comparison with experimental data. One can also envision an extension of this capability to other types of measurements, e.g., virtual laser diagnostics.

Once the problem is set up, computations can be started: total engine performance (thrust, impulse, mass, and eventually heat balance) is automatically measured, and the engine performs multiple cycles. A graphical display of engine performance is obtained in real-time; the user can decide whether to terminate the run after a number of cycles or let it run automatically. All information relevant to the engine performance is also recorded for postanalysis. One could also envision having the computations repeated automatically for a different set of initial and engine operation characteristics, specified from a "script" file. In that case, the performance could be computed automatically for a given trajectory and the optimal engine settings along this trajectory could be investigated. This capability will be implemented in the future.

14.4 APPROACH AND RESULTS

The most recent analysis looked at the effect of the overall equivalence ratio, ϕ. The configuration consists of a single cylindrical tube, with a nozzle of exit area 4 times the tube cross-sectional area. The ambient flow is stagnant at 1 atm, and the fuel is hydrogen. Figure 14.1 shows the variation of the specific impulse as a function of the "effective" equivalence ratio for two cases of filling strategy and two types of computations. The tube can be filled either with a uniform mixture or as a nonuniform charge, i.e., a stoichiometric ($\phi = 1$) region of variable size with the remainder being pure air. The single-pulse computations can be performed with the full chemical kinetics or by computing the chemical equilibrium at constant volume.

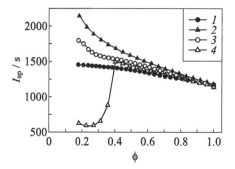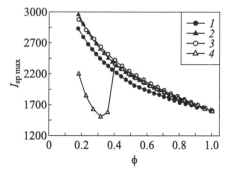

Figure 14.1 Variation of specific impulse as a function of "effective" equivalence ratio: *1* — nonuniform filling/equilibrium; *2* — uniform filling/equilibrium; *3* — nonuniform filling/kinetics; and *4* — uniform filling/kinetics.

Figure 14.2 Variation of maximum specific impulse as a function of "effective" equivalence ratio: *1* — nonuniform filling/equilibrium; *2* — nonuniform filling/kinetics; *3* — uniform filling/kinetics; and *4* — uniform filling/equilibrium.

The sudden performance drop at low ϕ for the uniform filling/kinetics case in Fig. 14.1 is due to the failure to achieve a detonation; the exact value of the critical ϕ is not necessarily reproduced here, since it would require a much higher spatial resolution. The flat curve of the nonuniform filling/equilibrium case appears to be a result of the moderate pressure increase due to constant-volume combustion, compared to the peak pressure of the detonation. In fact, the detonation pressure profile, as well as its kinetic energy, makes it ideal to induce momentum in the inert air charge at the end of the tube, in an ejector action (nonuniform filling/kinetics curve). The highest performance at low ϕ probably results from the fact that equilibrium combustion becomes more effective (less dissociation and more complete combustion) in this regime. Practicality of this mode is more questionable; it could be achieved through repeated shock reflection and relatively long confinement times, such as small nozzle-throat area, or side-wall ignition and transverse detonation propagation. However, these approaches imply other significant inefficiencies.

The performance curves of Fig. 14.1 are based on the final impulse. Corresponding performance results using the maximum impulse generated during blowdown are shown in Fig. 14.2; this time all profiles are similar, notwithstanding the detonation failure at low ϕ. The relative gain in I_{sp} is similar (80%–90%) to the best case of final impulse. The ability to base engine performance on I_{max} depends on whether the low-pressure, drag-producing phase of the blowdown can be avoided. This depends on the pressure at which injection of the fresh charge can proceed.

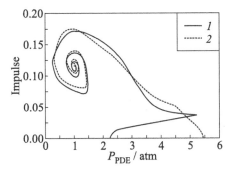

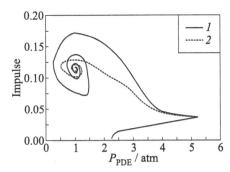

Figure 14.3 Impulse evolution vs. pressure at tube closed-end for $\phi = 0.5$ and uniform filling: *1* — kinetic and *2* — equilibrium.

Figure 14.4 Impulse evolution vs. pressure at tube closed-end for $\phi = 0.5$, uniform filling kinetics, without (*1*) and with (*2*) expanding nozzle.

The profile of the impulse as a function of the pressure inside the PDE tube (more specifically at the closed-end) provides a useful check on the potential for impulse maximization. As shown in Fig. 14.3 for a uniform mixture ($\phi = 0.5$), the pressure–impulse combination evolves from the lower-right corner towards the higher-left corner, at which time the maximum impulse is achieved. The point then enters a tightening spiral until steady state is achieved. If the pressure at the top-left corner is sufficiently low, one could initiate injection sufficiently early to avoid the start of the spiral, which results in lowering of the impulse and cycling frequency. According to Fig. 14.3, optimization can be achieved if the reservoir pressure is higher than 1 atm.

The evolution pattern can also depend on the nozzle configuration. Replacing the expanding nozzle by a straight tube extension of the same cross-sectional area, one finds from Fig. 14.4 that the spiral becomes smaller, i.e., the excursion between maximum and minimum impulses becomes smaller. In this case, there is no great loss in performance if the refilling cannot occur before the absolute minimum of tube pressure is achieved (leftmost point in the curve). On the other hand, the impulse achieved in a cycle can potentially be higher if one can refill from a higher pressure reservoir. This would seem to indicate that, although the expanding nozzle may appear detrimental if the drag-producing phase cannot be avoided, it could also significantly improve the performance if properly used. The ability to have higher injection pressure from a PDE at rest and at ambient condition may appear difficult, at least if the PDE is self-aspirated. However, this may also indicate how a pressure exchanger (e.g., turbocharger) could be used effectively with a nozzle configuration.

Of course, these results are preliminary, since they are based on single-cycle studies and for a single value of the stoichiometric fraction ϕ. Multicycle studies are, however, easily set up with the latest version of the code and have been initiated. This optimization problem will be further examined.

14.5 CONCLUDING REMARKS

The pure Java-based software for the design and evaluation of PDE technology is under further development. Recent code modifications have been implemented to facilitate multicycle studies. Systematic parametric studies of the engine operation and cycle optimization have been initiated and are being continued. Preliminary results confirm that PDE can dramatically increase its specific performance with an ejector action, thus lowering the effective equivalence ratio. It also appears that careful timing of the cycling can be coordinated with the geometric configuration design (i.e., nozzle) to increase the performance. Other optimization strategies will also be examined [6].

ACKNOWLEDGMENTS

The work was performed under ONR contract N00014-00-1-0474.

REFERENCES

1. Kailasanath, K., G. Patnaik, and C. Li. 1999. Computational studies of pulse detonation engines: A status report. AIAA Paper No. 99-2634.
2. Häuser, J., T. Ludewig, T. Gollnick, R. Winkelmann, R. Williams, J. Muylaert, and M. Spil. 1999. A pure Java parallel flow solver. AIAA Paper No. 99-0549.
3. Häuser, J., T. Ludewig, R. Williams, R. Winkelmann, T. Gollnick, S. Brunett, and J. Muylaert. 2000. A test suite for high-performance parallel Java. *Advances Engineering Software* 31(8–9):687–96.
4. Häuser, J. Private communication. See also: http://math.nist.gov/jnt.
5. Cambier, J.-L. 1996. Development of numerical tools for pulsed detonation engine studies. FFA TN-1996-50, FFA, Sweden.
6. Cambier, J.-L., M. R. Amin, and H. Z. Rouf. 2003. Parametric investigations of pulse detonation engine operation with an automated performance optimization software. AIAA Paper No. 2003-0890.

INDICES

Subject Index

A

absorption, 10, 11, 14, 19, 234, 285,
    366–368, 370, 372, 373
    infrared (IR), 9–11, 13
    line-of-sight (LOS), 12
acetylene, 355, 361
acoustics, 168, 201, 213, 214, 220,
    224, 227, 246, 248, 254, 256,
    313
    aero, 213, 214, 218, 221
    chamber, 159
    combustor, 216
activation energy, 71, 294, 372, 373
actuator, 9, 21, 91, 92, 159, 175,
    191–193, 234, 240–242
additive, 321–323, 326–328
    metallic, 331
afterburner, 225, 248
air-breathing propulsion, 170
algorithm
    conjugate, 196
    flux-corrected transport, 113
    Hooke and Jeeves, 196, 198, 199
    least-mean-square-based, 193–
    195, 198

noise-propagation, 249
recursive proper orthogonal de-
    composition (RePOD), 201,
    202, 204–206
Rosenbrock, 197
time-averaged-gradient (TAG),
    198
aluminum, 67, 144
analysis
    computational fluid dynamics,
    187
    dimensional, 218
    frequency domain, 203
    linearized perturbation, 40
    proper orthogonal decomposi-
    tion, 9, 19, 125
    quasi-steady, 390
    spectral, 215, 220
    stability, 202, 203, 210
    statistical energy, 220
analyzer
    Fourier transform infrared
    (FTIR), 140
    phase Doppler particle (PDPA),
    98

Author Index

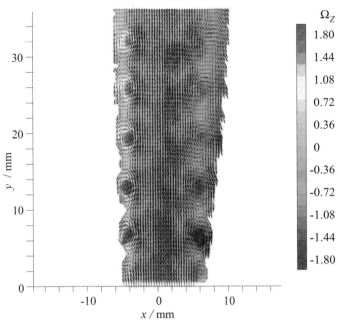

Figure 1.4 Flow field velocity color coded with the vorticity. (Refer Lourenco and Koc-Alkislar, p. 6.)

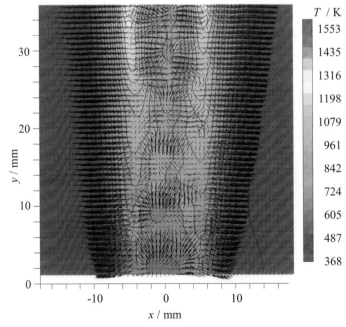

Figure 1.5 Temperature field for propane–air jet flame. (Refer Lourenco and Koc-Alkislar, p. 6.)

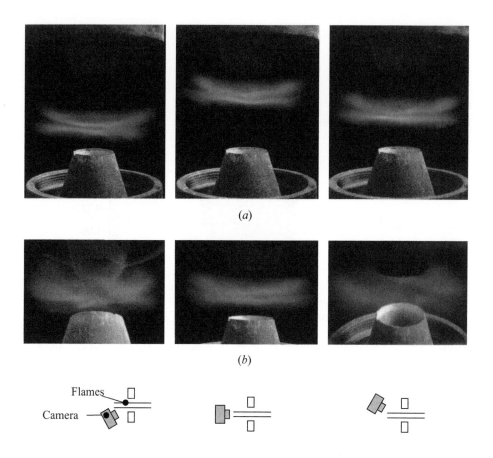

Figure 6.2 Unstable and stable twin-brush flames: (*a*) $\Phi = 0.7$, $U_b = 2$ m/s, and $H/D = 2.0$ (unstable, at three times); and (*b*) $\Phi = 0.7$, $U_b = 2$ m/s, and $H/D = 1.0$ (stable, at three angles shown below). (Refer Korusoy and Whitelaw, p. 53.)

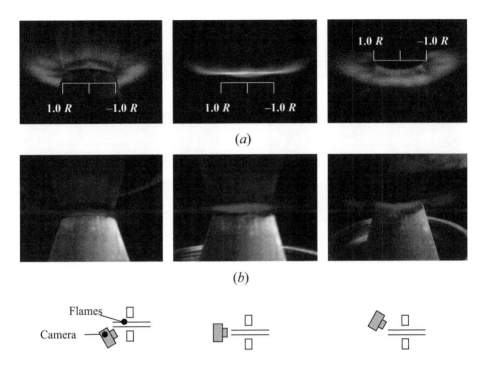

Figure 6.3 Stable single-brush flames and their extinction: (*a*) Φ = 0.7, U_b = 2.00 m/s, and H/D = 0.2 (stable); and (*b*) Φ = 0.7, U_b = 2.16 m/s, and H/D = 0.2 (partial extinction). Flames imaged at three different times and three angles shown below. (Refer Korusoy and Whitelaw, p. 54.)

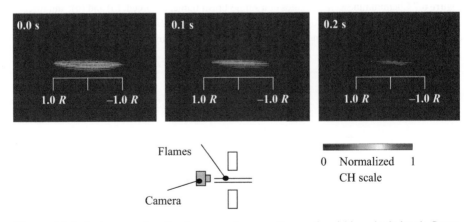

Figure 6.4 Instantaneous chemiluminescence images of an extinguishing single-brush flame: Φ = 0.7, U_b=2.26 m/s, H/D=0.2, and image intervals = 0.1 s. (Refer Korusoy and Whitelaw, p. 55.)

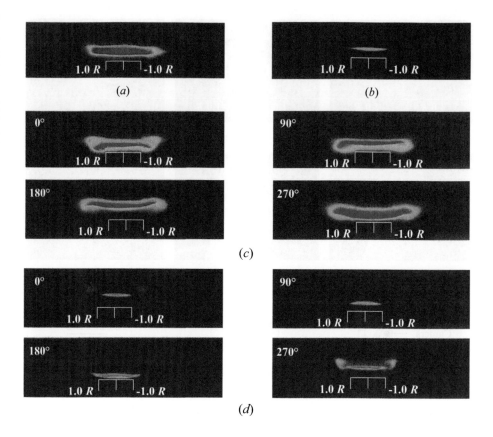

Figure 6.7 Chemiluminescence images of a single-brush flame, unforced: (*a*) and (*b*); and forced at 100 Hz: (*c*) and (*d*); (*a*) and (*c*): $\Phi = 0.7$, $U_b = 2.00$ m/s, and $H/D = 0.4$; (*b*) and (*d*): partially quenched, $\Phi = 0.7$, $U_b = 2.16$ m/s, and $H/D = 0.2$. (Refer Korusoy and Whitelaw, p. 60.)

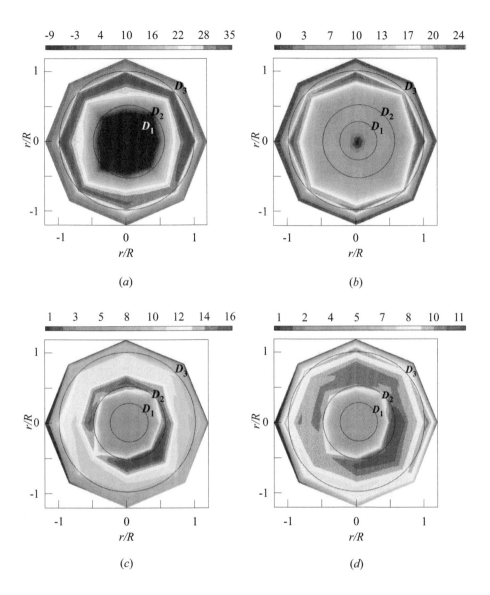

Figure 10.3 Velocity mapping of Swirler 504545 at $z/R = 0.1$ cross-sectional plane: (*a*) mean axial velocity (m/s); (*b*) mean tangential velocity (m/s); (*c*) rms axial velocity (m/s); and (*d*) rms tangential velocity (m/s). (Refer Gutmark *et al.*, p. 101.)

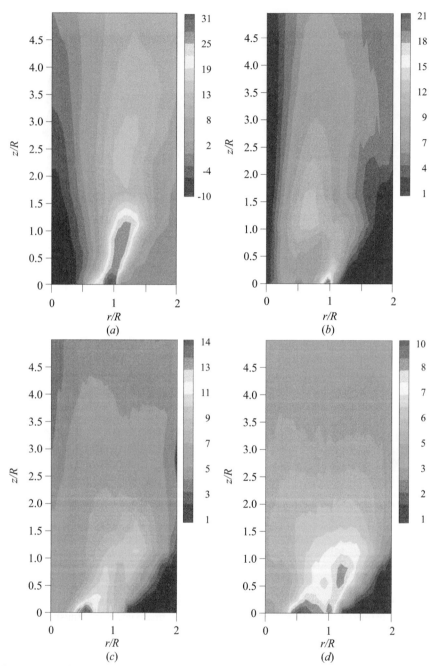

Figure 10.5 Velocity mapping for half streamwise plane of Swirler 504545: (*a*) mean axial; (*b*) mean tangential; (*c*) rms axial; and (*d*) rms tangential. (Refer Gutmark *et al.*, p. 103.)

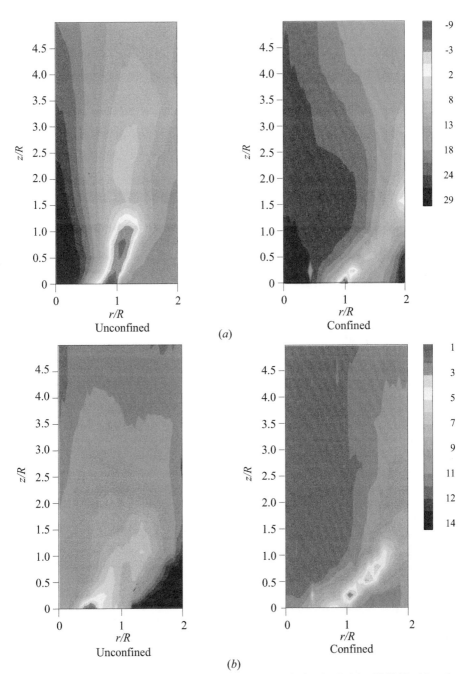

Figure 10.6 Comparison of axial velocity and rms axial velocity for Swirler 504545 with and without a confined tube: mean (*a*) and rms (*b*) axial velocities (m/s) on half streamwise plane. (Refer Gutmark *et al.*, p. 105.)

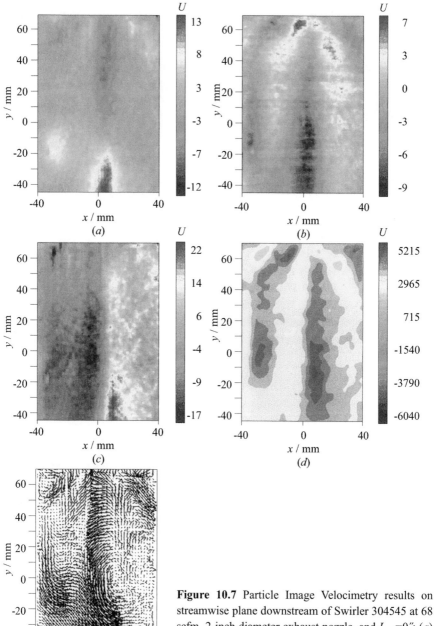

Figure 10.7 Particle Image Velocimetry results on streamwise plane downstream of Swirler 304545 at 68 scfm, 2-inch diameter exhaust nozzle, and $L_{mt}=0''$: (a) mean axial velocity component; (b) mean tangential velocity component; (c) mean radial velocity component; (d) vorticity; and (e) 3D vector. (Refer Gutmark et al., p. 107.)

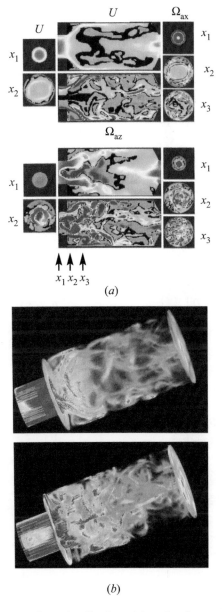

(a)

(b)

Figure 11.3 Instantaneous flow visualizations (a) and volume renderings of the vorticity magnitude Ω (b) for $S = 0.56$; LM-6000 (top), turbulent pipe flow (bottom). Instantaneous visualizations involve axial velocity levels between $-0.33U_0$ (dark blue) and U_0 (dark red) and vorticity values between $-R/(2U_0)$ (blue) and $+R/(2U_0)$ (red); more details are in [6]. (Refer Grinstein and Young, p. 117.)

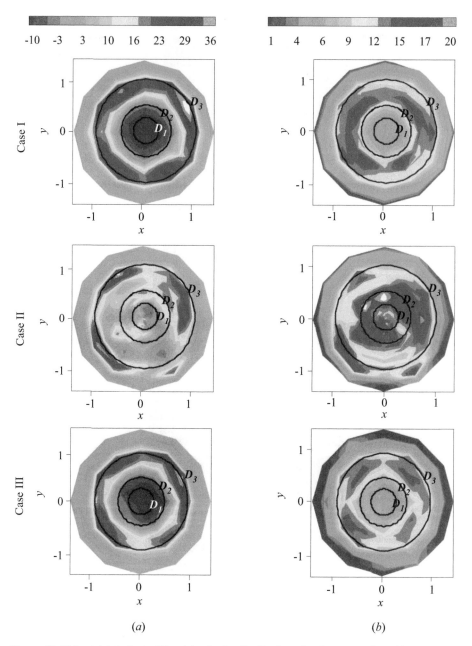

Figure 11.6 Mean (*a*) and rms (*b*) axial velocity distributions for three cases in Table 11.1 at a cross-stream plane near TARS outlet (UC LDV data [6]); levels: mean velocity between –10 m/s (blue) and 36 m/s (white); rms velocity between 1 m/s (blue) and 20 m/s (white); more details in [6]. (Refer Grinstein and Young, p. 121.)

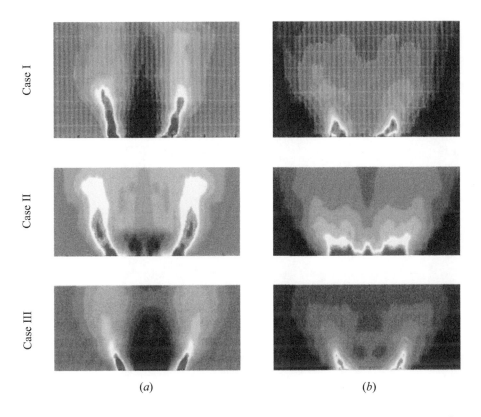

Case I

Case II

Case III

(a)   (b)

Figure 11.7 Caption is the same as in Fig. 11.6, at plane passing through the combastor axis; levels as in Fig. 11.6. (Refer Grinstein and Young, p. 122.)

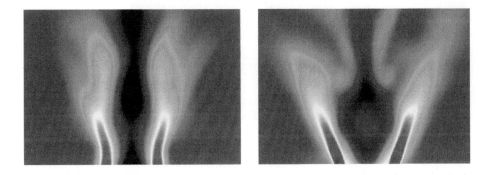

Figure 11.9a Mean axial velocity distributions in a streamwise plane from LES; levels chosen as in Figs. 11.6 and 11.7 for Case I (left) and Case III (right). (Refer Grinstein and Young, p. 123.)

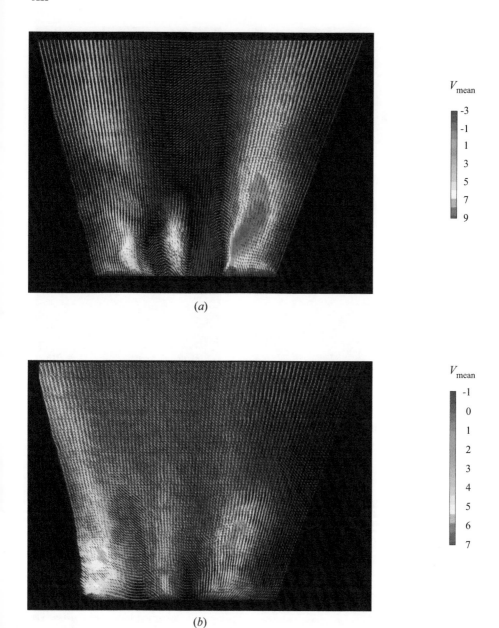

V_{mean}

-3
-1
1
3
5
7
9

(a)

V_{mean}

-1
0
1
2
3
4
5
6
7

(b)

Figure 12.3 Contours of mean axial velocity (m/s) for 65°/30° (a) and 50°/30° (b) swirl distribution in the burner with equal flow distribution under nonburning conditions. (Refer Gupta *et al.*, p. 134.)

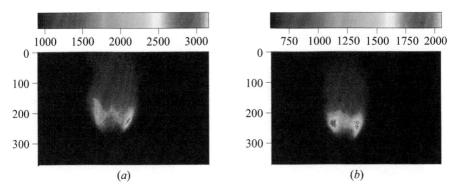

Figure 12.4 Measured OH distribution in the propane flames. Flow distribution: 25% inner and 75% outer annulus. Swirl combinations: 55°/30° (*a*) and 65°/30° (*b*). (Refer Gupta *et al.*, p. 135.)

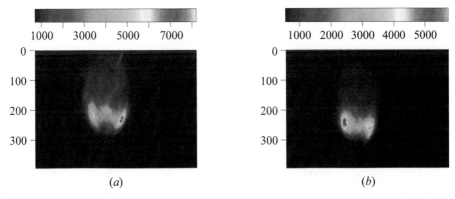

Figure 12.5 Measured CH distribution in the propane flames. Flow distribution: 25% inner and 75% outer annulus. Swirl combinations: 55°/30° (*a*) and 65°/30° (*b*). (Refer Gupta *et al.*, p. 135.)

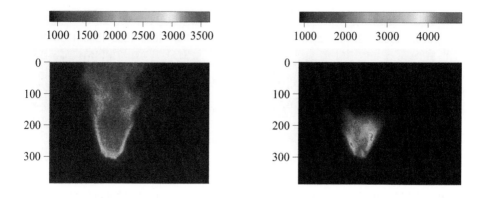

Figure 12.6 Measured OH distribution in the kerosene spray flames. Swirl combination: 55°/30°. Flow distributions: 25% inner and 75% outer annulus (*a*); and 75% inner and 25% outer annulus (*b*). (Refer Gupta *et al.*, p. 136.)

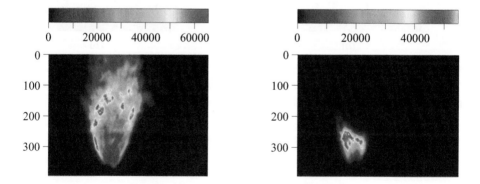

Figure 12.7 Measured CH distribution in the kerosene spray flames. Swirl combination: 55°/30°. Flow distributions: 25% inner and 75% outer annulus (*a*); and 75% inner and 25% outer annulus (*b*). (Refer Gupta *et al.*, p. 136.)

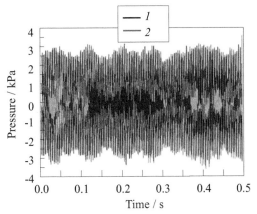

Figure 15.2 Time-trace of the pressure signals (*1* — combustor pressure and *2* — inlet pressure). Inner and annular airflow rates: 16 and 44 cfm, respectively. Primary and secondary fuel flow rates: 2.0 gph each. (Refer Acharya and Uhm, p. 159.)

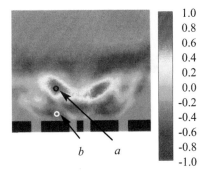

Figure 15.4 Rayleigh index at the same flow condition as that in Fig. 15.3. (Refer Acharya and Uhm, p. 162.)

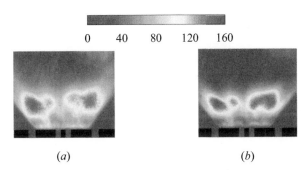

Figure 15.8 Time–averaged heat–release distribution of CH: (*a*) without control, and (*b*) with LQG-LTR control. (Refer Acharya and Uhm, p. 165.)

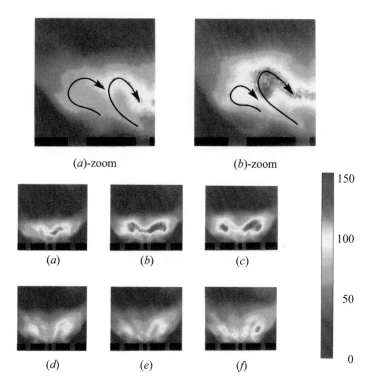

Figure 15.3 Pressure signal and spectrum (top row), and heat release distribution of CH radical intensity synchronized with the pressure signal of instability cycle (213 Hz). Secondary fuel supply of 2.5 gph. Primary and secondary air: 11 and 55 cfm. (Refer Acharya and Uhm, p. 160.)

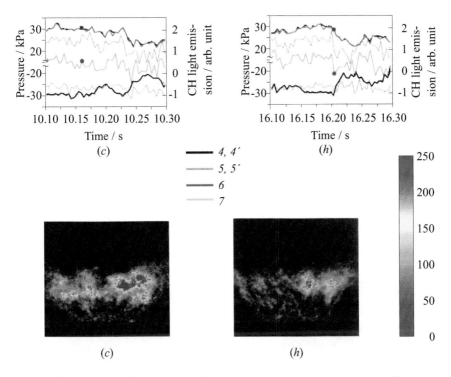

Figure 15.6 Top: normalized values of the peak pressures and corresponding CH signals and pixel-averaged CH image. Typical trigger location for CH imaging is denoted by (*a*) through (*h*). Inner and annular airflow rates: 16 and 44 cfm. Primary and secondary fuel flow rate: 2.0 gph. Expanded time trace and CH images for (*c*) and (*h*) instances: *4* and *4′* — maximum and minimum peaks of pressure; *5* and *5′* — maximum and minimum peaks of CH signals; *6* and *7* — pressure and CH values around trigger point. (Refer Acharya and Uhm, p. 163.)

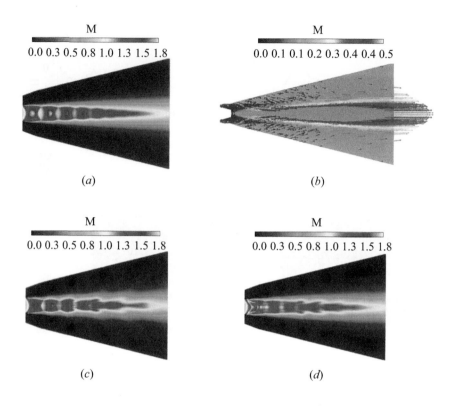

Figure 5.5 Mach number contours: (*a*) baseline, (*b*) velocity vectors, (*c*) LM case, and (*d*) HM case. (Refer Dash *et al.*, p. 263.)

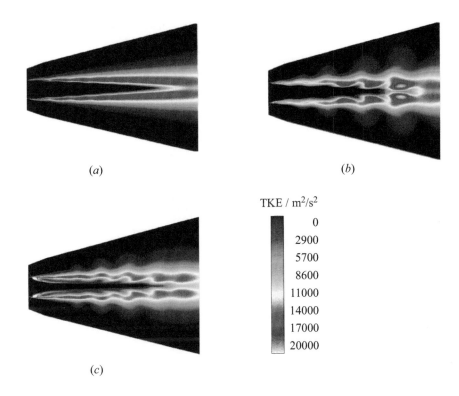

Figure 5.6 TKE contours: (*a*) baseline, (*b*) LM case, and (*c*) HM case. (Refer Dash *et al.*, p. 264.)

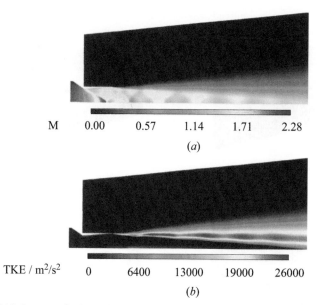

Figure 5.10 Contours for $M_e = 0$: (*a*) baseline Mach number and (*b*) baseline TKE. (Refer Dash *et al.*, p. 267.)

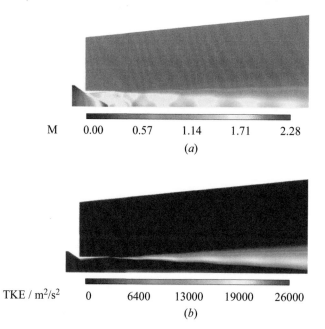

Figure 5.11 Contours for $M_e = 0.5$: (*a*) baseline Mach number and (*b*) baseline TKE. (Refer Dash *et al.*, p. 268.)

Emission magnitude

Low   High

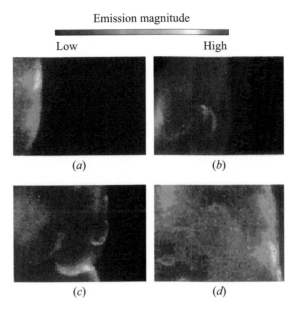

(a)   (b)

(c)   (d)

Figure 3.5 Time sequence CH*-chemiluminescence images of a successful diffraction of a locally overdriven C_2H_4–air detonation ($\varphi = 1.0$) from a 5.08- to 12.7-centimeter circular combustor; $t = 12$ μs (a); 19 μs (b); 26 μs (c); and 33 μs (d). (Refer Brophy *et al.*, p. 298.)

Emission magnitude

Low   High

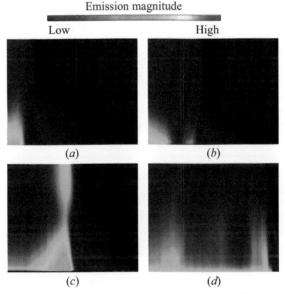

(a)   (b)

(c)   (d)

Figure 3.6 Images of the visible emission from a successful 2D diffraction of a locally overdriven C_2H_4–air ($\varphi = 1.0$) detonation in the square detonation-tube geometry $(Y/Y_0 = 1.33)$; $t = 15$ μs (a); 22 μs (b); 29 μs (c); and 36 μs (d). (Refer Brophy *et al.*, p. 299.)

Emission magnitude

Low · · · · · · · · High

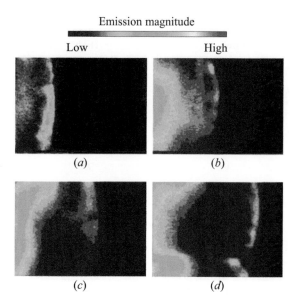

Figure 3.7 C_2H_4–air ($\varphi = 1.1$) detonation transmission in the 2D geometry ($Y/Y_0 = 1.33$); $t = 15$ μs (a); 22 μs (b); 29 μs (c); and 36 μs (d). (Refer Brophy *et al.*, p. 300.)

Emission magnitude

Low · · · · · · · · High

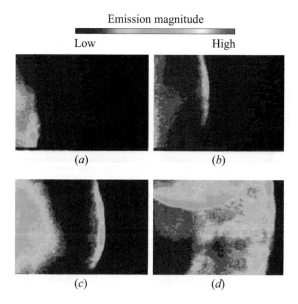

Figure 3.8 CH*-chemiluminescence images of a successful diffraction of a locally overdriven C_2H_4–air ($\varphi = 1.1$) detonation in the 2D geometry ($Y/Y_0 = 2.00$); $t = 15$ μs (a); 22 μs (b); 29 μs (c); and 36 μs (d). (Refer Brophy *et al.*, p. 301.)

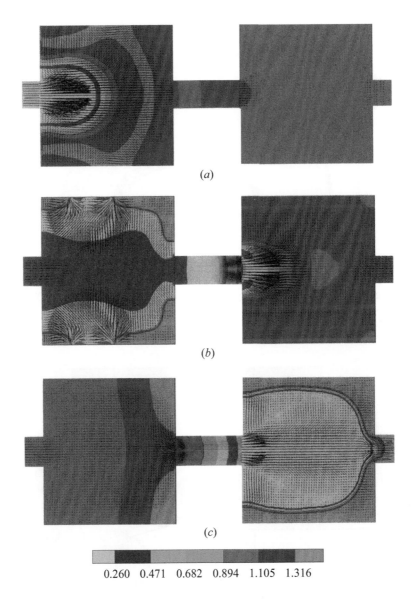

(a)

(b)

(c)

0.260   0.471   0.682   0.894   1.105   1.316

Figure 4.3 Density maps in the 6th and 7th chambers. Expansion ratio β_{ER}=96%, C_f=0.012: (a) t = 9.01 ms, (b) 9.82 ms; and (c) 10.60 ms. (Refer Smirnov *et al.*, p. 308.)

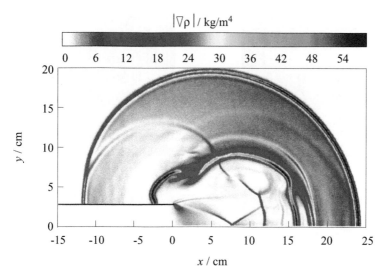

Figure 13.2 Snapshot of density-gradient field. (Refer Yang *et al.*, p. 399.)

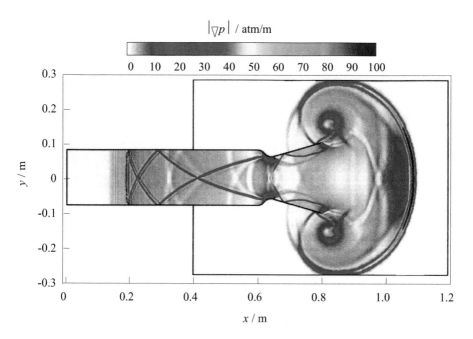

Figure 13.4 Snapshot of pressure-gradient field at $t = 0.80$ ms; $\tau_{cycle} = 3$ ms, $\tau_{close} = 2.1$ ms, and $\tau_{purge} = 0.1$ ms. (Refer Yang et al., p. 400.)

Printed and bound by CPI Group (UK) Ltd, Croydon, CR0 4YY

08/05/2025

01864818-0002